Radio Production

Radio Production is for professionals and students interested in under-standing the radio industry in today's ever-changing world. This book features up-to-date coverage of the purpose and use of radio with detailed coverage of current production techniques in the studio and on location. In addition there is exploration of technological advances, including hand-held digital recording devices, the use of digital, analogue, and virtual mixing desks and current methods of music storage and playback. Within a global context, the sixth edition also explores American radio by pro-viding an overview of the rules, regulations, and purpose of the Federal Communications Commission.

The sixth edition includes:

- updated material on new digital recording methods, and the develop-ment of outside broadcast techniques, including smartphone use;
- the use of social media as news sources, and an expansion of the sta-tion's presence;
- global government regulation and journalistic codes of practice;
- comprehensive advice on interviewing, phone-ins, news, radio drama, music, and scheduling.

This edition is further enhanced by a companion website featuring examples, exercises, and resources: www.focalpress.com/cw/mcleish.

After 33 years with BBC Radio, **Robert McLeish** was a consultant in the United States, Australia, Russia, India, and beyond. He has written and lectured in academia and served on the boards of international broadcast-ing organizations.

Jeff Link's work in broadcasting spans radio, television, and multimedia. His expertise lies in sound mixing, reporting, presenting, and productions, and he serves as an independent media consultant.

Radio Production

Sixth edition

Robert McLeish and Jeff Link

Focal Press
Taylor & Francis Group

NEW YORK AND LONDON

Sixth edition published 2016
by Focal Press
711 Third Avenue, New York, NY 10017

Simultaneously published in the UK
by Focal Press
2 Park Square, Milton Park, Abingdon, Oxon OX14 4RN

Focal Press is an imprint of the Taylor & Francis Group, an informa business

First edition published 1978 by Focal/Hastings House
Fifth edition published 2005 by Focal Press

Library of Congress Cataloging in Publication Data
McLeish, Robert.
Radio production / by Robert McLeish and Jeff Link. — Sixth edition.
pages cm
1. Radio—Production and direction. 2. Radio plays—Production and direction. I. Link, Jeff, 1946–
II. Title.
PN1991.75.M34 2016
791.4402'32—dc23
2015013102

ISBN-13: 978-1-138-81997-9 (pbk)
ISBN-13: 978-1-315-74404-9 (ebk)

Typeset in Times New Roman
by Swales & Willis Ltd, Exeter, Devon UK

Printed and bound by CPI Group (UK) Ltd, Croydon, CR0 4YY

Contents

Preface to the sixth edition

The Internet has been a game changer.

These words open Chapter 4 but I repeat them here because of the huge effect that this digital web has had on radio. It has radically affected both its transmission and its reception. It is possible now to do without complex studios and power-hungry transmitters, and the output of music and speech from almost anywhere in the world can be received on a small handheld device, so that refugees fleeing from their Middle-Eastern homeland, out of range of their country's transmitters, can still keep in touch on a cell phone.

The trend to ever smaller devices, with intuitive functions, providing more and more services, will continue. Consumers want user-friendly gadgets that will slip into their pocket or handbag, capable of delivering touch-screen, voice-activated Internet connections, GPS, SatNav, email, radio and TV. They want to take photos and video, dictate to it to record audio – and provide a visible transcript – and, with reliable coverage, send texts and make phone calls, using a tiny WiFi earpiece. It has to be theft-proof, or if stolen, remotely capable of data destruction.

All this has led to ever more radio channels. As Joseph Kebbie, a broadcaster from Ghana told me, 'Radio is key. In parts of the world, especially across Africa, radio is still the means whereby most people get their information – more than by television or the Internet which, like a reliable electricity supply, is by no means universal.' Radio is cheap to run and small stations can be really local, specific not only to the language of the region but to the dialect of the area they serve.

Thus the broadcaster gets ever closer to the listener – something which the use of social media helps to reinforce. One of the themes here is that of making programming more personal – getting alongside your listener, generating companionship, support and loyalty.

Once again, I must acknowledge my debt to so many who have contributed to this book – to the late Frank Gillard for his encouragement with the original draft and for Dave Wilkinson's ideas for the early illustrations. BBC colleagues were ever helpful in sharing their technical expertise with us – Geoff Woolf, Kevin Johnson, and Iain Meikle, who as our technical reader made many useful observations and amendments. I am grateful for help from programme innovators, Graham Dixon of Radio 3, Chris Harris, David Clayton, and Edd Smith at BBC Radios Solent and Norfolk. Community radio's Daniele Fisichella at Future Radio, and Kevin Potter at Hope FM Bournemouth, together with the children's Internet station, Abracadabra Radio, have all contributed. James Lacey of the Cloudbass media company took me over their OB vehicles, my nephew, Craig McLeish, wrote a special arrangement of 'Down by the Riverside' for p. 287, and Dr Graham Mytton, the international audience research consultant, very helpfully updated the section on programme evaluation.

We have covered much ground to include our international flavour with pictures and ideas from Roger Stoll of MediAfrique, Jos Verhoogen, Tim Hollingdale and Paul Vernon of Feba Radio, Jane Elliot of the Nova Radio Network in Australia, Kevin Keegan of Trans World Radio, Andrew Steele of the International Communications Training Institute, and Mike Adams of First Response Radio who, with Charles Randall, gave us his experience of the hurricane in the Philippines. I am grateful too for the colourful Paraiso School of Samba, who appear on p. 265, and also Ervin Munir for the use of his specialist music studio.

This edition comes with its own website and there are many who have helped there with images, recording and transcripts. I especially thank Mhairi Campbell for her work here, and also the advertising agencies for the use of their commercials – for the printed text and the audio on the website.

Finally, and above all, I owe a huge debt of gratitude to my co-writer, Jeff Link, whose special skills cover not only all things Internet and social media, but also for his photos and new writing throughout this edition. He has also designed the accompanying website. I am most grateful, and as we look forward to the further expansion of radio usage, we are certain of its development into unexpected areas, with new technologies coming to us in ways yet to be explored.

Robert McLeish

1

Characteristics of the medium

From its first tentative experiments and the early days of wireless, radio has expanded into a universal medium of communication. It leaps around the world on short waves linking capitals in a fraction of a second. It jumps to high satellites to put its footprint across a continent, and it streams through the Internet to reach every digital device around the globe. It brings that world to those who cannot read and helps maintain a contact for those who cannot see.

It is used by armies in war and by radio hams for fun. It controls air traffic, directs the taxi, and is essential for fire brigades and police. It is the enabler of business and commerce, and as WiFi feeds the smartphone, iPad and countless other devices, broadcasters pour out millions of words every minute in an effort to inform, educate and entertain, propagandise and persuade; music fills the air. Community radio makes broadcasters out of listeners and the Citizen Band gives transmitter power to the individual.

Whatever else can be said of the medium, it is plentiful. It has lost the sense of awe which attended its early years, becoming instead a very ordinary and 'unspecial' method of communication. To use it well we have to adapt the formal 'written' language that we learnt at school and rediscover our oral traditions.

To succeed in a highly competitive marketplace where television, a huge range of social media, Internet websites, lifestyle magazines, newspapers, cinema, theatre, DVDs, CDs and apps jostle for the attention of a media-conscious public, the radio student and professional programme makers must first understand the strengths and weaknesses of radio in order to decide how best to use it in their own context.

Radio makes pictures

It is a blind medium but one that can stimulate the imagination so that as soon as a voice is heard, the listener attempts to visualise the source of the sound and to create in the mind's eye the owner of the voice. What pictures are created when the voice carries an emotional content – interviews with witnesses of a bomb blast – the breathless joy of a victorious sports team.

Unlike television, where the pictures are limited by the size of the screen, radio's pictures are any size you care to make them. For the writer of radio drama it is easy to involve us in a battle between goblins and giants, or to have our spaceship land on a strange and distant planet. Created by appropriate sound effects and supported by the right music, virtually any situation can be brought to us. As the schoolboy said when asked about television drama, 'I prefer radio, the scenery is so much better.'

But is it more accurate? In reporting news, there is much to be said for seeing video of, say, a public demonstration rather than leaving it to our imagination. Both sound and vision are susceptible to the distortions of selectivity, and in reporting an event it is up to the integrity of the individual on the spot to produce as fair, honest and factual an account as possible. In the case of radio, its great strength of appealing directly to the imagination must not become the weakness of allowing individual interpretation of a factual event, let alone the deliberate exaggeration of that event by the broadcaster. The radio writer and commentator choose words with precision so that they create appropriate pictures in the listener's mind, making the subject understood and its occasion memorable.

Radio speaks to millions

Radio is one of the mass media. The very term 'broadcasting' indicates a wide scattering of the output covering every home, village, town, city and country within the range of the transmitter – and wider still when its output is streamed on a nearly universal Internet so that even a local station has a worldwide reach. A local station is now reaching its expatriate target audience – still interested in what is happening 'at home', especially the sports results – but broadcasters should remember that their words and meanings can now be heard in a very different culture from their local target audience.

Its *potential* for communication, therefore, is very great but the *actual* effect might be quite small. The difference between potential and actual will depend on matters to which this book is dedicated – programme relevance, editorial excellence and creativity, qualities of 'likeability' and persuasiveness, operational competence, technical reliability and consistency of the received signal. It will also be affected by the size and strength of the competition in its many forms. Broadcasters sometimes forget that people have other things to do – life is not all about listening to radio and watching television.

Audience researchers talk about *share* and *reach*. Audience share is the amount of time spent listening to a particular station, expressed as a percentage of the total radio listening in its area. Audience reach is the number of people who *do* listen to something from the station over the period of a day or week, expressed as a percentage of the total population who *could* listen. Both figures are significant. A station in a highly competitive environment might have quite a small share of the total listening, but if it manages to build a substantial following to even one of its programmes, let alone the aggregate of several minorities, it will enjoy a large reach. The mass media should always be interested in reach.

Radio speaks to the individual

Unlike television, where the viewer is observing something coming out of a box 'over there', the sights and sounds of radio are created within us, and can have greater impact and involvement. Radio on headphones happens literally inside your head. Here is a quote from some Ofcom research: 'With radio, you're walking down the street and just having a laugh to yourself – and nobody else knows what you're laughing at – it gives me a lift.'

Television is, in general, watched by small groups of people and the reaction to a programme is often affected by the reaction between individuals. Radio is much more a personal thing, coming direct to the listener. There are obvious exceptions: communal listening happens in clubs, bars, workshops, canteens and shops, and in the rural areas of less developed countries a whole village may gather round the set. However, even here, a radio is an everyday personal item.

The broadcaster should not abuse this directness of the medium by regarding the microphone as an input to a public address system, but rather a means of talking directly to the individual – multiplied perhaps millions of times.

Radio is private and personal

Radio is often private, heard behind closed doors or alone in the car. It can, therefore, offer help, support and advice to people who have nowhere else to turn. In many countries the so-called 'shameful subjects', especially those affecting women, can only be dealt with in this way. Forced marriage, female genital mutilation, domestic violence, child abuse, trafficking and slavery, the abortion of female foetuses, AIDS and HIV are real issues which, heard on radio in private, demonstrate that they *can* be talked about, dispelling rumours and myths and opening up the possibility of being able to initiate action – something that has actually been shown to save lives.

The speed of radio

The ancient Greeks would listen to the messenger who could run the fastest – who brought the news first. Technically uncumbersome, the medium is enormously flexible and is often at its best in the totally immediate 'live' situation. No waiting for the presses or the physical distribution of newspapers or magazines. A crisis report from a correspondent overseas, a listener talking on the phone, a sports result from the local stadium, a concert from the capital, radio is immediate. The recorded programme introduces a timeshift and, like a newspaper, may quickly become out of date, but the medium itself is essentially live and 'now'.

The ability to move about geographically generates its own excitement. Long since regarded as a commonplace, both for television and radio, pictures and sounds are sent electronically around the world, bringing any event anywhere to our immediate attention. Radio speeds up the dissemination of information so that everyone – the leaders and the led – knows of the same news event, the same political idea, declaration or threat. If knowledge is power, radio gives power to us all whether we exercise authority or not.

Radio has no boundaries

Books and magazines can be stopped at national frontiers but radio is no respecter of territorial limits. Its signals clear mountain barriers and cross ocean deeps. Radio can bring together those separated by geography or nationality – it can help to close other distances of culture, learning or status. The programmes of political propagandists or of Christian missionaries can be sent in one country and heard in any other. Sometimes met with hostile jamming, sometimes welcomed as a life-sustaining truth, programmes have a liberty independent of lines on a map, obeying only the rules of transmitter power, sunspot activity, channel interference, or receiver sensitivity. Furthermore, laptop, tablet and smartphone ownership makes us independent even of these constraints so that any studio can have an almost worldwide reach. Refugees fleeing from their homeland out of range of their own transmitters can still hear their radio via the Internet. Crossing political boundaries, radio can bring freedoms to the oppressed and enlightenment to those in darkness.

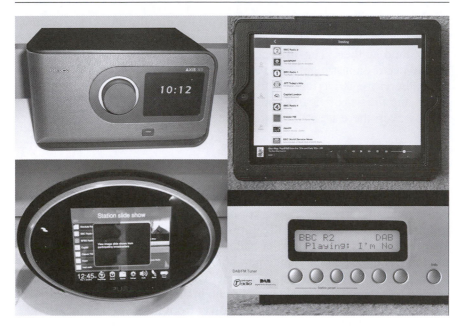

Figure 1.1 Radios for receiving conventional transmissions and via the Internet. Some are DAB compatible

The transient nature of radio

It is, in general, a very ephemeral medium and if the listener is not in time for the news bulletin, it is gone and it's necessary to wait for the next. Unlike the newspaper that a reader can put down, come back to, or pass round, broadcasting mostly imposes a strict discipline of having to be there at the right time. The radio producer must recognise that while it's possible to store programmes in the archives, they are only short-lived for the listener. This is not to say that they may not be memorable, but memory is fallible and without a written record it is easy to be misquoted or taken out of context. For this reason it is often advisable for the broadcaster to have some form of audio or written log as a check on what was said, and by whom. In some cases this may be a statutory requirement of a radio station as part of its public accountability. Where this is not so, lawyers have been known to argue that it is better to have no record of what was said – for example, in a public phone-in. Practice would suggest, however, that the keeping of a recording of the transmission is a useful safeguard against allegations of malpractice, particularly from complainants who missed the broadcast and who heard about it at second-hand.

The transitory nature of the medium means that the radio listener must not only hear the programme at the time of its broadcast, but must also understand it then. The impact and intelligibility of the spoken word should occur on hearing it – there is seldom a second chance. The producer must, therefore, strive for the utmost logic and order in the presentation of ideas, and the use of clearly understood language.

However, there are other more lasting ways of listening.

Radio on demand

Radio streamed on the Internet has the very great advantage of not only being available live and 'now', but of extending the life of previous programmes, news bulletins and features so they can be recalled on demand. By offering audio files of earlier material as 'podcasts', the station website overcomes the essentially ephemeral nature of broadcasting in that it provides specific listening when the listener wants it, not simply when it happens to be broadcast. Internet use of radio in this way radically changes the medium, transferring the scheduling power from the station to the listener. Once downloaded, programmes can be kept permanently, unless they are designed to expire after a given time.

Radio as background

Radio allows a more flexible link with its user than that insisted upon by television or print. The medium is less demanding in that it permits us to do other things at the same time – programmes become an accompaniment to something else. We read with music on, eat to a news magazine, or hang wallpaper while listening to a play. Radio suffers from its own generosity – it is easily interruptible. Television is more complete, taking our whole attention, 'spoon-feeding' without demanding effort or response, and tending to be compulsive at a far lower level of interest than radio requires of its audience.

Because radio is so often used as background, it frequently results in a low level of commitment on the part of the listener. If the broadcaster really wants the listener to *do* something – to act – then radio should be used in conjunction with another medium such as follow-up texts or email. Educational broadcasting, for example, needs an accompanying website, printed fact-sheets, booklet material, and tutor hotlines involving schools or universities. Radio evangelism has to be linked with follow-up correspondence and involve local churches or on-the-ground missionaries. Advertising requires appropriate recall and point of sale material. While radio can claim some spectacular individual action results, in general, producers have to work very hard to retain their part-share of the listener's attention.

Radio is selective

There is a different kind of responsibility on the broadcaster from that of the newspaper editor in that the radio producer selects exactly what is to be received by the consumer. In print, a large number of news stories, articles and other features are set out across several pages. Each one is headlined or identified in some way to make for easy selection. The reader scans the pages choosing to read those items of particular interest – using his or her own judgement. With radio this is not possible. The selection process takes place in the studio and the listener is presented with a single thread of material – it is a linear medium. Here are some more comments from the Ofcom research:

> 'Radio – you can't always have what you want.'
> 'I have lots of tracks on my tablet so that if I'm working, I can press the shuffle button and that's like having your own radio station.'

So radio choice for the listener exists only in the mental switching-off which occurs during an item which fails to maintain interest, or by tuning to another station. It is possible for the listener to scan the channels or search 'on demand' sources to find a missed item, but in general any one channel of radio or television is rather more autocratic than a newspaper.

Radio lacks space

A newspaper might carry 30 or 40 columns of news copy – a 10-minute radio bulletin is equivalent to a mere column and a half. Again, the selection and shaping of the spoken material has to be tighter and more logical. Papers can devote large amounts of space to advertisements, particularly to the 'small ads', and personal announcements such as births, deaths and marriages. This is ideal scanning material and it is not possible to provide such detailed coverage in a radio programme.

A newspaper is able to give an important item additional impact simply by using more space. The big story is run using large headlines and bold type – the picture is blown up and splashed across the front page. The equivalent in a radio bulletin is to lead with the story and to illustrate it with a voice report or interview. There is a tendency for everything in radio – including advertisements – to come out of the set the same size. An item might be run longer but this is not necessarily the same as 'bigger'.

Coverage described as 'in depth' might only be 'at length'. There is limited scope for indicating the differing importance of an economic crisis, a show-business item, a murder, the arrival of a pop group, the market prices and the weather forecast. It could be argued that the press is more likely to use this ability to emphasise certain stories to impose its own value judgements on the consumer – known perhaps as bias? This naturally depends on the policy of the individual newspaper editor. The radio producer is denied the same freedom of manoeuvre and this can lead to the feeling that all subjects are treated in the same way, a criticism of bland superficiality not infrequently heard. On the other hand, this characteristic of radio perhaps restores the balance of democracy, imposing less on the listener and allowing greater freedom of judgement as to what is important.

The personality of radio

A great advantage of an aural medium over print lies in the sound of the human voice – the warmth, the compassion, the anger, the pain and the laughter. A voice is capable of conveying much more than reported speech. It has inflection and accent, hesitation and pause, a variety of emphasis, lightness and speed. The information that a speaker imparts is to do with the style of presentation as much as the content of what is said. The vitality of radio depends on the diversity of voices that it uses and the extent to which it allows the colourful turn of phrase and the local idiom.

This is borne out by the Ofcom research:

'Radio provides company in a way that music alone can't.'
'Human voices are entertaining, comforting and give you a sense of security.'

It is important that all kinds of voices are heard and not just those of professional broadcasters, power holders and articulate spokesmen. The technicalities of the medium must not deter people in all walks of life from expressing themselves with a naturalness and sincerity that reflects their true personalities. Here radio, uncluttered by the pictures that accompany the talk of television, is capable of great sensitivity, and of engendering great trust.

The simplicity of radio

The basic unit comprises one person with a handheld recorder rather than even a small TV crew. This encourages greater mobility and also makes it easier for the non-professional to take part, thereby enlarging the possibilities for public access to the medium. In any case, sound is better understood than vision, with online radio, recorders, playback and stereo equipment found in most schools and homes. While the amateur video diary is an accepted format, far more complex programme ideas can be well executed by part-timers in radio. It is also probably true that whereas with television or print any loss of technical standards becomes immediately obvious and unacceptable, with radio there is a recognisable margin between the excellent and the adequate. Of course one should strive for the highest possible radio standards, but sound alone is easier to work with for the non-specialist. Hence, the growth in onlookers to a news incident filing their own reports – 'citizen journalism'.

For the broadcaster, radio's comparative simplicity means a flexibility in its scheduling. Items within programmes, or even whole programmes, can be dropped to be replaced at short notice by something more urgent.

Radio is low cost

Relative to the other media, both its capital cost and its running expenses are small. As broadcasters round the world have discovered, the main difficulty in setting up a station is often not financial but lies in obtaining a transmission frequency. Such frequencies are safeguarded by governments as signatories to international agreements, they are finite, a limited resource, and are not easily assigned. A free-for-all would lead to chaos on the air.

However, because the medium is cheap to use and can attract a substantial audience, the cost per hour – or more significantly the cost per listener hour – is low. Such figures have to be provided for advertisers, sponsors, supporters and accountants. But it is also important for the producer as well as the executive manager to know what a programme costs relative to its audience. This is not to say that cost-effectiveness is the only measure of worth – it most certainly is not – but it is one of the factors that inform scheduling decisions.

The relatively low cost once again means that the medium is ideal for use by the non-professional. Because time is not so expensive or so rare, radio stations – unlike their television counterparts – are encouraged to take a few gambles in programming. Radio is a commodity that cannot be hoarded, neither is it so special that it cannot be used by anyone with something interesting to say. Through all sorts of methods of listener participation, the medium is capable of offering a role as a two-way communicator, particularly in the area of community broadcasting.

Radio can reduce its costs still further by using an automated playout system whereby the station provides a full output schedule without anyone

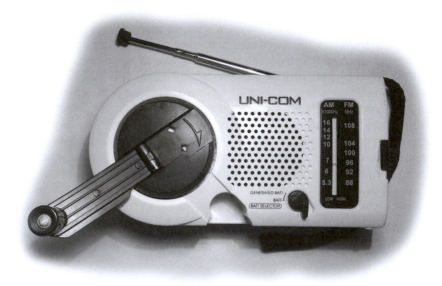

Figure 1.2 A small wind-up radio for use where batteries are expensive or unobtainable. It also has a torch light and will charge a cell phone

being present to oversee the transmission. This is particularly useful for regular half-hour blocks of programming, or to cover the night hours.

Radio is also cheap for the listener. Solid state technology allows sets to be mass produced at a cost which enables their virtual total distribution. Where batteries are expensive, a set may operate with a wind-up clock-work motor or built-in solar panel. More affordable than a set of books, good radio brings its own 'library' which is of especial value to those who, for whatever reason, are deprived of literature in their own language. Because of its relatively low cost, radio can serve a small community not only in its mother tongue, but in the local dialect. The broadcaster should never forget that while it's easy to regard the technical installations (studios, transmitters, etc.) as expensive, the greater part of the total capital cost of any broadcasting system is borne directly by the public in buying receivers.

Radio for the disadvantaged

Because of its low cost and because it does not require the education level of literacy, radio is particularly well suited to meet the needs of the poor and disadvantaged. The UK Government's Department for International

Development sets special store in using radio. Their booklet, *Media and Good Governance*, says

> 'The news media can be used to convince the poor of the benefits of having and realizing rights – and can help them assert these rights in practice. In many countries this is best achieved through radio which reaches a wide audience, rather than television which may be accessible to only a small minority.'

Radio teaches

Radio works particularly well in the world of ideas. As a medium of education it excels with concepts as well as facts. From dramatically illustrating an event in history to pursuing current political thought it has a capability with any subject that can be discussed, taking the learner at a predetermined pace through a given body of knowledge. With musical

Figure 1.3 Her prized possession. A listener in the Democratic Republic of the Congo (Photo: Roger Stoll)

appreciation and language teaching it is totally at home. Of course it lacks television's ability to demonstrate and show, it does not have charts, maps and graphs – as a medium it is more literate than numerate – but backed up by a website with teacher's notes even these limitations can be overcome, and a booklet helps to give memory to the understanding. Add a texting or correspondence element and you have the two-way questioning process which is at the heart of all personal learning.

From Australia's School of the Air to the UK's Open University and Africa's many educational programmes, radio effectively reaches out to meet the formal and informal learning needs of people who want to grow.

Radio has music

Here are the Beethoven symphonies, the Top 40, tunes of our childhood, jazz, opera, rock and favourite shows. From the best on CD or download to a quite passable local church organist, radio provides the pleasantness of an unobtrusive background or the focus of our total absorption. It relaxes and stimulates, inducing pleasure, nostalgia, excitement or curiosity. The range of music is wider than the coverage of the most comprehensive record library and can therefore give the listener a chance to discover new or unfamiliar forms of music.

DAB (Digital Audio Broadcasting) and RDS (Radio Data System) enhance the experience by supplying a range of additional information such as station name, music title, artist and any special messages the station wants to send. DAB stations' content is also deliverable through a digital TV, tablet or laptop, so multiplying the outlets for a programme. This also has the advantage of the easy recording of radio using an associated digital recorder.

Radio can surprise

Unlike the music we download, the CD we choose to play, or the book we pick up at home, selected to match our taste and feelings of the moment, music and speech on radio is chosen for us and may, if we let it, change our mood and take us out of ourselves. We can be presented with something new and enjoy a chance encounter with the unexpected. Radio should surprise – make us laugh. Here is a quote, again from the Ofcom research: 'I love interesting speech and classic comedy. Radio has the capacity to surprise.'

Broadcasters are tempted to think in terms of format radio where the content lies precisely between narrowly defined limits. This gives consistency and enables the listener to hear what is expected, which is probably why the radio is switched on in the first place. But radio can also provide the opportunity for innovation and experiment – a risk that producers must take if the medium is to surprise us in a way that is both creative and stimulating.

Radio can suffer from interference

While a newspaper or magazine is normally received in exactly the form in which it was published, radio has no such automatic guarantee. Short-wave transmission is frequently subject to deep fading and co-channel interference. Medium wave too, especially at night, may suffer from the intrusion of other stations. The quality of the sound received is likely to be very different in its dynamic or frequency range from the carefully produced balance heard in the studio. Even FM can be temperamental – liable to a range of distortions, from the flutter caused by a passing aircraft, to ignition interference from cars and other electrical equipment. Web listening can be subject to gaps and extraneous noises with a poor or slow Internet connection, or where the electricity supply is intermittent. Reception in a moving vehicle can also be difficult as the signal strength varies.

Digital transmission – DAB, or HD in the United States – and direct broadcasting by satellite overcome most of these problems – at a cost – but it is as well for the producer to remember that what leaves the studio is not necessarily what is heard in the possibly noisy environment of the listener. Difficult reception conditions require compelling programmes in order to retain a faithful audience.

Given these basic characteristics of the medium – 19 of them – how is radio to be used? What are its possibilities? Details vary across the world but broadly it functions in two main ways – it serves the individual, and operates on behalf of society as a whole.

Radio for the individual

- It diverts people from their problems and anxieties, providing relaxation and entertainment. It reduces feelings of loneliness, creating a sense of companionship.
- It is the supplier or confirmer of local, national or international news, and establishes for us a current agenda of issues and events. It makes government explicit and satisfies our personal curiosity about what is going on around us.
- It offers hope and inspiration to those who feel isolated or oppressed within a restrictive environment, or where their own media are manipulated or suppressed.

- It helps to solve problems by acting as a source of information and advice. It can do this either through direct personal access to the programme or in a general way by indicating sources of further help, especially in times of crisis.
- It enlarges personal 'experience', stimulating interest in previously unknown topics, events or people. It promotes creativity and can point towards new personal activity. It meets individual needs for formal and informal education.
- It contributes to self-knowledge and awareness, offering security and support. It enables us to see ourselves in relation to others, and links individuals with leaders and 'experts'.
- It guides social behaviour, setting standards and offering role models with which to identify.
- It aids personal contacts by providing topics of conversation through shared experience – 'Did you hear the programme last night?'
- It enables individuals to exercise choice, to make decisions and act as citizens, especially in a democracy through the unbiased dissemination of news and information.

Radio for society

- It acts as a multiplier of change, speeding up the process of informing a population, and heightening an awareness of key issues.
- It provides information about jobs, goods and services and so helps to shape markets by providing incentives for earning and spending.
- It holds leaders to account, acting as a watchdog on power holders, providing contact between them and the public.
- It helps to develop agreed objectives and political choice, and enables social and political debate, exposing issues and options for action.
- It acts as a catalyst and focus for celebration – or mourning – enabling individuals to act together, forming a common consciousness, encouraging community uniqueness.
- It contributes to the artistic and intellectual culture, providing opportunities for new and established performers of all kinds.
- It disseminates ideas. These may be radical, leading to new beliefs and values, so promoting diversity and change – or they may reinforce traditional values, so helping to maintain social order through the status quo.
- It enables individuals and groups to speak to each other, developing an awareness of a common membership of society.
- It mobilises public and private resources for personal or community ends, particularly in an emergency.

Some of these functions are in mutual conflict, some are applicable more to a local rather than a national community, and some apply fully only in conditions of crisis. However, the programme producer should be clear about what it is he or she is trying to achieve. Lack of clarity about a programme's purpose leads to a fuzzy, ineffective end product – it also leads to arguments in the studio over what should and should not be included. We shall return to the point, but it is not enough for the producer to want to make an excellent programme – you might as well want merely to modulate the transmitter. The question is why? What is to be its effect – on the listener, that is? Before looking at some possible personal motivations for making programmes at all, we should examine the meaning of the much-used phrase – broadcasting as a public service.

The public servant

Public service broadcasting is sometimes regarded as an alternative to commercial radio. The terms are not mutually exclusive, however; it is possible to run commercial radio as a public service, especially in near-monopoly conditions or where there is little competition for the available advertising. It is a matter of where the radio managers see their first loyalty. To put this in perspective here are ten attributes of service, of being community-minded, of doing what is required and being useful, using the analogy of the perfect employee, someone servicing our car, or family helper. He or she:

- is loyal to the employer and does not try to serve other interests or use this position of trust for personal gain. Genuine helpers are people who are clear about their purpose and demonstrate this in everything they do;
- understands the culture, nuances and foibles of the people being served, and in accepting them also becomes fully accepted. A helping friend is not critical or judgemental of those he or she serves but may challenge them, advise them, or even restrain them. But the friend is not there to act as a policeman; there is no corrective mandate;
- is available when needed, and for whoever in the group needs help, the young and the old as well as the boss. It may be tempting to pay court to the power holders, to keep in with the paymaster – but it could be the sick or disadvantaged who need most attention;
- is actually useful, meeting stated requirements and anticipating needs and difficulties – ready to offer original solutions to problems, as or even before they arrive. This means looking ahead – being creative;

- is well informed, knows what is going on, offers good advice and is able to relate unpalatable truth, risking unpopularity in the process. This requires courage as well as a well-stocked mind;
- is hard working, technically expert and efficient. Does not cut corners or waste resources, but is honest and open, and if called to account for particular behaviour, responds willingly. There is nothing underhand – integrity is a key attribute;
- is witty and companionable, courteous and punctual. The radio servant inhabits our house, comes with us in the car, and is someone with whom we can have a relationship that is enjoyable as well as professional;
- has to realise that it's not possible to do everything – to be in two places at once – there are priorities. The users of the service must understand this too;
- works smoothly alongside others and does not crave recognition or status. When the number to be served is very large, other helpers may need to be brought in. We may want more than one kind of service;
- is affordable. My helper, informer, and friend must not bankrupt me. However much I value this service, long-term cost-effectiveness is crucial.

Each of these characteristics of service has its equivalent in public service broadcasting. Its hallmarks can be drawn from our ideas of what a real helper is and does. Such a station is certainly not arrogant, setting itself up as a power in its own right. It is responsive to need, making itself available for everyone – not simply the rich and powerful; indeed its universality makes a point of including the disadvantaged. It is wide ranging in its appeal, competent and reliable, entertaining and informative. Its programmes for minorities are not to be hidden away in the small hours but are part of the diversity available at prime time. It is popular in that over a period of time it reaches a significant proportion of the population. It does not 'import' its programmes from 'foreign' sources but is culturally in tune with its audience, producing most of the output itself. It provides useful and necessary things – things of the quality asked for, but also unexpected pleasures. Above all, it is editorially free from interference by political, commercial or other interests, serving only one master to whom it remains essentially accountable – its public.

But there is an immediate dilemma – such perfection may be too expensive. As in everything else we get the level of service we pay for and it might not be possible to afford a 24 hours a day, seven days a week output catering for all needs. Live concerts, stereo drama and world news are expensive commodities, and radio managers need to make judgements about what can be provided at an acceptable price.

The concept works well for a service that is adequately paid for by its listeners – by public licence or by subscription. But if this is not the case, can a public service station do deals with third parties to raise additional revenue? Can a government, commercial or religious station be run as a genuine public service? Yes it can, but the difficulties are obvious.

First, a publicly funded service that makes arrangements with commercial interests is putting its first loyalty, its editorial integrity, at risk. Any producer making a programme as a co-production, or acting under special rules, must say so – not as part of some secret or unstated pay-off, but to meet the requirements of public service accountability.

Second, there is a strong tendency for 'piper payers' to want to call the tune. A government does not like to hear criticism of its policy on a station that it regards as its own. Authority in general does not wish to be challenged – as from time to time political interviewers must. Ministries and departments are highly sensitive to items which 'in the public interest' they would rather not have broadcast. Officials will avoid or delay 'bad news', however true. Similarly, a commercial station often needs to maximise its audience in order to justify the rates – certainly at peak times – so pushing sectional interests to one side to satisfy the advertisers' desire for mass popularity. A Christian fund-raising constituency may press for the gospel *it* wants to hear, forgetting the need to serve people in a multiplicity of ways. The need to survive in a harsh political climate, or in a fiercely competitive one, exacerbates these pressures. The fact is that a station dedicated to public service but controlled or funded by a third party, having its own agenda and interests to consider, is almost certain to weaken in its commitment.

Personal motivations

So what is *our* purpose for being in radio? Do you want to be a public servant, or is radio attractive because it has some appearance of power – able to sway public opinion and make people do things? If so, it has to be said that this is very rare and most unlikely to be achieved by this medium alone. Is it to be the protagonist mouthpiece of someone else – or are there reasons that meet my own needs?

It is as well for the producer to understand what some of these personal motivations are:

- to inform people – the role of the media journalist;
- to educate – enabling people to acquire knowledge or skill;
- to entertain – making people laugh, relax or pass the time agreeably;
- to reassure – providing supportive companionship;
- to shock – the sensation station;

- to make money – a means of earning a living;
- to enjoy oneself – a means of artistic expression;
- to create change – persuading for a new society;
- to preserve the status quo – resisting change, maintaining the established order;
- to convert to one's own belief – proselytising a faith or political creed;
- to serve people, to present options – assisting citizenship, allowing the listener to exercise choice.

Each programme maker must ask, 'Why do I want to be in broadcasting?' Is it just a job? Certainly it may be in order to earn a living, but also because he or she has something to say. It may be out of a genuine desire to serve one's fellow men and women – to allow other people to have their say, to provide options for action by opening up possibilities, entertaining people, making them better informed, or something to do with 'truth'. Perhaps it's a more complicated amalgam than any of these.

One's success criteria are perhaps best summed up as a response to the question, 'What do you want to say, to whom, and with what effect?' The 'to whom?' is important for, in the end, radio is about a relationship. Much more than on television, the presenter, DJ, or newsreader can establish a sense of rapport with the listener.

The successful station is more than the sum of its programmes: it understands and values the nature of this relationship – and the role which it has for its community as both leader and servant.

2

Structure and regulation

No radio station – and therefore no producer – exists in a vacuum. It has a context of connections, each one useful and necessary but also representing a source of potential pressure. Figure 2.1 illustrates some of these. At the top is the national regulator – the body with overall responsibility to its government for the supervision of all that country's broadcasting services. There may be additional advisory bodies, sponsors and advertisers, community and educational interests, sources of information like the police and fire services, local councils, programme suppliers and supporters' clubs. The station may be part of a large organisation or a media chain with other affiliate stations and a headquarters office. Every station

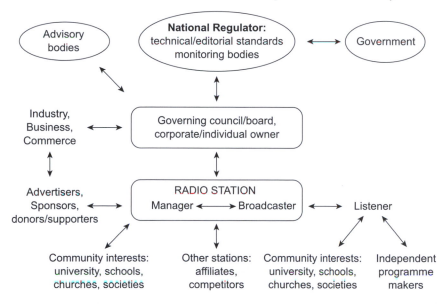

Figure 2.1   Stakeholders and interested parties around a typical station

is likely to have its own governing council, board, or owner, to whom the manager reports.

Each of these bodies is linked here by two-way arrows – what do they represent? Is money changing hands – or information, or advice, or control? It is a good exercise for broadcasters to get to know their own situation.

The national regulator

Starting at the top, every country has an authority for the overall regulation of broadcasting. Their roles are very similar – they allocate station frequencies and set limits for transmitted power in order to avoid interference. They impose certain programme standards, set rules and monitor the output of stations to ensure these are kept. They do listener research. Here are three examples.

1   *The Federal Communications Commission.* Responsible to the US Congress, the FCC regulates all radio and television broadcasting, including inter-state and international satellite, broadband and cable services. It is the US's primary authority for communications law and technical innovation. It states its vision as:

> To promote innovation and investment, competition and consumer empowerment for the communications platforms of today and the future – maximizing the power of communications technology to expand our economy, create jobs, enhance U.S. competitiveness and unleash broad opportunity for all Americans.

Based in Washington DC, the Commission comprises a number of specialist committees. The Technological Committee deals with matters such as interference issues, security and industry standards. As far as the Internet is concerned the FCC says its aim is to:

> Encourage broadband deployment and preserve and promote the open and interconnected nature of the public Internet; Consumers are entitled to access the lawful Internet content of their choice; Consumers are entitled to run applications and use services of their choice, subject to the needs of law enforcement; Consumers are entitled to connect their choice of legal devices that do not harm the network.

There are specific rules on broadcasting telephone conversations, broadcast hoaxes and equal opportunities. It prohibits programming that is obscene, indecent or profane. All sponsored material must be explicitly identified at the time of broadcast. The FCC has much to say about political programming. It has the power to inspect stations and insists that broadcasters keep station logs and records. It undertakes research into listening and viewing to determine how best to use the available frequency spectrum and adjust to technological development.

The FCC sets out its work in considerable detail on a well signposted website: www.fcc.gov.

2   *The UK Office of Communication.* Ofcom regulates all radio and television in the UK. Currently the BBC is responsible to its own Board of Trustees. However, this may change – in any case the Corporation must abide by the Ofcom Code. This 'Broadcasting Code' is required reading for all UK stations. The Code sets out to:

> protect listeners from harmful and offensive content but also ensures that broadcasters have the freedom to make challenging programmes. For example, broadcasters can transmit provocative material – even if some people consider it offensive – provided it is editorially justified and the audience is given appropriate information.

As well as harm and offence, the Code covers other areas like impartiality and accuracy, fairness, privacy, protecting the under-18, elections and referendums, crime, religion, sponsorship and product placement.

It is a crucial aim of a commercial station to make a profit for its shareholders by selling advertising. However, their licence conditions require them to provide a certain amount of local programming. This is why Ofcom sets out its duties as:

- promoting the interests of citizens and consumers;
- securing a range and diversity of local commercial services which (taken as a whole) are of high quality and appeal to a variety of tastes and interests;
- ensuring for each local commercial station an appropriate amount of local material with a suitable proportion of that material being made locally.

Pursuing these objectives, Ofcom says that community stations receiving some advertising, sponsorship, and local grants must provide 'social gain' – that is, community information, news, weather, sport, events, etc. Ofcom carries out research into the way people use radio which shows that small-scale stations are 'highly valued, fostering a real sense of belonging for which listeners have a unique sense of affection'. It showed that while people may not listen for great lengths of time, their level of engagement was high. Ofcom issues regular bulletins about its work. The website for details is: www.ofcom.org.uk.

3   *The Australian Communications and Media Authority.* ACMA is similar to the others on matters of frequency allocation and interference, has a list of prohibited programme content – transmitted or online – and has developed the Australian Internet Security Initiative to help address the problem of the surreptitious sending and installation of malicious software.

The Authority regulates advertising, which must be clearly distinguishable from other content. It has Codes of Practice concerning betting and gambling, and the incitement of violence, or hatred against a particular individual or group. It rules against anything that presents as desirable the misuse of alcohol, illegal drugs or tobacco. Programmes must not simulate news events in a way that could mislead or alarm, and 'must not offend generally accepted standards of decency, having regard to the demographic characteristics of their audience'. Its equality policies require the use of terms such as 'fire-fighter' not fireman, similarly – police officer and sales rep., etc. Full details can be found at: www.acma.gov.au.

These examples serve to highlight the nature of the overall control which all national regulators have. Programme makers everywhere should make themselves aware of the requirements which apply to them – they are in general not overly restrictive but are in tune with the culture of the country.

The local board

Immediately in charge of the operation of the station is the manager or managing editor and his or her staff. This person is likely to be responsible to a local governing council or board.

To avoid constant misunderstandings and argument, the board's terms of reference should be very precisely defined. Its job will depend on how the station is set up – its legal constitution. It may have a controlling function, deciding how the programming or finance should be determined, or its role may simply be advisory in assisting the station to meet its declared aims. It may play a part in appointing the staff; it should be ready to negotiate with outside bodies and always to protect the station against unfair criticism. If the station is run as a charity, the board will form the group of trustees responsible to the Charity Commission or other ruling authority.

It is a good idea for the working producer to know at least two members of the organisation's board or council. Not only does this help the programme makers to appreciate the context within which the station is operating, but it also acquaints board members with the difficulties and encouragements experienced by those who actually create the product. Such introductions are best made through the manager.

Station finance

Radio is often categorised and structured not so much by what it does as by how it is financed. Again, there is a wide variation in how a station gets its money, but each method of funding has a direct result on the programming which a station can afford or is prepared to offer. This, in turn, is affected by the degree of competition which it faces.

The main types of organisational funding are as follows:

- public service, funded by a licence fee and run by a national corporation;
- commercial station financed by national and local spot advertising or sponsorship, and run as a public company with shareholders;
- government station paid for from taxation and run as a government department;
- government-owned station, funded largely by commercial advertising, operating under a government-appointed board;
- public service, funded by government funds or grant-in-aid, run by a publicly accountable board, independent of government;
- public service, subscription station, does not take advertising and is funded by individual subscribers and donors;
- private ownership, funded by personal income of all kinds, e.g. commercial advertising, subscriptions, donations;
- community ownership, not-for-profit station often supported by local advertising and sponsors;
- institutional ownership, e.g. university campus, hospital or factory radio, run and paid for by the organisation for the benefit of its students, patients, employees, etc.;
- radio organisation run for specific religious or charitable purposes – sells airtime and raises income through supporter contributions;
- restricted Service Licence – RSL stations, on low power and having a limited lifespan to meet a particular need, e.g. one-month licence to cover a city festival.

The ability to charge premium rates for advertising depends either on having a large audience in response to popular appeal programming, or an audience with a high purchasing power for the individual advertiser. Competing for income from a limited supply is quite different from the competition which a station faces to win an audience.

Around the world, stations combine different forms of funding and supplement their income by all possible means, from profit-making events such as concerts, publishing, and the sale of programme downloads – both to the public and to other broadcasters – to the sale of T-shirts and CDs, community coffee mornings and car boot sales.

Station structure

Finance directly affects a station's structure, but given the vast range of audience requirements, from huge national and international organisations to very small community needs, it is not surprising that there is no such thing as a 'typical station'. It is possible, however, to give some examples which generalise how things work – how the boss, manager, director, editor, advertising people, producers, reporters, operators, engineers, administrative staff and freelances work together to create a coherent output.

Here is what a commercial station might look like.

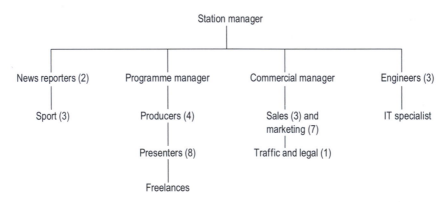

The station manager would be a member of the board of directors, or might be one of a group of station managers overseen by a senior programme controller or general manager.

The programme head is the deputy manager. The engineers look after the equipment and the IT aspect of the website, and even though it's a self-op station, they may do studio operation for some of the freelance

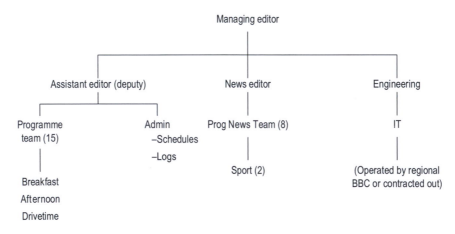

Typical local BBC station

presenters. There's generally a front of house receptionist, and the larger stations would have secretaries for the administrative logs and legal work, plus perhaps catering and cleaning staff. Security is often contracted out.

A BBC local station typically has a very flat management structure integrating the news output with the lengthy Sequence or Strip programmes across the day. They share the same office space or newsroom (see Figure 7.2).

The producers and broadcast assistants exist within the programme teams, together with the freelance contributors. There is a receptionist and secretarial staff as required. The engineering resource is often shared between several stations.

Community radio

Like other non-BBC stations in the UK, community radio comes under Ofcom and has its own advisory and funding bodies (see Figure 2.2).

The purpose of a station such as this is not only to serve the community as listeners, but to involve local people in creating the output and to provide educational courses and music events. A fairly typical station – Future Radio in Norwich – has three full-time paid staff and some 120 volunteers. The full-time staff are: a station manager, an advertising coordinator (who is the deputy station manager) and a broadcast assistant. The advertising

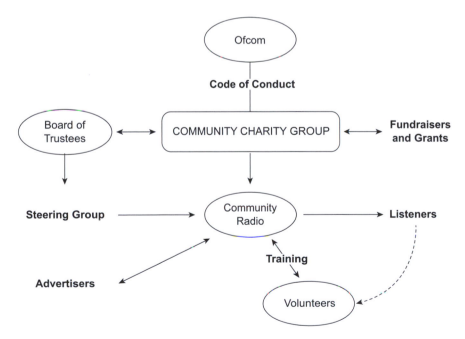

Figure 2.2 The typical community radio context

coordinator is in charge of commercial sales and promotional events, also supervising the production of trailers and commercials for broadcasting. The broadcast assistant looks after everything related to the studio, including scheduling, managing playlists, station IDs, studio bookings and helping the volunteers with the daily operations of the station. Because the radio station is part of an umbrella charity group it also shares the expertise of a salaried IT manager, who deals with IT and buildings, and a business support officer who helps with fund raising. There are also opportunities in a community station such as this for young people to apply for work-experience posts.

The Community Media Association is the UK representative and supporting body for this sector of broadcasting (www.commedia.org.uk). The manager of Future Radio talks about his job on the website.

Station life

A radio station should be a creative place. Programme makers are always wanting to win prizes and awards for what they do, but more important is the winning of an audience – and keeping it. As this book says more than once, it is about being relevant and engaging – knowing about your listeners and what motivates them. Not simply waiting for people to come to you, but reaching out to them – an idea never bettered than in the catchline of CBS station WINS in New York:

> 'You give us twenty-two minutes – we'll give you the world.'

Whether you are in a city area, an African township, a huge expanse of rural countryside, or across the Internet, you have to care about what concerns people – or makes them laugh – your listener is the be-all and end-all of your endeavour. Meeting those needs is a serious business requiring insight, accuracy and truth. But it should also be fun – an enjoyable pursuit to which you give much of your time and energy. Many are the broadcasters who have said, 'It's amazing, I actually get paid for doing what I like best.' Of course there are rows and different opinions, competing for ratings, arguments over ideas and budgets. But the ideal working atmosphere has a certain amount of mutual support, of appreciating that the worth of a station is greater than the sum of its individual parts or programmes, and that everyone who works there, in whatever capacity, has a share in the final outcome.

Getting a job

There are hundreds of opportunities, so you'd think it would be easy to find a place somewhere in local radio or in hospital, student or community radio.

The key is to acquaint yourself with what's going on – the debates and issues on broadcasting. Read the trade magazines and reviews in newspapers, look at station websites, download recordings, get tickets for a show if you can, and of course listen to as much radio as possible – of all different kinds so that you know what suits you best.

Getting a job at a station means first of all understanding its structure, what its aims are, what kind of output it provides, what it needs, and how it works. Does the work involve television and website coverage as well as radio? The most likely way in is to offer to work as a freelance helper, or contributor of ideas and items and, if successful, move on to the staff as a programme assistant. Many successful broadcasters have started by making the tea – and doing so willingly. A station may offer an attachment to begin with, or an internship – paid or unpaid. If so, it is essential to get to know the equipment really well, understanding the software technicalities and processes of the studio. This can be very helpful to some of the established programme people, especially the older contributors. Station staff appreciate newcomers who are keen and enthusiastic, arrive on time, ask questions, come up with ideas and stay late to get things finished.

If given a programme brief, however small, it is important to make sure your name is attached to it – programme credits are crucial to your subsequent CV. These count hugely if you are interviewed for a job – for which the key attributes are enthusiasm, and having done your homework.

3

The radio studio

Here is an introduction to the crucial operational process of creating the sounds that the listener will hear – sounds that must be clear and accurate. Anything that is distorted, confused or poorly assembled is tiring to understand and will not retain the listener's interest.

The quality of the end product depends in part directly on the engineering and operational standards. It scarcely matters how good the ideas are, how brilliant the production, how polished the presentation, because all will founder on poor operational technique. Whether the broadcaster is using other operational staff, or is in the 'self-op' mode, a basic familiarity with the studio equipment is essential – it must be second nature.

Taking the analogy of driving a car, the good driver is not preoccupied with how to change gear, or with which foot pedal does what, but is more concerned with road sense, i.e. the position and speed of the car on the road in relation to other vehicles and people. So it is with the broadcaster. The proper use of the tools of the trade – the studio mixer, microphones, computers, recorders – these must all be at the command of the producer who can then concentrate on what broadcasting is really about, the communication of ideas in music and speech.

Good technique comes first, and having been mastered, should not be allowed to impinge on the programme. The technicalities of broadcasting – editing, fading, control of levels, sound quality and so on – should be so good that they do not show. By being invisible they allow the programme content to come through.

In common with other performing arts such as film, theatre and television, the hallmark of the good communicator is that the means are not always apparent. Basic craft skills are seldom discernible except to the professional who recognises in their unobtrusiveness a mastery of the medium.

There are programme producers who declare themselves uninterested in technical things; they will leave the mixer or computer operation to others so that they can concentrate on 'higher' editorial matters. Unfortunately, if you do not know what is technically possible, then you cannot fully realise the potential of the medium. Without knowing the limitations there can be no attempt to overcome them. You simply suffer the frustrations of always having to depend on someone else who does understand. So here, we will explore the studio and discuss some of the equipment.

Studio layout

Studios for transmission or rehearsal/recording may consist simply of a single room containing all the equipment, including one or more microphones. This arrangement is designed for use by one person and is called a self-operation or self-op studio.

Where two or more rooms are used together, the room with the mixer and other equipment is often referred to as the control room or cubicle, while the actual studio – containing mostly microphones – is used for interviewees, actors, musicians, etc. If the control cubicle also has a mic it may still be capable of self-operation. In any area, when the mic is faded up and becomes 'live', the loudspeaker is cut to avoid 'feedback' or 'howl-round', and monitoring must be done on headphones.

Figure 3.1 A typical analogue cubicle with digital add-ons. 1. Analogue mixing desk. 2. Racks panel with jackfield and power supply. 3. CD players. 4. Screens connected to digital add-ons. 5. Gram deck. 6. Loudspeakers

The studio desk, mixer, control panel, console, or board

Most studios will include some kind of audio mixer – analogue, digital or fully computerised. What it is called depends on which country you are in, such as panel, console, board, mixer, desk, etc. It is essentially a device for mixing together the various programme sources, controlling their level or volume, and sending the combined output to the required destination – generally either the transmitter, a recorder or streamed to the Internet. Traditionally, it contains three types of circuit function:

1 *Programme circuits:* a series of differently sourced audio channels, their individual volume levels controlled by separate slider faders. In addition to the main output, a second or auxiliary output – generally controlled by a small rotary fader on each channel – can provide a different mix of programme material typically used for public address, echo, foldback into the studio for contributors to hear, a clean feed, or separate audio mix sent to a distant contributor, etc.
2 *Monitoring circuits:* a visual indication (either by a programme meter or an LED bargraph) and an aural indication (loudspeaker or headphones), to enable the operator to hear and measure the individual sources as well as the final mixed output.
3 *Control circuits:* the means of communicating with other studios or outside broadcasts by means of 'talkback' or telephone.

In learning to operate a mixer there is little substitute for first understanding the principles of the individual equipment, then practising until its operation becomes second nature. The following are some operational points for the beginner.

The operator must be comfortable. The correct chair height and easy access to all necessary equipment is important for fluid operation. This mostly calls for a swivel chair on castors.

The first function to be considered is the *monitoring* of the programme. Nothing on a mixer, which might possibly be on-air, should be touched until the question has been answered – what am I listening to? The loudspeaker is normally required to carry the direct output of the desk, as, for example, in the case of a rehearsal or recording. In transmission conditions it will often take its programme feed off-air, although it may not be feasible to listen via a receiver when transmitting on short wave or on the web. As far as possible, the programme should be monitored as it will be heard by the listener, i.e. after the transmitter, not simply as it leaves the studio. With a digital desk and transmission chain, the inherent delay and latency in the signal often means that one cannot monitor much beyond the desk output.

'TRADITIONAL' LAYOUT - STUDIO/CUBICLE

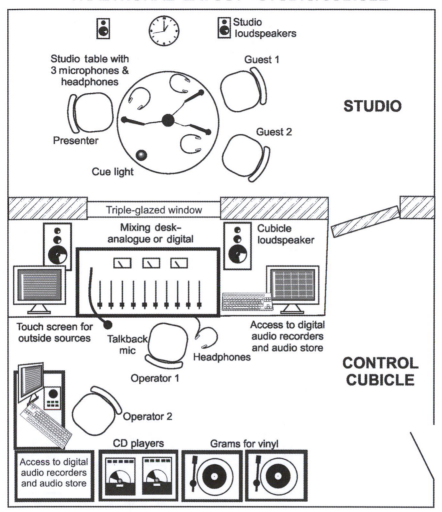

Figure 3.2 Layout of a traditional studio, controlled or 'driven' by its cubicle. This arrangement is primarily designed for complex demands such as lengthy news and current affairs programmes with many guests and recorded inserts. The studio area has a table and chairs with two or more microphones. Monitoring is by loudspeaker or headphones, with the speakers in the studio area being muted when the mic channels are open. The headphones in the studio also carry talkback from the cubicle. The size of the mixing desk can vary – in this illustration there are 10 channels, which can be assigned by the operator. Depending on the complexity of the programme the cubicle may be run by one or more operators. The first operator will control the desk, with a second operator at the back of the cubicle controlling media playout, recordings and outside sources

The volume of the monitoring loudspeaker should be adjusted to a comfortable level and then left alone. It is impossible to make subjective assessments of relative loudness within a programme if the volume of the loudspeaker is constantly being changed. If the loudspeaker has to be turned down, for example for a phone call, it should be done with a single key operation so that the original volume is easily restored afterwards. If monitoring is done on headphones, care should be taken to avoid too high a level, which can damage the hearing.

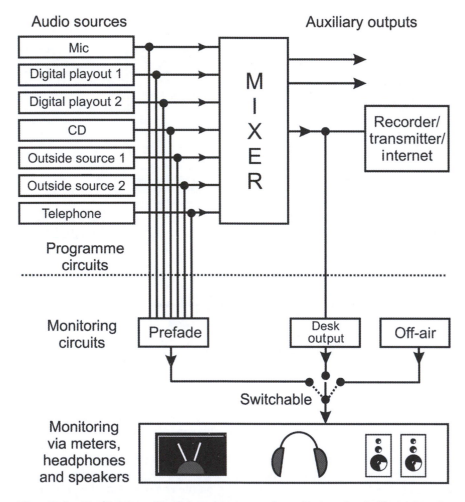

Figure 3.3 Studio mixer. Typical programme and monitoring circuits illustrating the principle of main and auxiliary outputs, prefade listening and measurement of all sources, desk output and off-air monitoring

Loudspeakers should also be kept to reasonable levels if a risk of hearing loss is to be avoided.

In *mixing* sources together – mics, computer playout, etc. – the general rule is to bring the new sound in before taking the old one out. This avoids the loss of atmosphere which occurs when all the faders are closed. A slow mix from one sound source to another is the 'crossfade'.

In assessing the relative *sound levels* of one programme source against another, either in a mix or in juxtaposition, the most important devices

Figure 3.4 A versatile mixer that can be used for permanent installation in a small studio or as a portable mixer for outside broadcasts. There are comprehensive mixing and monitoring controls, including left and right LED columns for stereo level indication. This mixer has eight faders for mono inputs, which can be panned left or right, plus two faders for stereo inputs. Each channel has rotary controls for gain, EQ and auxiliary output which could be routed to reverberation devices. The last two faders on the right are master faders for the left and right outputs

are the operator's own ears. The question of how loud speech should be against music depends on a variety of factors, including the nature of the programme and the probable listening conditions of the audience, as well as the type of music and the voice characteristics of the speech. There will certainly be a maximum level that can be sent to the line feeding the transmitter, and this represents the upper limit against which everything else is judged. Obviously for the orchestral concert, music needs to be louder than speech. However, the reverse is the case where the speech is of predominant importance or where the music is already dynamically compressed, as it is with rock or pop. This 'speech louder than music' characteristic is general for most programming or when the music is designed for background listening. It is particularly important when the listening conditions are likely to be noisy, for example at busy domestic times or in the car.

In a situation of fiercely competing transmitters, maximum signal penetration is obtained by sacrificing dynamic subtlety. The sound level of all sources is kept as high as possible and the transmitter is given a large dose of compression. It is as well for the producer to know about this, otherwise it is possible to spend a good deal of time obtaining a certain kind of effect or perfecting fades, only to have them overruled by an uncomprehending transmission system!

Probably the most important aspect of mixer operation is *self-organisation*. It does require multitasking so it is essential to have a system for handling any physical items: that is, the running order, scripts, CDs, etc. The second requirement is *accurate reading of the computer screens*. The good operator is always one step ahead, knowing what has to be done next, and having done it, setting up the next step.

Digital mixers

The general change from analogue mixers to digital has not been without its problems.

In the digital cubicle, the audio signal does not come into the desk but is controlled by the faders remotely with the advantage that levels can be preset and 'remembered', so that any given setting can be restored by the touch of a 'recall' button – the faders being motorised to reset themselves. A digital desk offers a large amount of processing – EQ, compression, echo, etc., often by means of a touch-screen. Also since the audio signal is remote from the desk – and is digital – it is less prone to noise. Digital desks are often more expensive than analogue types and are ideal for the more complex mixing required for outside broadcast (OB), orchestral or theatrical work. However, if their advantages are not applicable to the smaller station, there are many alternative analogue desks that come with a USB connection to provide the advantages of computer playback or recording.

Figure 3.5 A typical digital cubicle. The mixing is digitally controlled by the fader banks, with the signals being routed through a main apparatus room in another part of the building. 1. Digital mixing faders. 2. Racks panel with power supplies and additional XLR inputs and outputs. 3. CD players. 4. Screens connected to digital playout and production systems. 5. Gram deck. 6. Loudspeakers

There are also forms of virtual mixer where the 'faders' are operated by a touch-screen and all the controls are visual. Given the right software a laptop becomes a mixer – the advantage is the low cost. Mixers like this are mostly used for small Internet stations.

Studio software

Capable of recording, editing, storing and replaying audio material in digital form, the software offers very high-quality sound, and immediate access to any part of the programme held. The computer in this context is used in two different ways:

1   As part of an integrated network system where all the material is stored in a central server and can be accessed or manipulated by any one of a number of terminals. Individual programmes or items can be protected by allowing access only through a password.
2   As a terminal capable of editing, storing or broadcasting material, used essentially as a production computer.

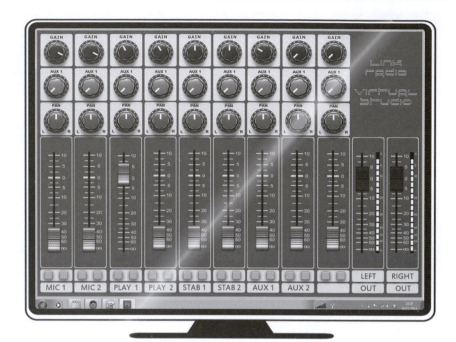

Figure 3.6 A self-contained virtual mixer on a computer screen. The graphics simulate a traditional physical mixer, giving the software a familiar look which makes the device very user-friendly. The faders and rotary controls can be operated by a mouse/keyboard combination or by using the touch-screen monitor. The inputs are connected to the computer hardware either directly, or through an add-on called a 'break-out box' which allows for additional inputs and outputs

The advantage of an integrated system is that after editing an item – a news report, for example – on a terminal in the newsroom, it is then immediately available to the on-air broadcaster in the studio. The studio terminal can have access to an immense range of programme items – it simply depends on the presenter understanding the system well enough to know what is available.

As a programme presenter, producer or journalist operating in a live studio, the basic operations are:

- to transfer into the database an item from another source, e.g. a recorder;
- to retrieve an audio item from the computer database;
- to edit, rename and save an item;
- to open a programme running order;
- to insert material into a running order;
- to move items within a running order;

Figure 3.7 A versatile 'portable' digital mixer. Inputs, outputs, auxiliary sends and other channel settings can be configured using the touch-screen and subsequently saved to internal memory to be recalled as required

- to delete material from a running order;
- to play out an item through the desk on-air.

The range of playout software for the recording, editing, storing, scheduling and playback of audio material is very extensive, including SCISYS dira! – which includes the StarTrack, Highlander and Orion applications – SADiE, RCS Master Control, Radioman, DALET, Myriad P Squared, Soundforge, Soundscape, Simian, Pro-Tools, VoxPro, Zetta, Prisma, D-Cart, Adobe Audition (formerly Cool Edit Pro), Audacity, etc. Some of these systems are capable of virtually running the station, maintaining a database, recording audio, storing, editing, linking together, scheduling and playing back music tracks, voice links, commercials and promos, etc. in any desired order.

It is absolutely key for any operator to know the software in use really well – what it is designed to do, its details and shortcuts. Suffice it to say that no system should be handed over to the operational users until it is thoroughly tested and proved by the installation engineers. Once operational, the software must be totally reliable. Even so, computers can crash and newsrooms in particular will produce bulletin scripts as back-up.

Figure 3.8 Community station 'Future Radio', showing an analogue mixer connected to a digital media playout system. It also has facilities for DJs who prefer to stand while broadcasting their 'mix sets', either from vinyl or CD. 1. Analogue mixing desk. 2. Racks panel with power supply. 3. CD players. 4. Screens connected to digital media playout. 5. Gram decks. 6. Loudspeakers. 7. Adapted CD players providing vari-speed and scratch facilities for DJ 'mix sets'

Experience has shown that it is also wise to have a CD player with pre-recorded music easily and quickly available on one of the mixer channels.

Digital compression

Digital recorders and computers store an audio signal in the form of digital data. A recognised standard was set in 1983 by the compact disc, storing music by using 16-bit samples taken 44,100 times per second, creating a digital file of about 10 MB per minute of audio. Newer formats provide options to record digital audio at higher sample rates, e.g. 48 kHz and 96 kHz. Higher sample rates will involve much more data, and although this does not pose problems for computers with their fast processors and huge storage, large amounts of data are difficult and time-consuming to send on the Internet or by email.

It is possible to reduce the size of an audio file by halving the sampling rate to 22.05 kHz. This limits the frequency response of the audio to around 11 kHz which, while not good enough for music, is quite acceptable for most speech. Here are some sampling rates and the kind of quality they give. The figures are for dual channel audio.

8,000 kHz	telephone quality
11,025	poor AM radio quality
22,050	near FM radio quality
32,000	better than FM
44,100	CD quality
48,000	DAT quality
96,000	DVD – audio (typical)

The 'Moving Pictures Experts Group' – MPEG for short – was established with the aim of developing an international standard for encoding full-motion film, video, and high-quality audio. They came up with the MP3 format, an abbreviation of MPEG-1 audio layer 3. This shrinks down the original sound data by a factor of 12 or more by cleverly discarding the faint and extreme sounds that the ear is unlikely to hear. The perceived audible effect is therefore minimal. Different rates of MP3 compression can be heard on the website.

Digitally compressed audio is ideal for a reporter sending material from a remote location back to the studio. A report recorded, for example, on a smartphone or digital recorder, which itself introduces compression, is transferred to a laptop computer where it is edited and compressed to MP3 format. This file is then sent via a local connection to the Internet or email, and downloaded by the radio station ready for broadcast. One minute of high-quality audio, equivalent now to only 1 MB of data, will take relatively few minutes to upload via the Internet, or even a mobile data connection.

Digital development

Following the installation of digital studios, systems are being introduced that are based around central digital hubs supporting more than one radio station in a corporate group.

The BBC's ViLoR system – Virtualisation of Local Radio – is designed to have 39 stations connected to one of two digital hubs where all the audio files – speech and music – are held. This means that the same uncompressed linear audio signal is used all the way to the transmitter, or online audience, so avoiding transcoding between formats. This results in reduced cost – less digital equipment on-station, better quality and less noise.

Music playout

Music reproduction employs a wide variety of methods. Many stations have all their music either remotely or locally held on a central hard drive computer system with instant access to any track. A few may rely on a library of vinyl records built up over the years. Some use CDs in one form or another, either on individual players or through a jukebox system. It is not surprising, therefore, that studio equipment differs widely from station to station.

In playing a music item, most control desks have a 'prefade', 'pre-hear' or 'audition' facility which enables the operator to listen to the track and

Figure 3.9 A typical digital playout screen, providing options for manual or automatic playout to transmitters or for Internet radio streaming. This screen displays a Beatles special – a previously formatted playlist is already playing out. The title shown top left is 'On the air', and the 'Next' track is preloaded and shown top right. The upcoming tracks are listed in the main panel, and last minute adjustments to the running order can be made as required. The software usefully displays the time remaining on each track, in addition to current time and weather information. The operator must take manual control for live items such as travel and news bulletins. The small fader (top centre) controls the level of the music output

adjust its volume before setting it up to play on air. This provides the opportunity of checking that it is the right piece of music, even though listening only to the beginning may give a false idea of its volume throughout.

The fully digital station, using only a central server as a music source, is wise to keep at least one CD player or conventional turntable for vinyl records – even the occasional 78 rpm disc can be brought in by a listener.

A possible argument against the remote operation of music is that presenters often prefer to feel in control of their programme through physical contact with their discs. The disadvantage here is that the disc is then separated from its inlay or sleeve notes. However, this information can always be made available from a data store and brought up on a computer screen as required.

Recording formats

From spools of wire, reels of magnetic tape and cassettes, through minidiscs and DAT, to SD cards and solid state storage, recording formats have changed immensely. The trend has always been to store longer and longer durations in smaller and smaller spaces. Reel-to-reel recordings were mostly made on ¼-inch width tape, but they may also be found on

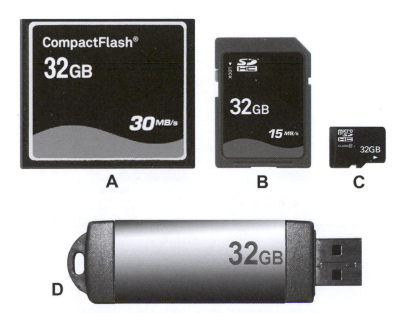

Figure 3.10 Illustration comparing the physical size of four common types of solid state storage: (A) Compact Flash card; (B) SDHC card; (C) Micro SDHC; (D) USB stick

1- and 2-inch sizes for multi-track recordings. Huge stocks of archive and library material exist on tape, which is why many studios keep a reel-to-reel machine in the corner.

Digital audio workstation

This facility is likely to be in a small cubicle area separate from the studio, often adjacent to the newsroom. It could be a 'stand-alone' terminal complete in itself, or be fully integrated with a network computer system. The example here is a typical DAW (Figure 3.11). It comprises an audio mixer (one source of which is a microphone), computer and keyboard, and perhaps a unit for picking up incoming phone lines. In addition, the mixer could be supplied with feeds from the local newsroom, a remote studio or outside broadcast. This arrangement can therefore be used for local voice recording, interviewing someone at a remote source, editing material or correcting levels of an already recorded mix. It is an ideal arrangement for mixing items to create a short self-op news package. In this example we are making a programme opening, mixing together three separate sources (Figure 3.12).

Given the appropriate software, different studio items are recorded on the computer to appear as separate tracks. Here is the script we are working on.

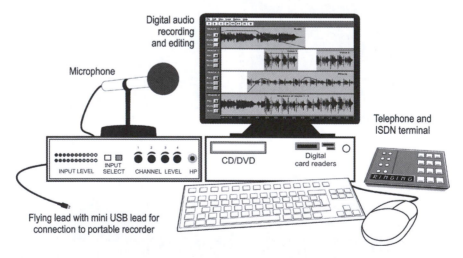

Figure 3.11 Digital audio workstation or DAW capable of downloading from a portable recorder, or recording from outside sources or mic and mixing or editing. Suitable for making finished packages

```
MUSIC:          Locally recorded guitar rhythm
                   (fade after 4 seconds and mix with Fx)
Fx:             Surf waves on seashore
                   (fade up, peak and hold under)
NARRATOR:       The Comoros Islands. On the map they look like tiny
                specks in the Indian Ocean
                   (slowly fade down music)
                But walking along the white sandy beaches you're
                overwhelmed by the mass of lush green vegetation
                climbing up the huge mountain peaks. But they're not
                just mountains – they're volcanoes. Ask any passing
                fisherman.
                   (music out)
Fx:             Briefly up, down, hold under)
FISHERMAN:      Oh yes – the last real eruption we had was in 1977 –
                lava poured down into the sea . . .
```

Here, the music is put on Track 1, the speech on Track 2, and the effects on Track 3. The resulting mix appears on Track 4. At the bottom is a time-scale marked off in seconds from the left.

Each of the tracks can be played independently and its level adjusted by clicking the mouse on to the volume line, or envelope. This is the solid line on each track indicating its relative level at any one time – commonly known as the 'rubber band'. The level is altered by moving it up or down i.e. 'rubber banding'.

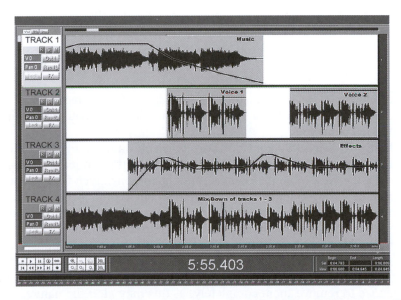

Figure 3.12 Mixing the programme opening. Music starts on Track 1, then effects on Track 3 faded up, both down for Voice 1 on Track 2. Music faded out. Effects up and down for Voice 2 – effects held under. The solid black lines or 'rubber bands' show the sound level of each track and are manipulated by the mouse. The final mix appears on Track 4

The music begins on its own and after 3 seconds the seawash is faded up to mix with it. After 5 seconds this mix is faded down and held under the narrator's voice. The music continues to be faded and is completely out by the time the seawash is briefly peaked before the fisherman – Voice 2 – comes in. Note that the fisherman was recorded at a slightly lower level than the narrator's voice so that Track 2 is brought up to compensate.

The advantage of this facility is that it allows the mixing to be repeated with 100 per cent accuracy while perhaps changing one of the levels or cues to get the desired result. It provides for split-second timing, it leaves all the original recordings intact, it doesn't tie up a whole studio, and takes only one person to achieve the final programme. If that person is the narrator who is also the producer, the production method is not only very efficient but can also create a high level of personal job satisfaction.

Editing principles

The purpose of editing can be summarised as:

1   to rearrange recorded material into a more logical sequence;
2   to remove the uninteresting, repetitive or technically unacceptable;
3   to reduce the running time;
4   for creative effect to produce new juxtapositions of speech, music, sound and silence.

Editing must not be used to alter the sense of what has been said – which would be regarded as unethical – or to place the material within an unintended context.

There are always two considerations when editing, namely the editorial and the technical. In the editorial sense it is important to leave intact, for example, the view of an interviewee, and the reasons given for its support. It would be wrong to include a key statement but to omit an essential qualification through lack of time. On the other hand, facts can often be edited out and included more economically in the introductory cue material. It is often possible to remove some or all of the interviewer's questions, letting the interviewee continue. If the interviewee has a stammer, or pauses for long periods, editing can obviously remove these gaps. However, it would be unwise to remove them completely, as this may alter the nature of the individual voice. It would be positively misleading to edit pauses out of an interview where they indicate thought or hesitation.

The most frequent fault in editing is the removal of the tiny breathing pauses which occur naturally in speech. There is little point in increasing the pace while destroying half the meaning – silence is not necessarily a negative quantity.

Editing practice

Once audio material has been transferred on to a hard disk, e.g. from a Flashmic, it can be manipulated, cut, rearranged or treated in a variety of ways depending on the software used.

This may be Adobe Audition, Audacity, Quick Edit Pro, Sound Forge, WavePad, etc. A general point of technique is almost always to cut on the beginning of a word, as it makes a definite edit point, rather than at the end of the preceding word – although occasionally the most definite point is in the middle of the word. However, while it is tempting to edit visually using the waveform on the screen, it is essential to listen carefully to the sounds, distinguishing between the end of sentence breath and the mid-sentence breath. These are typically of different length and getting them right gives a naturalness to the end result.

An advantage of this form of editing is that it leaves the original recording intact – non-destructive editing. It is therefore possible to do the same edit several times, or to try alternatives, to get it absolutely right. This is a valuable facility for editing both speech and, especially, music. If a skilled music editor, faced with the master recording and a couple of retakes, can assure the producer that after editing, the show will be perfect, then the musicians can go and a lot of money will be saved.

Microphones

The good microphone converts acoustic energy into electrical energy very precisely. It reacts quickly to the sudden onset of sound – its transient response; it reacts equally to all levels of pitch – its frequency response; and it operates appropriately to sounds of different loudness – its sensitivity and dynamic response. It should be sensitive to the quietest sounds, yet not so delicate as to be easily broken or susceptible to vibration through its mounting. It should not generate noise of its own. Add to these factors desirable qualities in terms of size, weight, appearance, good handling, ease of use, reliability and low cost, and microphone design becomes a highly specialised scientific art.

To the producer, the most useful characteristic of a microphone is probably its directional property. It may be sensitive to sounds arriving from all directions – omni-directional – and such a microphone is useful for location recording and interviewing, audience reaction, and talkback purposes. Alternatively a directional mic is essential in most types of

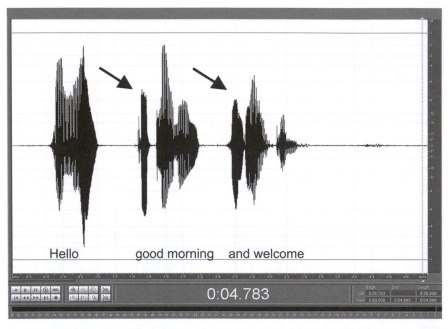

Hello   good morning   and welcome

0:04.783

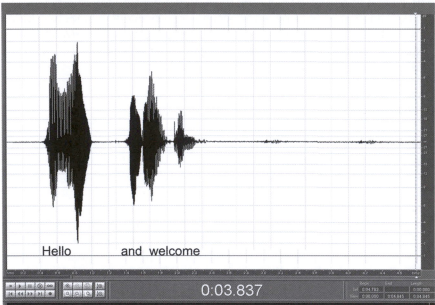

Hello   and welcome

0:03.837

Figure 3.13 Editing. Using the mouse, the arrows are set at the edit points in order to remove the 'good morning' for the afternoon repeat

music balance, quiz shows, and where there is any form of public address system.

The choice of mic for a particular job requires some thought and although it might be possible to rely on the expertise of a technician, it will pay a producer to become familiar with the advantages and limitations of each type available. For example, some mics include on/off switches, or a switch to start a recorder. Some incorporate an optional bass-cut facility. Some require a mains unit or battery pack, or have a directivity pattern that can be changed while in use. Some operate better out of doors than others, some will make a presenter sound good when working close to it, others just distort. Producers will need to decide whether a radio mic is necessary, when clip-on personal mics are appropriate, or if a highly directional 'rifle' mic is required. The more one knows about the right use

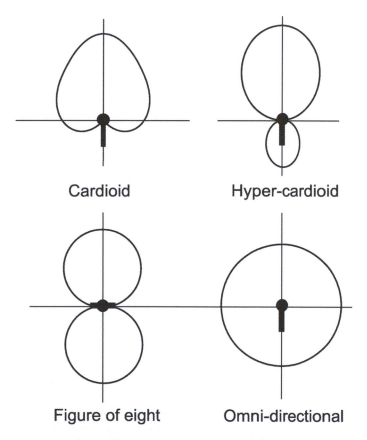

Figure 3.14 *Microphone directivity patterns. A microphone is sensitive to sounds within its area of pick-up. In selecting one for a particular purpose, consideration must be given to how well it will reject unwanted sounds from another direction*

of the right equipment, the more the technicalities of programme making become properly subservient to the editorial decisions.

Stereo

Simply stated, the stereo microphone gives two electrical outputs instead of one. These relate to sounds arriving from its left and its right. This 'positional information' is carried through the entire system via two transmission channels, arriving at the stereo receiver to be heard on left and right loudspeakers. When stereo was introduced in the late 1950s, the BBC introduced a metering formula which has continued to be followed and has also been adopted by some other broadcasters. Here, the left channel is generally referred to as the 'A' (red) output and the right channel is the 'B' (green) output. The meter monitoring the electrical levels may have two needles – red and green (following navigational rules), or there may be two meters – left and right. The signal sent to a mono transmitter – the 'M' signal – is the *combination* of both left and right, i.e. 'A + B', while the stereo information – the 'S' signal – consists of the *differences* between what is happening on the left and on the right, i.e. 'A – B'. Sometimes a second monitoring meter is available to look at the 'M' and 'S' signals. Again, it has two needles conventionally coloured, respectively, white and yellow. Vertical columns of LEDs are an alternative way of indicating the signal level. What does the producer need to know about all this?

First, that if a programme is to be carried by both monaural and stereo transmitters some thought has to be given to the question of compatibility. Material designed for stereo can sound pointless in mono, or even technically bad. For example, speech and music together can be distinguished in stereo purely because of their *positional* difference; in mono the same mix may be unacceptable since a difference in *level* is needed. The producer will optimise the programme for the primary audience, but also ensure it is compatible for both. It is all too easy to fall in love with the stereo sound in the studio and forget the needs of the mono listener.

Second, it is not necessary to use a stereo mic to generate a stereo signal. Two directional mono mics (or a 'co-incident pair') connected to a stereo mixer in such a way as to simulate left and right signals, for example through 'pan-pots', will give excellent results. This technique is useful in an interview or phone-in when the voices can be given some additional left/right separation for the stereo listener.

Third, a pan-pot on a mixer channel can give a mono source both size and position. For example, a mono recording of a sound effect can be placed across part, or all, of the sound picture. Two mono recordings, for example of rain, can give a convincing stereo picture if one is panned to the left and the other to the right, with some overlap in the middle. When

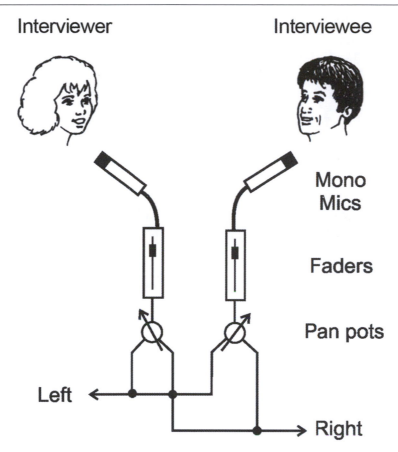

Figure 3.15 Mono mics and faders create a stereo effect when their pan-pots are set to give a left and right placing to each sound source

recording music, special effects can be obtained by the deliberate mislocation of a particular source – hence the piano 10 metres wide at the front of the orchestra, or the trumpeter whose fanfare flies around the sound stage simply by the twirl of the pan-pot!

And fourth, that working in stereo is a challenge to the producer's creativity. To establish distance and effect movement in something as simple as a station promo gives it impact. To play three tracks simultaneously – one left, one centre and one right – as a 'guess the title' competition is intriguing. A spatial round-table discussion that really separates the speakers has a much more live feel than its mono counterpart. The drama – or commercial – in which voices can be made to appear from anywhere, dart around or 'float', literally adds another dimension. For the listener, a stereo station should do more than keep the stereo indicator light on.

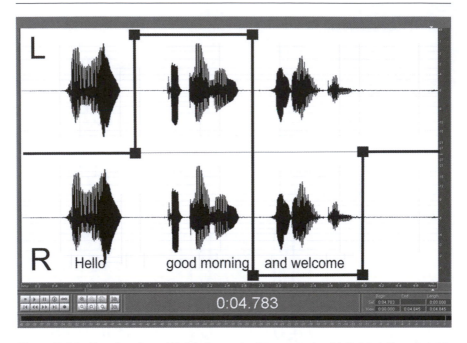

Figure 3.16 Panning for artificial stereo. In this example, 'Hello' is left and right, 'good morning' is taken to the left, and 'welcome' is moved right. The centre is then restored

Equipment faults

Studios are complex places. There is a lot to go wrong – a knob or key on the mixer works loose, the software seems unreliable, the clock is not absolutely accurate, a small indicator light may blow or a headphone lead break. Perhaps it is something physical like a squeaky door or an unsafe mic stand. Whatever the problem, it is likely to affect programmes and must be put right. It is up to every studio user to report any problem and the easiest way of doing this is to have a special routine in the computer system for tracking faults and errors. Alternatively, hang a small notebook permanently in a conspicuous place. It is the responsibility of the person who discovers the fault to make a note of its symptoms and the time it happened. Every morning the system is routinely checked by a maintenance technician. Intermittent faults are particularly troublesome and may require a measure of detective work. It is in everyone's interest to record any technical incident, however slight, in order to maintain high operational standards.

4

Using the Internet and social media

The Internet has been a game changer, introducing new ways to listen to radio and bringing about a revolution in the way the listener and the broadcaster can communicate and share content. In previous decades, the revolution has been led by the radio industry, driving and controlling the innovation, by embracing scientific discovery, technical developments and innovation in production techniques. In contrast, the 'Internet revolution' has been largely driven by the audience. From the 'selfie' to YouTube videos, from tweets to messages on Facebook, the public has driven and satisfied an insatiable desire to 'share'.

The gradual expansion of broadband and mobile capability has compelled the radio industry to question some of its preconceptions about audiences, established production practices and distribution methods. Broadcasters initially found it difficult to accept that no one person owns the Internet and there is no 'gatekeeper', for this methodology is very different from the original model of the radio industry. Now it is possible for anyone to set up a radio station without the usual restrictions of spectrum and wavelength agreements. After a short period of playing 'catch-up', the industry has, more recently, eagerly embraced the Internet, delighted that radio is being brought to new audiences through ever expanding platforms.

An online presence

The Internet has changed the way we listen and watch, the way we read and learn, the way we communicate and the way we interact with each other. But it is important to clarify a common misconception: the Internet and the World Wide Web are not synonymous. The *Internet* is a networking infrastructure that connects together millions of computers across the world. Using this network any computer can communicate with any other computer, as long as they are both connected to the Internet. It is the

Internet that is used for email, instant messaging and the transfer of multimedia files from one location to another via FTP (File Transfer Protocol).

The *World Wide Web* (www or 'the Web') is a way of accessing and sharing information over the Internet, using a language, or protocol, called Hypertext Transfer Protocol (HTTP). The web resides on the Internet, and is responsible for a large amount of data traffic that makes up the Internet. Our personal computers make use of a software application called 'a browser' to access web pages. Popular browsers include Internet Explorer, Firefox and Chrome. The browser is designed to search for information on the web, with search engine results generally being presented in the form of 'results pages' containing information in the form of text, images, audio, video and other types of data. Popular search engines include Google and Bing, and searching for the most topical keywords may bring forth many thousands of results pages.

An online presence is vital for all radio stations who wish to have any chance of competing for audience reach and share in the twenty-first century. It reinforces the presence of a station and reinforces the identity of its presenters. For a smaller station, Internet presence may be limited to a simple website incorporating a few pages with contact information and a basic schedule. The 'Home' page will often have an option to 'Listen Live', perhaps displaying information regarding the current audio item playing. Larger stations will often support a website that extends to many pages, some of which are being frequently updated to reflect changes to news and events. There will be sections for breaking news stories, weather, sport, events and entertainment news. Every day more pages will be added so the site will steadily grow, quickly creating a wealth of archive material. Within a very short time the station may have a website that extends to many hundreds, even thousands of pages.

Internet radio

The global roll-out of faster Internet connections and the growth in the number of mobile- and Internet-capable devices has led to a revolution in the distribution of radio stations. Many thousands of stations are now available via the Internet. Some make use of the Internet as an auxiliary distribution option in addition to transmitters, while others use it as the sole distribution method. The first station to make its programmes available via the Internet was in 1994 in the United States, and in the space of just a few years *live streaming* has become a worldwide phenomenon, with even small stations choosing to increase their audience reach in this way. Every radio station with an Internet feed is now an international broadcaster, with the potential for listeners anywhere in the world where an Internet connection is available.

In its simplest form, live streaming is a straightforward process that requires an audio source, a computer and an Internet connection. The audio is processed and compressed via a soundcard and software application into a continuous data feed, uploaded as a continuous stream to the Internet. This is available to the listener as a continuous playback stream, without the need to download before listening. For a 24-hour streaming service the data use will be considerable, and this must be considered when costs are estimated. The popularity of unlimited data packages, which are 'uncapped', makes this less of a problem today than it was a few years ago.

Any individual or group can, for a modest investment, set up a global Internet-only radio station with a potential for many thousands of listeners. Without listeners it is hardly a radio station, so it will be necessary to find ways to tell your potential listeners about the service you are offering. Fortunately there are many services on the Internet which will help to promote Internet radio stations and assist in distribution and marketing – for a modest fee.

In addition to live streaming, larger stations will often make their programmes available as an 'on-demand' service. This will enable those who have missed a programme when it was broadcast live to catch up at a later time, often for a period of seven to 30 days after the original broadcast. On-demand services may also be available as downloads. Depending on intellectual property rights, these downloads may be protected by a Digital Rights Management System (DRM) which controls how long the audio file can be played before it 'expires'. The time may vary from a few days to a month or more, and this expiry information is encoded in the audio file.

In the case of larger stations, the administrator of the website will need to be aware of how quickly the storage requirements of the site will grow in quite a short space of time. Whereas many thousands of pages of text can be stored in a few megabytes of storage space, audio and video can soon consume many terabytes of storage.

In addition to a personal computer, there are multiple ways for a listener to receive and play Internet radio – mobile phones, laptops, netbooks, iPods, tablets, Internet radios, smart televisions, smart watches and other mobile devices. To aid listening on a computer the originating station will arrange a loading page on their website so that a listener can simply click and play the audio stream. The audio will then be available via the listener's web browser and will play live via the media player on the computer. Web browsers also come installed on some televisions, and inexpensive add-ons are available for those who wish to extend the capability of an older television receiver.

Tablets and smartphones make use of client application software – or 'apps' – as a convenient way to access Internet radio. In addition to offering ways to search through the thousands of Internet radio streams available, apps also provide options to save favourite stations, and also access recently played stations.

Portable radios which are Internet capable often have additional DAB and FM tuners. Some also add solid state memory, usually in the form of a removable SD memory storage card, so that the user can record programmes for later listening. A 'pause audio' option may also be available. This allows a listener, who experiences an unexpected interruption, to press the pause button, and resume listening from the same point when the intervention has passed.

Radio podcasts

A podcast is a digital audio file downloaded through the Internet or streamed online to a computer or mobile device, featuring a pre-recorded radio programme. The word is derived from 'broadcast' and 'pod' from the success of the iPod, as audio podcasts are often listened to on portable media players.

There is often an option for the listener to 'subscribe' to the podcast feed to enable the download process to be automated. When the publisher adds new podcasts, the files are downloaded automatically as soon as they become available. Podcast files are then stored locally on the user's device ready for use offline, even when an Internet connection may not be available – for instance during a daily commute to work. Downloaded podcasts are usually free of DRM and can usually be played indefinitely, although some may omit or shorten music inserts for copyright reasons.

Other names for podcasting exist, such as 'netcast', which is a vendor-neutral term avoiding a reference to the Apple iPod. In the first few years of development, enhanced podcasts which displayed images synchronised with the audio found great popularity. These have largely given way to audio-only podcasts, largely because faster connections have made video podcasts possible where graphics and images are desirable.

A station may provide podcasts from a range of its popular programmes. These may vary from full programmes which have previously been broadcast to compilations of 'best bits'. These podcasts of entertaining extracts are extremely popular with those who might have missed their favourite show or wish to do a quick catch-up by listening to bite-sized chunks. Daily compilations of news also provide the opportunity for the listener to catch up on local, national and international developments they might have missed. Podcasts are also made by celebrities who might not necessarily have regular radio broadcast shows. The British actor and comedian, Ricky Gervais, is one popular exponent of the podcast, and at one time held the record for the most downloaded podcast. In addition, publishers of newspapers and magazines create well-subscribed weekly podcasts.

Whether these podcasts, without a supporting radio station, could technically be called 'radio' is doubtful – what is clear is that, providing you have a microphone and a device that is capable of connecting to the Internet, anyone can become a 'podcaster'.

Internet research

Internet research is now part and parcel of everyday life and, for the radio producer or presenter, searching is a skill that is well worth perfecting. The Internet not only provides instant access to news and current affairs but also can help with more general research. Key words are used to search for just about anything, but that does not always mean finding the answer you need. It is essential for efficient research to learn the specific syntax of your preferred search tool. Carrying out focused and defined searches will allow for at least some of the material to be filtered out. For example, it is useful to check out Google's own help pages.

Figure 4.1 A USB microphone which connects directly to a computer and, using software which is provided with the mic, has a headphone socket built in to the body of the mic. Used for simple podcasts and Internet streaming

While carrying out research online it is necessary to be aware of some of the possible pitfalls. These can range from digging up too much information to relying on non-trusted information and even hoax sites. Remember, anyone can write anything they like and put it on a website! It is best to find several sources so that research findings can be cross-checked.

Concerns about material on a website can be verified with the owner of the site by using a generic tool such as 'whois.com' or 'coolwhois.com'. Also, if you are unsure about the provenance of an image that has been sent by a listener you can use 'Tineye.com'. This site allows you to check where an image has come from and whether other versions of it exist.

For investigative journalists online safety is paramount. It may be necessary to keep your email secure by using an anonymous tool, and also to be able to use programs to hide your IP address. This can be accomplished by using a web-based proxy tool, several of which are available free of charge.

Making the best use of social media

If any justification were needed to support the use of the word 'revolution' in describing the changes the Internet has brought to the broadcast industry, then all that is required is the mention of two sites, *Facebook* and *Twitter*. Although Facebook was created in 2004 for an academic community, it was not until two years later that it was available to all. The same year, 2006, saw the birth of Twitter.

People have swarmed to embrace the idea of sharing content online and, for many, social media is a daily routine that has become a vital part of their lives. As of 2015, statistics from Facebook quote 864 million daily active users, while Twitter is said to have 284 million active users.

The broadcast industry made early attempts to encourage listeners to join a 'community' and contribute to an online conversation by creating 'message boards'. Like-minded listeners could join in a shared community discussing programmes, presenters, news stories and so on. The broadcasters hosted the content themselves – and here was the problem. Because what was written on the message boards could be construed as being the voice of the broadcaster, the radio station had to take full responsibility for the content created by its listeners. As a result 'moderation' became necessary – the role of the moderator was to monitor comments being uploaded to the site and act as editor, removing any comments that could be considered offensive or libellous. The users did not like the intervention of a moderator, viewing the role as a form of censorship; and the broadcasters did not like it much either, as it used valuable editorial resources. Then along came Facebook, Twitter and other popular social media sites which the broadcasters were keen to embrace, as these provided a sense of community and content sharing, but without the burden of hosting and editorial responsibilities.

Different types of social media have different uses, strengths and advantages. Sites such as Tumblr, Flickr, Beebo, Instagram, Myspace, Google+, YouTube and Pinterest, for example, provide ways for listeners to get in touch, stay in touch and share content. The post or tweet is there for all the community to see – listeners and broadcasters alike. Everyone is sharing the same space on an equal footing and the communication process is instant and, unlike a phone call, text or email, a tweet or a post does not require a personal response.

So what are the advantages to a radio station of embracing social media? There are two main aims – to create content and to increase the audience and reach and share, i.e. gain more listeners and encourage existing listeners to listen longer. Each post or message is a marketing opportunity for a radio show, or the station as a whole, remembering that each post may be shared on Facebook or re-tweeted on Twitter. Thinking of the way fruit flies multiply, from a pair to millions in a few days, it is easy to comprehend how a post can 'go viral' overnight. Using social media, it is possible for an audience to a daily show to create content even when the programme is not on the air. The 'Breakfast Show' may only be on air from, say, six o'clock till nine, but with the aid of social media the show lives on 24/7.

The producer of the 'Breakfast Show' arriving at 3.30 in the morning can find the gift of a dozen new items which have been suggested by listeners overnight. Is this a reason for the production team to be lazy and complacent? Certainly not – but in addition to the leads and exclusives that may be found in this way, the producer can also make use of social media to reinforce the brand of the station and launch topics for future shows. By encouraging listeners to be 'followers', a tweet will be brought to their attention at any time, even when the show is not on the air. Careful timing of tweets can greatly increase the effectiveness and volume of responses, and a simple well-timed tweet can remind followers when the programme is back on air or alert them to special subjects.

Listeners are now very capable at collecting good-quality audio (and video) via their mobile phones. This 'citizen journalism' can make a major contribution to breaking stories. Listeners are likely to be first on the scene and it is often far less cumbersome to upload multimedia to a Facebook account than to do the same using a station's own website. Audio material which has been collected by a listener who is 'on the spot' can be used as actuality for a news story. Even the uploading of images in the form of video and photographs is actively encouraged. They bring a new dimension to the radio station which is receptive to social media, as they can be shared with their own followers and added to the station's own website. This rich media content will be especially important during severe weather conditions or at times of other natural disasters. Permissions for use may have to be sought, and questions regarding ownership and intellectual property rights might need to be discussed; and of course it will be necessary to treat all posts with an element of caution and to check each one for authenticity and accuracy before broadcast. Apart from the risk of spurious, malicious or joke posts, what is often overlooked is that businesses and commercial organisations will put much time and effort into creating and crafting posts, writing tweets and filling news feeds with content that marketing departments want you to see – and broadcast!

There are other dangers and pitfalls associated with an over-reliance on social media. It is necessary for presenters to remember that not every listener is riding high on the wave of new technology and social media, and it is easy to disenfranchise this section of the audience. There are still many 'old fashioned' listeners who just want to listen to the radio without contributing content. These listeners do not wish to communicate with other listeners, do not want to share content or tweet their views on anything and everything to the programme presenter. Such listeners may easily tire of hearing the presenter frequently imploring listeners to 'Join us via Twitter', 'Visit our Facebook page', 'Contact us on email at so and so', 'Text us on xxx' or 'Call us on xxx'. Reading out a list like this can take up to 30 seconds, and it can easily sound like desperation. The key is in the balance – it might be enough to mention the website and perhaps one other method of contact which can be varied at each junction. The full contact information is listed on the station website and the tech-savvy will be quite familiar with using it as the first port of call for details.

It is fairly safe to assume that listeners to a programme or station aiming for a younger audience will be more familiar with the Internet and social media. For instance, in the UK a station like BBC Radio 6 Music, with a diet of music aimed at adults under 35 years old, was conceived as a digital station. When its future was in doubt because of financial cutbacks it was largely a campaign on Facebook that saved the station from closure.

Personal Twitter accounts

Presenters are often keen to encourage listeners to follow them via their own personal social media accounts. There is a danger here that any personally posted messages may be considered to be views and comments of the radio station – particularly difficult if the comments are 'off message', perhaps with a political bias or with the possibility of causing offence. There could be conflicts regarding commercial interests. What if a presenter tweets about a movie he/she has just seen and considers the plot was 'rubbish', and the radio station is running a series of commercials for the same movie? Similarly, if a presenter endorses a product in a social media posting, when the station is running a lucrative commercial campaign for a product that is a major competitor. In either case the commercial director is not going to be pleased! When a personal profile is promoted, and even flaunted online, there is also the danger that a presenter could come under attack via social media trolling.

The divergence and expansion of social media has meant that customer service needs to be very good, and seen to be very good. A few years ago people would write a letter of complaint to a radio station and await a reply. Now a complaint is posted on social media and a conversation happens instantly – and in public.

Another risk the station might have to face regarding presenters who promote their own social media accounts, is that when a contract comes to an end, and that presenter moves to another competing radio channel, the management could find that a large section of the presenter's followers will jump ship too, taking a large chunk of the audience with them. For this reason many stations have agreed policies and standards regarding personal emails, tweets and posts. The advice may be to keep personal accounts quite separate, or to have a consistent pattern to the Twitter names, via a station account, such as @firstnameradiostation. Although this arrangement may not be universally popular with presenters, they will be less likely to deviate from agreed policies when their name is seen to be inextricably linked with the station.

Instant messaging via social media can be a real guide to the way a programme or subject is engaging the audience. A hot topic will bring forth many instant tweets and comments, providing a way for the station to gauge and understand its actual influence on the audience. When reading out messages, tweets, emails and texts it is important for producers and presenters to be aware of their responsibility regarding the privacy of personal information. The detail the listener provides might be of a highly confidential nature, and it is easy for too much to be disclosed on air. It is important to remember that people get support from communicating with like-minded people, but also to be fully aware of the station's responsibilities regarding the prevailing data protection laws.

By its very nature this instant communication via social media necessitates that interaction is broadcast live; judgements have to be made instantly – and this is where experience and good staff training are vital.

Blogs

The dilemmas and conflicts of interest that may be associated with personal posts are an even greater risk in *blogs*. Formed from two words 'web' and 'log', a blog is a personal discussion or expression of views published on the World Wide Web. The posts are usually published via a web publishing tool or 'blogger'. The advantage of a blog is that the writer is not restricted to the 140-character limit of a Twitter message. Blogs often provide quite lengthy commentaries and debate on particular subjects, usually offering interactivity so that users are able to leave comments and even message each other. Typically a blog might include text, images and links to other blogs. Editors, producers and presenters may all engage in either a daily or weekly blog. The same editorial standards that apply to a broadcast should also apply to all posts, whether via a blog or other social media tools. Crossing the line of acceptable taste and ignoring legal liabilities can lead to serious consequences. Exact legislation varies from country to country, but courts are now bringing actions against posts that are considered to be defamatory, and charges have led to successful prosecutions. The safest approach is to consider every tweet and post as a broadcast, and it is sensible advice not to write anything on a social media account that you would not actually say on air.

A case study

Here is an illustration of social media in action during a developing story.

David Clayton is Editor of BBC Radio Norfolk, a local station covering the east coast of England. The station broadcasts across the large rural county of Norfolk which, by area, is the fifth largest in England. With nearly 100 miles of coastline exposed to the North Sea the risk of dramatic coastal erosion and flooding is high. In the region there have been major flooding incidents over the past century, most notably in 'the 1953 flood', when over 300 people died. More recently, in December 2013, Britain was once again battered by a tidal surge and Norfolk was particularly at risk with 20-foot waves smashing against a string of east coast towns. No one could predict how serious the outcome would be, but some months later the radio coverage of the night won the station a gold award in the journalism category at the annual Radio Academy ceremony in London. The Judges said the radio station's coverage was:

> Vivid, dramatic but always responsible reporting that placed this talented news team right at the heart of a developing story. This is local news-gathering at its finest – when radio becomes the essential 'go to place' for information. Story telling that was always in touch with its community, often moving, and always compelling.

Clayton admits that social media made a big contribution to the Radio Norfolk coverage. The station and its team of presenters had already

Figure 4.2   A mobile phone showing how a local radio station was able to use social media to good advantage on the night of a storm emergency

established a good rapport with listeners via social media and in addition to the team of regular reporters this was a great opportunity for 'citizen journalism'. Over the previous year, the station's listeners had been encouraged to participate and contribute via emails, Facebook posts and Twitter tweets, in addition to the traditional methods of phone and email (letters are something of a rarity these days). The station had already established its own Facebook and Twitter accounts, and on the night the feed was almost unstoppable, with people contacting the station with updates on the state of the sea defences and contributing dramatic pictures. Houses were going over the cliff and seaside cafés were being washed into the sea. On the night of the floods it was shown that Twitter was much more effective with quick two-way communications, whereas Facebook was used for more considered posts. A member of the team was responsible for managing social media, disseminating images, tweets and texts from

Figure 4.3   The aftermath of the storm (Courtesy of the BBC)

reporters and listeners and uploading them to the station's accounts. When re-posting images and messages contributed by listeners Clayton stressed it was important to attribute each one and put it in context. He felt listeners were now accustomed to differentiating between contributions from a member of the public and official announcements from a reporter or the emergency services.

The website has an audio interview with David Clayton, with illustrative clips from BBC Radio Norfolk's storm coverage.

5

Ethics

If you put 'media ethics' into an Internet search engine such as Google, you are likely to be confronted by nearly 4 million results. Such is the concern, interest and serious intent of the overwhelming number of the world's journalists and broadcasters that there is a right and a wrong way of going about these things, that the practicality of 'truth' is paramount, and that a properly informed public is a highly desirable objective. This leads to a widespread continuing debate to codify best practice in achieving these ends. These sets of rules, standards and Codes of Practice regulate how journalists do their work and are far from theoretical ideas. They are the practical application of a series of ethical principles – generally defined as 'a system of morals or rules of behaviour, recognised and approved professional standards of conduct'. But who recognises or approves such conduct – the morality of what is done in practice – and the values which inform those decisions? As we shall see, such rules are decided by journalists and broadcasters themselves, yet it is often the lack of proper reflection by media practitioners into what they do that leads to much unethical behaviour – journalists making too many decisions, too quickly and without the appropriate time to think things through. However, professional journalism is essentially connected to the real – and far from ideal – world.

Governments, institutions, commercial interests, and individuals have things to hide, things they would rather not make public because personal reputations, share prices, votes, individual wealth or power would suffer. The desire for total truth is not all that universal, hence the need for news people in particular to have a set of rules to govern what they do, and for everyone else to know what these are. This is why in 2008 the International Federation of Journalists launched a global Ethical Journalism Initiative aimed at strengthening awareness of the key issues within professional bodies.

Declarations of intent

On 10 December 1948, the General Assembly of the United Nations pro-
claimed the Universal Declaration of Human Rights, to which every UN
member has, in theory at least, agreed. Article 19 states:

> Everyone has the right to freedom of opinion and expression; this right includes
> freedom to hold opinions without interference and to seek, receive and impart
> information and ideas through any media regardless of frontiers.

Here is a basic human belief in the value of the freedom of expression
and in the giving and receiving of ideas. Such freedom is subsequently
constrained by national laws that, for example, make it illegal for such
expression to incite racial hatred or undermine the State.

One of the UN's specialist bodies, UNESCO (the United Nations'
Educational Scientific and Cultural Organization), produced the
International Principles of Professional Ethics in Journalism. Published on
20 November 1983 and representing 400,000 working journalists worldwide, the first of these states:

> People and individuals have the right to acquire an objective picture of reality
> by means of accurate and comprehensive information as well as to express
> themselves freely through the various media of culture and communication.

The document goes on to set out:

- the journalist's dedication to objective truth and reality;
- the journalist's responsibility to society;
- matters of professional integrity;
- the need for public access to, and participation in, the media;
- respect for privacy and human dignity;
- respect for the public interest and democratic institutions;
- concern for public morals, and cultural diversity, etc.

Of the more important websites to appear under Media Ethics, are those
of the Press Councils (see p. 386). Maintained by the Independent Press
Councils, it is a forum listing over 300 Codes of Practice established by
broadcasting and press interests from Australia to Zambia. While differing in length from half a page to 50 pages, their underlying aspirations are
remarkably similar and form the basis of how we should treat our material, the people involved, the stakeholders, our colleagues and the public.
They set out the essential Code of Journalistic Principles as follows:

- to be professionally competent;
- to be independent (from political, economic, intellectual forces);
- to have a wide and deep definition of news (not just the obvious, the interesting, the superficial);
- to give a full, accurate, fair, understandable report of the news;
- to serve all groups (rich/poor, young/old, conservative/liberal, etc.);
- to defend and promote human rights and democracy;
- to work towards an improvement of society;
- to do nothing that might decrease the public's trust in media.

Another code for journalists was set as The Munich Charter of 1971 (available online) and specialist Codes of Practice exist for advertising (p. 229), and for general programming (pp. 351, 386).

The job of the media is to mediate – between those involved in making news and the public who consume it – and to provide that conduit with as much clarity, transparency and honesty as possible. This means separating fact from opinion, issuing corrections when mistakes are made, and being as open as possible regarding the *process* of journalism. It is very much in the public interest that producers, reporters, interviewers – broadcasters in general – should constantly challenge their own standards and ethical assumptions. It helps, therefore, to take time out to establish a culture of values and clarify local practice. There is much to be learned from the sharing of ideas and concerns.

Objectivity, impartiality, and fairness

Some declare it to be impossible, that we are inevitably creatures of our own age and environment, seeing the world through the lens of a particular time and culture. In this sense only God is truly objective. But broadcasters must be concerned with truth – even when quite different perceptions and beliefs are held to be true. Objectivity here means recounting these truths accurately and within their own context, even when they conflict with our own personal values. The difficulty is that professional news judgements must, in the end, rely on personal decisions. This is why the question of individual motivation is so important: *why* do I wish to cover this story in this way? To tell the truth, or to make a point?

What the editor, producer or reporter must not do is to introduce a partiality as a result of conscious but undisclosed personal convictions and motivation, even for the best of reasons. Making decisions based on one's own political, religious or commercial views is to put oneself before the listener. The impartiality of chairmanship is an ideal to which the producer must adhere, for any bias will seriously damage one's credibility for honest reporting. Yet in a world when one man's 'terrorist' is another man's 'freedom fighter', the very language we use in imparting the facts is itself a matter of dispute and allegiance. In this example one needs to use either more neutral words such as 'guerrilla', 'gunman' and 'insurgent', or to concentrate purely on the factual outcome of an act – 'the bombing killed six, caused many serious casualties and created an atmosphere of shock and fear'.

Here are further examples where the words used betray a partisanship or bias in the case being presented. *The Guardian* newspaper analysed some of the terms used in the British press during the Gulf War of 1991. The findings still give us cause for thought.

We – take out	They – destroy
neutralize	kill
dig in	cower in their foxholes
We – launch first strikes	They – sneak missile attacks
Our missiles cause –	Their missiles cause –
collateral damage	civilian casualties
We have –	They have –
Army, Navy, Air force	A War machine
reporting guidelines	censorship

Objectivity becomes more difficult and more crucial as society becomes less ordered in its deliberations and more divided in its values. This is something that many countries have witnessed in recent years. The crumbling of an established code of behaviour alters the precepts for making decisions.

A useful definition of objectivity, quoted in *Journalism: Principles and Practice* by Tony Harcup (Sage, 2004), comes from J. H. Boyer:

- balance and even-handedness in presenting different sides of an issue;
- accuracy and realism in reporting;
- presenting all main relevant points;
- separating fact from opinion, but treating opinion as relevant;
- minimising the influences of the writer's own attitude, opinion or involvement;
- avoiding slant, rancour or devious purposes.

This means that nothing is added to truth, and nothing significant taken away. An issue is presented impartially with all its constituent parts and differences, unaffected by the presenter's opinion. Nevertheless, broadcasters are thinking human beings and are not meant to be without views – simply not to let them affect their work.

Democracy cannot be exercised within a society unless its individual members are given an unobstructed choice on which to make their own moral, political and social decisions. That choice does not exist unless the alternatives are presented in an atmosphere of free discussion. This in turn cannot exist without freedom – under the law – of the press and broadcasting. The key to objectivity lies in the avoidance of secret motivation and the broadcaster's willingness to be part of the total freedom of discussion – to know that even editorial judgement, the very basis of programme making, is open to challenge. Keep the listener informed about what you are doing and why you are doing it – that is the public interest.

Limitation of harm

During the normal course of an assignment a reporter might go about gathering facts and details, conducting interviews, doing research, making background checks, taking photos, videotaping, recording sound, and so on. Harm limitation deals with the question of whether everything learned should be reported and, if so, how. This principle of limitation means that some weight needs to be given to the negative consequences of full disclosure, creating a practical and ethical dilemma. The Society of Professional Journalists' code of ethics offers the following advice, which is representative of the practical ideals of most professional journalists. Quoting directly:

Journalists should:

- have compassion for those who may be affected adversely by news coverage. Use special sensitivity when dealing with children and inexperienced sources or subjects;
- be sensitive when seeking or using interviews or photographs of those affected by tragedy or grief;
- recognize that gathering and reporting information may cause harm or discomfort. Pursuit of the news is not a license for arrogance;
- recognize that private people have a greater right to control information about themselves than do public officials and others who seek power, influence or attention. Only an overriding public need can justify intrusion into anyone's privacy;
- show good taste. Avoid pandering to lurid curiosity;
- be cautious about identifying juvenile suspects or victims of sex crimes;
- be judicious about naming criminal suspects before the formal filing of charges;
- balance a criminal suspect's fair trial rights with the public's right to be informed.

In other words, there is a need to consider what harm may be caused by publishing news, as well as withholding it.

Watchdog

As an independent monitor of power, the media serves a useful function when it can watch over the few in society on behalf of the many, to guard against the abuse of power, the unfair treatment of individuals and, in the extreme, to counter tyranny. This is where an independent media goes hand in hand with a democratic society. This monitoring role can certainly go wrong, for the press and broadcasting itself wields power that must also be supervised by its own publicly accountable control bodies. In Britain, this means the Independent Press Standards Organisation and, for broadcasting, the National Regulator. Properly overseen and following ethical guidelines – which are also in the public domain – the media should be able to supply information, comment and amplification on matters that enable society to understand and perhaps to work through its disagreements.

Whistleblowers

Another dilemma arises when those matters that a government's legitimate intelligence services wish to keep secret in the interests of national security, are suddenly exposed by an internal whistleblower. Or data which is the property of a national institution or private company, is made public by someone who has hacked into their computer. Should the media collude with the perpetrator in these cases and publicise the information, even if it is placed on the Internet in the public domain? The answer has to be 'no, not necessarily', in the same way that television does not show something, like a hostage murder atrocity, simply because it has been placed on a public website. The broadcaster must always ask the question, 'Is this in the best interests – overall short-term and long-term – of my listener or viewer?' Sometimes there are no easy answers.

Bad practice

Having determined those ethical principles that support truth, accuracy, objectivity, impartiality, fairness and balance – to which we shall return in the next chapter – we should look at some of those activities which lie outside these parameters and may therefore be said to be wrong.

- payment, actual or in kind, offered for the promotion or suppression of particular issues or interests. Commercial and subscriber stations in particular must have clear rules as to what is the legitimate buying of advertising time and what constitutes the greyer areas of 'favourable editorial'. Payments to or by contributors may constitute bribes instead of proper fees or expenses;

- reporting may become biased because of a strongly held personal belief. This is the more difficult when that belief can be justified on religious or moral grounds, e.g. the advocacy of war, the practice of human cloning or abortion, the treatment of 'other' religions or races, the coverage of crime and anti-social behaviour;

- the deliberate misrepresentation of a person's views in the way recorded interviews are edited, or by the context in which their contribution is used;

- an understandable desire for stations to want to maximise ratings or profits, which in itself may affect staff in reporting objectively. Priorities have to be weighed when there is an apparent conflict of interests;

- the prevention of any segment of public opinion from being included in the output. What is the basis for deciding whether a view is insignificant, too extreme or from too small a minority?;

- to be deceitful with the listening public about how material was obtained, especially when using secretive or surreptitious methods (see p. 90);

- getting the news out of proportion – magnifying the less important but possibly colourful story at the expense of the significant, but more difficult.

Sometimes it is easy to veer towards bad practice without any fault being intended. It is often expensive to cover an issue as fully as one would like; there might not be the resources to take on a government department, or a powerful corporation, or even to provide the necessary staff training to do the job well. The sin of omission, of not doing something because of an outside pressure, is often the one that doesn't show – except to the person who made the decision knowing it to be ethically wrong, or made it because of undisclosed personal motives.

It is natural to have loyalties when your own country is at war and one of the greatest moral dilemmas is how to report fairly in these circumstances. Is it even possible? It took courage for the late Sir Richard Francis, then Managing Director of BBC Radio, arguing for even-handedness back in the Falklands War, to declare that 'the grief of the widow in Buenos Aires is no less than the grief of the widow in Portsmouth'. This was not a popular sentiment with the government of the day.

It is easy to be cynical about the moral standpoint of others – 'well, they would say that, wouldn't they?' – and this is what tends to generate words of a particular colour. But cynicism itself eats away at all belief and starts to destroy the ability to perceive objective balance. It is right for news interviewers to be sceptical, otherwise they could not ask the probing questions needed to challenge assumptions and arrive at a deeper truth. But there is no need always to believe the worst of people and reject all opinion but one's own – that's cynicism.

> *Cynicism*: sarcastically to doubt or despise human sincerity and merit; to be incredulous of human goodness or integrity.

> *Scepticism*: to be inclined to suspend judgement, given to questioning the truth of stated facts and the soundness of inferences.

Not only should journalists be sceptical about what they see and hear, but they should be sceptical about their ability to know what it means – and this requires another essential quality, humility.

The status of the media

Given some pretty high-sounding aspirations by editors in general, they quickly founder if the climate in which they operate is repressive or unduly censorious. The relationship between the media and government and other major power-holders is therefore crucial. What if they do not share the ethical concerns of the journalist? The DFID booklet *Media and Good Governance* says:

> Free, independent and plural media (radio, TV, newspapers, internet etc.) provide a critical check on state abuse of power or corruption. Further, they enable informed and inclusive public debate on issues of concern to people living in poverty and give greater public recognition to the perspectives of marginalised citizens. Engaged citizens need information that allows them to exercise democratic choices.

The key questions in the relationship of media and government are, therefore:

- How pluralistic is media ownership?
- How independent are the media of government and political parties?
- What pressures might be brought by major multinational corporations?
- How representative are the media of different opinions, and how accessible are they to different sections of society?
- How effective are the media in investigating government and other powerful bodies?
- To what extent do the media report public and political events within acceptable bounds of accuracy and balance?

- How free are journalists from restrictive laws, harassment and intimidation?
- How free are private citizens from intrusion and harassment by the media?

One is wary of situations where the media are owned exclusively by government, or by a narrow elite so as to exclude other interests. We should be watchful when opposition or minority parties are denied a public hearing – especially in times of national stress.

The best and most crucial support for the ethical position adopted by a broadcaster comes from the public, who recognise that it is their best defence against deceit and manipulation. At no time is this more important than when there are strong forces that would seek to undermine it. Stations are closed down and journalists imprisoned and sometimes killed for attempting to tell the truth. The UN Charter notwithstanding, 'the freedom of the press' is by no means universal.

6

Writing for the ear

Writing words to be heard by the ear is quite different from words to be read by the eye. The layout of sentences, their order and construction has to be thought through in order to be totally clear and unambiguous at their first hearing. The listener does not automatically have the possibility of re-hearing something. It must make sense first time, and this places a special responsibility on the radio writer. So whether we are writing a 15-minute talk, a one-minute voice piece, or a cue to a recorded interview, the basic 'rules' of radio writing – and the pitfalls – need to be simply stated.

Who are you talking to?

The listener comes first. Decide who it is you are talking to. Is this for a specialist audience – such as children, doctors or farmers – or is it for the general, unspecified listener? In passing, it could be argued that there is no such thing as the 'general listener', since for consumer research purposes, we are all categorised by one or more of a number of criteria, e.g. socio-economic group, gender, age, religion, ethnicity, demographic location, social habits and so on. Even so, the style for the 'Morning Breakfast Show' will be tighter and punchier than the more relaxed 'Afternoon Show'. The language will be different but will nevertheless be appropriate when you know and visualise who you are writing for – the one person, the individual who is listening to you. Are they busily dashing about? Getting a meal, or quietly lying in bed? Forget the mass audience, as if talking in a hall full of people. Radio is not a PA system – 'some of you may have seen . . . '. Be personal, write directly for the person you want to talk to, seeing them as you write. It's then more likely to come out right – 'you may have seen . . . '.

We generally speak in quite short sentences – they are not long convoluted mouthfuls, not if we want to get through. So keep the written sentence short – 15 or 16 words. Try it.

Finally, avoid talking *about* your listener – not 'listeners who want to contact us should . . . ', but *to* the listener, 'if *you'd* like to contact us . . . '. Only when questions of what we want to say, and to whom, are answered can we properly start on the script.

What do you want to say?

Having decided who you are talking to, what thought do you want to leave with him or her? It may be that the script is simply to entertain – a lighthearted afternoon show – but it still needs thinking about unless it is to be just waffle. So, what do you want to say? What jokes, stories, anecdotes to tell? What information to include? Will you comment on news of the day, or local gossip? Start by listing the points to be made and put them in a logical order. Visualise the effect that each will have on the listener – bringing about a smile, or causing them to think about an issue, or to wonder at a particular fact. Build on the effect you create, leading from one point to the next so that you have connected strings of thought – so much more satisfying than everything being isolated and on its own.

It is important to have a strong opening – get the listener's attention at the start. In scripting a piece this is the part that often gets written last. It is difficult to set down a really good first couple of sentences on a blank page – much easier to come back to it when you know the shape of the whole and you can have an interesting, teasing or dramatic opener. Use an interesting metaphor, paint a picture, get me, the listener, to do some work by anticipating what you are on about – the issue, problem, or story. Grab my attention in the first sentence, tell me something in the second.

The storage of talk

You are talking, and writing a script is essentially the storing of that talk on paper or in a computer. So we need to write conversational language – quite different from the starchy prose we mostly learned at school. We want to use good colloquial words such as 'wasn't', 'didn't', and 'shouldn't've'. The easiest way of doing this, having listed out your points, is to say *out loud* your way of expressing them and writing down what you hear. In other words you record in script form an already spoken stream of thought. The sentences are generally shorter and simpler. You might need to check it and alter it later, but essentially it is the stuff of talk – audible communication – so much better than the more formal and often stilted language of the written essay or letter. A newsroom where scripts are being written should be a noisy place!

Even so, the best script is a fairly crude and imperfect form of storage. It gives no indication of emphasis and inflection, which can be very important in communicating sense. Neither does it say anything about speed or pause. Some people underline words to help with where the stress should go, but the danger here is that when reading it you concentrate so much on the underlining that the end result sounds artificial. Much better to communicate meaning by first understanding it.

For example, how should the following be said?

'You mean I have to be there at ten tomorrow?'

Put the emphasis on the 'I', the 'there', or the 'ten', and the implied meaning is altogether different. In other words, a script will tell you *what* to say, but not *how* to say it.

Delivering a script properly is presentation – the art of retrieving talk out of storage.

Words

These are the building blocks of our meaning and need to be used with a little care if that meaning is to be recreated in the mind of the listener. For example, a news story about a collapsed building included the sentence: 'Questions are being asked about whether there were flaws in the building.' It looks alright on paper, but on the air might not 'flaws' be confused with 'floors'?

Again, 'The government warned of attacks on foreigners.' What was that about 'a tax on foreigners'? Special care has to be taken over words that sound the same but have different meanings – like:

oral/aural	story/storey	sole/soul	two/too
draft/draught	hoard/horde	council/counsel	and so on.

In any language there are many such homophones and we have to beware of them. We cannot always rely on the context to make their meaning clear, and our purpose is to avoid misunderstanding and ambiguity, especially for the preoccupied, half-hearing listener. We look at this further in Chapter 7.

A final point on words is to prefer the simple to the complex. They generally communicate more directly, being easier to understand. So we would use 'begin' or 'start' rather than 'commence', 'find out' rather than 'ascertain', 'ask' rather than 'enquire', 'buy' instead of 'purchase', 'try' instead of 'endeavour'.

Words with Latin roots might sound 'educated', but our aim is to communicate in the most direct way that avoids mishearing. Our job is not to impress but to express.

Structure and signposting

After your strong opening, it is still true in a talk to say what you are going to say, say it, and then tell me what you said. Otherwise, be logical in the order you put things, and be interesting – that means being relevant, funny, useful or unusual.

With the printed page, a book or newspaper, it is possible to look back to clear up a point or to check how the writer got to the present assumption. With radio that's not possible, which is why the structure of our talk is important. So not only are our words simple, and our sentences shorter, but things must be in the right order – cause comes before effect. For example:

> 'Jim, who is about to leave school, where he's been for five years, which included a time as head boy, is looking for a job.'

Avoiding those relative clauses, this is better as:

> 'Jim has been at school for five years. This included a time as head boy. He's now about to leave and is looking for a job.'

Other 'joining' words like 'and' and 'well' can be added to ease the flow. Long convoluted sentences are difficult to follow and it is to no one's benefit to use polysyllabic words in complex phrases. Keep it simple and straightforward.

Signposting is the very useful technique in any oral communication of saying where you are in a talk, and where you are going next. For example, 'So much for the selection of staff, let's now look at their training.' Phrases like 'let me explain' are there to clarify structure, to help the listener follow your train of thought – 'and now, the weather'. Signposts, without being overdone, make listening easier.

As a matter of style we avoid using the same word twice in the same or adjacent sentences.

> 'The hurricane swept across the Florida coastline at midday bringing 120 mile an hour winds. By this evening the hurricane will be well inland.'

It sounds better if the second 'hurricane' is replaced by 'storm' or 'storm centre'.

The ending of the talk is what you will leave with the listener, so typically: repeat a main point, finish with a story that illustrates your theme, or look forward to the future. Give the listener something to hold on to – make it memorable.

Pictures and stories

Remember the visual nature of radio and illustrate what you are saying with pictorial colour. Appeal to the sense of smell and touch too if you can. Blue seas and white surf, red buses and silver cars, black umbrellas and grey stone streets, yellow bananas and the dimpled skin of shiny oranges. Recall the smell of new-mown grass or a Chinese meal, the reddish brown earth after rain and the blue haze of a warm camp fire. These all help the listener to be there, to share the experience – we shall meet this again in Chapter 19, on commentary.

Instead of strings of facts or concepts, be creative and turn them into evocative anecdotes and metaphor. Who said what to whom, how they responded, why they disagreed, and how it turned out in the end. Real life or made up, stories and pictures are memorable. This is why we turn a 'new 100,000 ton cruise liner' into a 'ship as tall as Big Ben with enough power to light a city the size of Southampton', or describe the solar panels for the latest space station as 'the size of two football pitches'. Such images, like any relevant visual aid, stay in the mind and your listener will remember better what you said.

Double meanings

In the same way that word sounds can have more than one meaning, so can phrases and sentences. Here are some real examples (that was a signpost).

> 'Our Reporter spoke to Mr X on the golf course, as he played a round with his wife.'

You have to avoid the double entendre, unless of course you intend to be funny.

> 'The Union said the Report was wrong.'
> 'The Union, said the Report, was wrong.'

The meaning depends on whether 'said the Report' is in parentheses. Two little commas alter the whole sense as to who was wrong.

'At first, supplies of the new car would be restricted to the home market.'

Does this mean that the car is being supplied only to the home market, or is it that home supply is restricted so it's all going for export? The meaning as written is surely unclear.

'The man was found lying on the pavement by his wife.'

What picture does that give you? Of a man lying alongside his wife? No, the story was of a shooting incident where the victim was found by his wife, lying on the pavement – and that's how the sentence should have been structured.

'The Prime Minister would say if he is to send more troops in a few days.'

To avoid confusion over what the 'few days' refers to, this is better as:

'The Prime Minister would say in a few days if he is to send more troops.'

And finally this dreadful example of double meaning in a story meant to calm the fear that Hormone Replacement Therapy treatment might increase the likelihood of cancer.

'The Report published today said that with the use of HRT there was no greater risk of contracting cancer.'

Do we mean that the risk was no greater, or that the risk could not be greater?

There are two ways of minimising errors of this sort. First, get the punctuation right. Punctuation shows you how to read it. Capital letters for the beginning of sentences and proper names, commas or dashes in the right places to indicate pauses or the subordinate clause. Second, speak *out loud* as you write, preferably to someone else. Many times a double meaning is avoided because, 'As soon as I said it, I knew it was wrong.'

Finally there are words that just won't cross the culture gap. My favourite is the use of the word 'momentarily'. In an American aircraft approaching touchdown you may get the announcement, 'We will be landing momentarily.' This can be disconcerting to a Brit for while to an American it means '*in* a moment', to the English it means '*for* a moment'!

Scriptwriting cross-culturally needs special care and advice.

The script

We speak at about 180 words a minute – three words a second is a good guide for a bulletin or scripted talk. A single typed line is 3–4 seconds, making a double-spaced page of A4 – 27 lines or 270 words – about one and a half minutes. Thus, a 30-second voice report needs about 90 words, and a three-minute piece for the breakfast programme about 540. The computer counting of words is extremely useful.

A script on the page or on the screen should, above all, be clear and easy to read. Double- or triple-spaced, with wide margins for any notes or alterations. Difficult words, foreign or unusual names may be given their phonetic pronunciation in brackets. Numbers can also be written in words if this helps, e.g. '10,700 (ten thousand seven hundred) tons of food aid were supplied'. Where possible such statistics are simplified, so this becomes, 'almost 11,000 tons'.

Clear paragraphs should be used to separate distinct thoughts or items. We use one side of the paper, and good quality paper at that, because it's quieter to handle.

Computers don't care about the ends of pages – they just go on printing on the next. However, best practice doesn't break a sentence at the bottom of a page, hoping that the reader will follow on. Each page ends with a full stop.

Finally, why have a written script at all? Whether on paper or on screen, its purpose is to tell us what to say, in what order, so that nothing gets left out and it runs to time. It is also a safety net, reducing the stress of having to remember. Essential for news, but even informal spontaneous programmes will have scripted notes – even the best ad-libbers are better with an aide-mémoire for names, points to make, stories to tell. More than this, all producers must learn that preparing a script provides the opportunity for thinking more deeply and creatively, adding substance, expressing ourselves more accurately, and developing the well-crafted memorable phrase.

7

News – policy and practice

The best short definition of news is 'that which is new, interesting and true'. 'New' in that it is an account of events that the listener has not heard before – or an update of a story previously broadcast. 'Interesting' in the sense of the material being relevant, or directly affecting the audience in some way. 'True', because the story as told is factually correct.

It is a useful definition not only because it is a reminder of three crucial aspects of a credible news service but because it leads to a consideration of its own omissions. If all news is to be really 'new' a story will be broadcast only once. Yet there is an obvious obligation to ensure that it is received by the widest possible audience. At what stage then can the news producer update a story, assuming that the listener already has the basic information? What do we mean by 'interesting' when we speak not of an individual but of a large, diversified group with a whole range of interests? Do we simply mean 'important'? In any case, how does the broadcaster balance the short-term interest with the long? And as for the *whole* truth – there simply is not time. So how should we decide, out of all the important and interesting events which confront us, what to leave out? And concerning what is included, how much of the context should be given in order to give an event its proper perspective? And to what extent is it possible to do this without indicating a particular point of view? And if the broadcaster is to remain impartial, do we mean under *all* conditions? These are some of the questions involved in the editorial judgement of news. Let's first admit that 'interesting' is by no means the same as 'important'.

'Interesting'

What is it that actually interests our fellow men and women? We noted that it was anything being relevant or directly affecting people – relating to them – or which might do.

Figure 7.1   Part of the main BBC newsroom at Broadcasting House, London. It is the largest newsroom in the world, supported by some 3,000 journalists. It provides 24-hour news for all the BBC's national and regional television and radio services, 41 local radio stations, World Service radio and television – and 27 language services, together with hundreds of thousands of Internet web pages. It is linked to a network of BBC correspondents around the world

Broad areas of interest are:

- *conflict:* wars, strikes, political argument, business take-overs, success and failure;
- *disaster:* tragedy, plane/train crashes, missing persons, epidemic illness, refugees, cyclone;
- *development:* building plans, new discovery, exploration, job creation, medical advance;
- *crime:* murder, corruption, theft, violence, fraud, drugs, trafficking, child abuse;
- *money:* lottery wins, the budget, stock market gains or losses, buried treasure, price increases;
- *sex:* how people behave, marriage and divorce, rape, paedophilia, research findings;
- *famous people:* celebrity lifestyle, achievement and awards, scandal, obituary;
- *sport:* league winners and losers, results (betting), records set – 'firsts', who is 'best'?

Just take one example – a cyclone that blew itself out over the Pacific wouldn't be all that interesting, but one that devastated the Philippines with huge loss of life was (see pp. 255–6). Quantity makes a story significant – how many people were killed? How much money was lost? By how much was the new record better? How long was the prison sentence?

Distance also affects our judgement – a mining disaster in China's Bei Shan Province is unlikely to have the same impact as a smaller one within our own borders.

We will come back to the detail of news values, but a key question is always – how will this affect my listeners' lives?

Starting with the consumer, what does the news listener expect to hear? Certainly in a true democracy the public has a general right to know and discuss what is going on. Most people will find that interesting. However, there will be limitations, defined and maintained by law – matters of national security, confidences of a business or private nature, to which the public does not have rightful access. But these reasons can be used to cloak the genuine interest of the individual. Caught in such a conflict, the broadcaster is faced with a moral problem – the not-unfamiliar one of deciding the greater good between upholding the law and championing the rights and freedom of the individual. At such times, those involved in public responsibility should consider three separate propositions:

1   Broadcasters are not elected: they are not the government and as such are not in a position to take decisions affecting the interest of the State. If they go against the practice of the law they do so as private citizens, with no special privileges because they have access to a radio station.
2   Associated with the public right to know is the private right not to divulge. A society that professes individual freedom does not compel or allow the media to extract that which a person wishes lawfully to keep to himself.
3   What harm or discomfort may the journalist be causing by reporting – or not reporting – events? It's important to show good taste and avoid pandering to salacious curiosity.

Thus the listener has a right to be informed; but although the constraints may be few and the breaches of it comparatively rare, the right is not total. Broadcasters must know where they stand and on what basis the lines of editorial demarcation are drawn. Journalists in particular will remember that they are accountable to the listener. That is the purpose of having a by-line or credit attached to their work.

Codes of Practice

We have already seen that newsrooms often have their own stringent ethical standards. These are often given the force of law – in Britain, the

Broadcasting Acts of 1990 and 1996, updated by the Communications Act of 2003 establishing Ofcom as the regulatory body to issue and uphold the professional Codes of Practice. The BBC has its own guidelines (see p. 386). These have to be fully understood and implemented by any working journalist. Some are self-evident – the need to be fair in political reporting, with particular constraints applying to election coverage, the care taken not to misrepresent an interviewee, the correction of factual errors, the dangers of simulated newscasts, interviewing people wanted by the police, any abusive treatment of the beliefs of adherents of any religion, and so on.

But others need special vigilance, such as the reporting of terrorism. The UK's Terrorism Act 2000 sets out to ensure that reports of terrorism do not amount to incitement. It broadly defines terrorism as 'the use or threat of action where the threat is designed to influence the government or intimidate the public . . . for the purpose of advancing a political, religious, or ideological cause'. Interviewing people who advocate violence therefore needs special care. It becomes a criminal offence to invite support for any proscribed organisation. Again, violence must not be described for its own sake and it must not be 'glorified or applauded'. The Official Secrets Acts also constitute an area requiring special expertise.

Objectivity

In the case of the BBC, the basis of news and current affairs broadcasting has always been – and still is – first, to separate for the listener the reporting of events (news) from the discussion of issues and comment (current affairs), and second, to give both, or more likely all, sides of an argument. This is best achieved from the position of being informed but independent. Of course there are journalists who see broadcasting as a means of indulging their own attempts at public manipulation, just as there are governments that see news purely in terms of propaganda for their own cause. But to quote the DFID booklet *Media in Governance*:

> If the media are puppets of government, they will soon lose credibility with the public. Without credibility, government manipulation of the media may fool politicians but will not ultimately fool the people. Even true messages may be disbelieved.

People sheltered from unpalatable truths cannot decide, and cannot grow. Arising from the broadcaster's privileged position as the custodian of this form of public debate, the role of a radio service, even one under government or commercial control, is to allow expression to the various components of controversy but not to engage itself in the argument nor to lend its support to a particular view, i.e. to be impartial. However, such a policy is by no means universal and stations in some countries are encouraged to take an editorial line. In Britain, radio was for many years the monopoly of the BBC, and as the only provider it had to be as objective as possible. Where there are several broadcasting sources, each may develop its own attitude to political and other controversial issues and, like a newspaper, attempt to sway public opinion.

Where news is written by a government office, there is always the emphasis on supporting its own achievements – 'The government has performed another miracle today . . . '. However, objectivity requires a news channel always to distance itself, even slightly, from its sources. This is better as: 'The government says it has performed another miracle today . . . '. The difficulty stems from the fact that many governments of new nations see their essential task as nation building, so they believe they have to say how good everything is. Their broadcasting is not about challenging the mediocre or wrong, so you can't publish anything detrimental if you want to keep your job. Nevertheless, ways should be found of expressing truth so that broadcasting is fully credible.

What is the meaning of impartiality when covering a complex industrial dispute involving official and unofficial representatives, breakaway groups, vocally militant individuals and separate employers' and government views and solutions?

Even more difficult situations were those such as Northern Ireland or post-war Iraq where there has been a 'limited' civil war. Do we give equal time for those who would uproot society – for those who oppose the rule of law? These are not easy questions since there is a limit to the extent to which anyone may be impartial. When one's own country is involved in armed conflict it is extremely difficult, perhaps not even desirable, to be neutral – but one must, as far as possible, remain truthful. While society may be divided and changing in its regard for what is right and wrong, it is less so in its more fundamental approach to good and evil. No public medium of communication can function properly and without critical dissent unless society is agreed within itself on what is lawful and unlawful. It is possible to be impartial in a peaceful discussion on attempts to bring about changes in the existing law, but such impartiality is not possible in reporting attempts to overthrow it by force. One can be objective in reporting the activities of the man with the gun, but not in deciding whether to propagate his views.

A former Director General of the BBC, Sir Hugh Greene, said:

> I do not mean to imply that a broadcasting system should be neutral in clear issues of right and wrong, even though it should be between Right and Left. I should not for a moment admit that a man who wanted to speak in favour of racial intolerance has the same rights as a man who wanted to condemn it. There are some questions on which one should not be impartial.

There are those who disagree that race relations is a proper area for show-ing partiality, just as there are those who oppose the underlying accept-ance of the Christian faith as a basis for conducting public affairs. This is not an abstract or purely academic issue, it is one that constantly faces the individual producer. Decisions must be made as to whether it is in the public interest to give voice to those who would challenge the very sys-tem of democracy which enables that freedom of expression. On the one hand, to give them a wider currency – 'the oxygen of publicity' – may be interpreted as a form of public endorsement, on the other, to expose them for what they are could result in their total censure. What is important is the maintenance of the freedom to exercise that choice, and ultimately to be accountable for it to an elected authority. Sir Geoffrey Cox, former Chief Executive of ITN (Independent Television News), has said of the broadcaster's function:

> It is not his duty, or his right, to editorialise on the question of democracy, to advocate its virtues or attack its detractors. But he has a firm duty to see that society is not endangered either because it is inadequately informed, or because the crucial issues of the day have not been so probed and debated as to establish their truth. A good broadcast news service is essential to the function-ing of democracy. It is as necessary to the political health of society as a good water supply is to its physical health.

Legality

To stay within the law demands a knowledge of the legal process and of the constraints that the law imposes on anyone, individual or radio station, to say what they like. In Britain no one, for example, is allowed to pre-judge a case that is to come before a court, to interfere with a trial, influence a jury or anticipate the findings. Thus there are considerable restrictions on what can be reported while a matter is *sub-judice.* To exceed the defined limits is to run the very severe risk of being held in contempt of court – an offence that is viewed with the utmost seriousness, since it may threaten the law's own credibility.

Under present British law the outline of what is permissible in reporting a crime falls under four distinct stages:

1 *Before an arrest is made* it is permissible to give the *facts* of a crime but the description of a death as 'murder' should only be used if the police have made a statement to that effect. Witnesses to the crime can be interviewed but they should not attempt to describe the identity of anyone they saw or speculate on the motive.

2   After an arrest is made, or if a warrant for arrest is issued, the case is said to be 'active'. This continues *while the trial is in progress* and it is not permissible to report on committal proceedings in a magistrate's court, other than by giving the names and addresses of the parties involved, the names of counsel and solicitors, the offence with which the defendant is charged, and the decision of the court. The reporting of subsequent proceedings in the higher court is permitted but no comment is allowed. The matter ceases to be 'active' on sentence or acquittal.

3   Responsible comment is permissible *after the conviction and sentence are announced*, so long as the judge is not criticised for the severity or otherwise of the sentence, and there is no allegation of bias or prejudice.

4   *If an appeal is lodged*, the matter again becomes *sub-judice*. No comment or speculation is allowed and only factual court reports should be broadcast.

Complications can arise if the police are too enthusiastic in saying that 'they have caught the person responsible'. This is for the court to decide and broadcasters should not collude with police in pre-judging a case. There are special rules that apply to the reporting of the juvenile and matrimonial courts. The key question throughout is whether what is broadcast is likely to help or hinder the police in their investigation or undermine the authority of the judicial process.

Such matters are the stock in trade of the journalist, and producers unacquainted with the courts are advised to proceed carefully and to seek expert advice.

The second great area of the law of which all programme makers must be aware is that of libel. The broadcaster enjoys no special rights over the individual and is not entitled to say anything that would 'expose a person to hatred, ridicule or contempt, cause him/her to be shunned or avoided, lowered in the estimation of society, or tend to injure him/her in their office, profession or trade'. To be upheld, a libel can only be committed against a clearly identifiable individual or group. In civil law, it is not possible to defame the dead. The most damaging accusation that can be brought against a broadcaster standing under the threat of a libel action is that he or she acted out of malice. This is not an unknown hazard for the investigative journalist working, for example, on a story about the possibility of corruption or dubious practice involving well-known public figures. The broadcaster's complete real defence against a charge of libel is that what was broadcast, i.e. published, was true, and that this can be proved to the satisfaction of a trial judge, and in some cases perhaps a jury as well. Again, we have the absolute necessity of checking the facts and using words with a precision that precludes a possibly deliberate misconstruction.

A second defence is that of 'fair comment'. This means that the views expressed were honestly held and made in good faith without malice. This attaches particularly to book reviews, or the critical appraisal of plays and films, but may also apply to comments made about politicians or other public figures. Such a defence also has to show that the remarks are based on demonstrable facts not misinformation.

In British law, to repeat a libellous statement made by someone else is no defence unless that person enjoys 'absolute privilege', as in a court of law or in parliament. Even so, reports of such proceedings have to be fair as well as accurate and if the statement made turns out to be wrong and an apology or correction is issued, this too is bound to be reported. A defence of 'qualified privilege' is available to reports of other public proceedings such as local authority council meetings, official tribunals, company annual general meetings open to the public, and other meetings to do with matters of public concern. The same defence may be used in relation to a fair and accurate report of a public notice or statement issued officially by the police, a government department or local authority. Where no 'privilege' exists, the broadcaster is as guilty as the actual perpetrator of the libel. Producers and presenters of the phone-in should be constantly on their guard for the caller who complains of shoddy workmanship, professional incompetence, or worse, on the part of an identifiable individual. An immediate reference by the presenter to the fact that 'well, that's only your view' may be regarded as a mitigation of the offence, but the broadcaster can nevertheless be held to have published the libel.

The law also impinges directly on the broadcaster in matters concerning 'official' secrets, elections, consumer programmes, sex discrimination, race relations, gaming and lotteries, reporting from foreign courts and copyright.

The individual producer, in whatever country he or she is working, should remain aware of the major legal pitfalls and must have a reliable source of legal advice. Without it the station is likely, sooner or later, to need the services of a good defence lawyer.

News values

From all the events and stories of the day how does the broadcaster decide what to include in the news bulletin? A decision to cover, or not to cover, a particular story could, itself, be construed as bias. The producer's initial selection of an item on the basis of it being worthy of coverage is often referred to as 'the media's agenda-setting function', and is a subject for much debate. People will discuss what they hear on the radio and are less likely to be concerned with topics not already given wide currency. So is a radio station's judgement as to what is significant worth having? If so, the process of selection, the reasons for rejection, and the weight accorded to each story (treatment, bulletin order and duration) are matters that deserve the utmost care.

There is sufficient evidence to support the significance of the *primacy* and *recency* effects in communication. This means that items presented at the beginning of a bulletin have greater influence than those coming later – also that the final statements exert an inordinate bearing on the total impact – probably because they are more easily recalled. These principles are made much use of in debates and trials but clearly apply also to bulletins, interviews and discussions. Who speaks first and who is allowed the last word is often a matter of some contention.

The broadcaster's power to select the issues to be debated – and their order of presentation – represents a considerable responsibility. Yet, given a list of news stories, a group of editors will each arrive at broadly similar running orders for a news bulletin designed for a particular audience. Are there any objective criteria on this matter of news *values*?

The first consideration is to produce a news package suited to the style of the programme in which is it broadcast, answering the question, 'What will my kind of listener be interested in?' A five-minute bulletin can be a world-view of 20 items, superficial but wide ranging; or it can be a more detailed coverage of four or five major stories. Both have their place, the first to set the scene at the beginning of the day, the second to highlight and update the development of certain stories as the day proceeds. The important point is that the shape and style of a bulletin should be matters of design and not of chance. Unlike a newspaper with its ability to vary the type size, radio can only emphasise the importance of a subject by its placing and treatment. A typical five-minute bulletin may consist of eight or nine items, the first two or three stories dealt with at one minute's length, the remainder decreasing to 30 seconds each or less. The point was made earlier that, compared with a newspaper, this represents a very severe limitation on *total* coverage.

Having decided the number and the length of items, the news producer has to select what is important as opposed to what is of passing interest. When time is short, it is easier to gain the interest of the listener with an item on the latest scandal than with one on the state of the economy. The second item is more significant for everyone in the long term, but requires more contextual information. The producer must not be put off by such difficulties, for it is the temptation of the easy option that leads to some justification in the charge that 'the media tends to trivialise'. An effect of the policy that news must always be available at a moment's notice is that stories of long-term significance do not find a place in the bulletin.

It is, after all, easier to report the blowing up of an aircraft than the development of one.

As already noted, a key criterion for selection is to favour items to do with people rather than things. The threat of an industrial dispute affecting hundreds of jobs will rate higher than a world record price paid for a painting. 'How could this event affect my listener?' is a reasonable question to ask. For the listener to a small station in Britain, 10 deaths following an outbreak of the Ebola virus in Cameroon would possibly be regarded with less significance than a serious local road accident. But should it? You don't have to wait for local doctors to volunteer to go to West Africa before you see the significance of the story. Particularly in local radio there is a tendency to run something because of its association with immediate mayhem and disaster rather than its wider relevance. A preoccupation with house fires and traffic accidents, otherwise referred to as 'ambulance chasing', is to be discouraged.

News values resolve themselves into what is of interest to, or affects, the listener. More significantly, 'interesting' is determined by what is:

- important – events and decisions that affect the world, the nation, the community, and therefore me;
- contentious – an election, war, court case, where the outcome is yet unknown;
- dramatic – the size of the disaster, accident, earthquake, storm, robbery;
- geographically near – the closer it is, the smaller it needs to be to affect me;
- culturally relevant – I may feel connected to even a distant incident if I have something in common with it;
- immediate – events rather than trends;
- novel – the unusual or coincidental as they affect people.

On a different scale, sport can be all of these.

News has been called 'the mirror of society'. But mirrors reflect the whole picture, and news certainly does not do that. Radio news is highly selective; by definition it is to do with the unusual and abnormal but the basis of news selection must not be whether a story arouses curiosity or is spectacular, but whether it is significant and relevant. This certainly does not mean adopting a loftily worthy approach – dullness is the enemy of interest – it requires finding the right point of human contact in a story. This may mean translating an obscure but important event into the listener's own understanding. A sharp change in the money markets will be readily understood by the specialist, but radio news must enable its significance to be appreciated by the man in the street. The job of news is not to shock but to inform. A broadcasting service will be judged as much by what it omits as what it includes.

Investigative reporting

The investigation of private conduct and organisational practice – and malpractice – is an important part of media activity. This is the watchdog function referred to earlier, particularly keeping an eye on those in positions of public trust. The role of the *Washington Post* in the Watergate exposé is a well-documented example. Radio too recognises that it is not enough to wait for every news story to break of its own accord – some, of genuine public concern, are protected from exposure simply because of the vested interests that work to ensure that the truth never gets out. It is therefore sometimes necessary to allocate newsroom effort to the process of research enquiry into a situation that is not yet established fact. The story may never materialise because insufficient fact comes to light. This will involve the station in some loss through unproductive effort, but it is nothing like the loss that will be suffered if the newsroom proceeds to broadcast a story of accusations that turn out to be false.

Government departments or commercial businesses involved in underhand dealings, public officials or others with power engaged in questionable financial practice, the rich and famous called to account for their sexual immorality – these are the most common areas of investigation. But who is to say what is underhand, questionable or immoral? While it might be possible to remain impartial in the reporting of news fact, the exposé inevitably carries with it an assessment of a situation against certain norms of behaviour. Such values are seldom purely objective. An investigation into the payment of a bribe in order to secure a contract may provoke a public scandal in one part of the world while in another it is simply the way in which business is conducted. In other words, investigation requires a judgement of some malpractice – of right and wrong. The reporter must therefore be correct on two counts – that the facts as reported stand up to later scrutiny, and that his or her own judgement as to the morality of the issue is subsequently endorsed by the listener, that is by society.

To enable the reporter's own values to remain largely outside the investigation, the most fruitful approach is often to use the stated values of the organisation or person being investigated as the basis of the judgement made. Thus, a body claiming to have been democratically elected but which subsequently was shown to have manipulated the polls lays itself open to criticism by its own standards. The same is true of governments which, while happy to be signatories of agreements on the treatment of prisoners, also allow their armed forces to practise beatings and torture;

or business firms which promise refunds in the event of customer complaint but which somehow always find a loophole to evade this particular
responsibility. The radio station may have to represent the listener in cases
of personal unfairness, or pursue the greater interests of society in the
face of public corruption. But the broadcaster must be right. This takes
patience, hard, wearying research, and the ability to distinguish relevant
fact from a smokescreen of detail.

Occasionally, outside pressure will be brought to bring the enquiry
to a halt. This could be the signal that someone is getting uncomfort-
able and that the effort is beginning to bear results. It is surprising how
often malpractice breeds dissatisfaction. Once the fact of an investigation
becomes known, a person with a grudge is likely to provide anonymous
information. Such 'leaks' and tip-offs of course need to be checked and
treated with the utmost caution. A story that is told too soon will fall
apart as surely as one that is wrong. Further, a station must resist the
temptation to get so involved with a story that it falls prey to the same
malpractice – although perhaps on a much smaller scale – as it is attempting to expose. So does it pay for information? Does it go in for surreptitious recording, to get the words of an illegal arms dealer or dubious
estate agent – does it impersonate a potential buyer, for instance? Yes,
this might be necessary, but should be sanctioned by senior management
given all the facts. Does it jeopardise its own integrity by giving false
information or staging events in the hope of laying a trap for others?
Again, perhaps so, but investigative reporters should not work alone, but
in twos or threes – to argue through the methods, develop theories and
assess results. Guidelines suggest:

- the story should be sufficiently vital to justify deception;
- this should be the only way of getting the story;
- the listener must be told how the story was obtained.

Reporters will be wise to stay in close contact with their management –
whose backing and continuing financial support are crucial. When it
works, the effects are immediate and considerable. The reputation gained
by the programme and station are incalculable. A 'scoop' puts competitors
nowhere. The public at large wants wrongs put right. People respect a moral
order, especially for others, and in the end prefer justice to expediency.

Campaigning journalism

Programmes, and their station or network, cease to be objective or impartial when they wholeheartedly promote a particular course of action. The
extent to which any such bias will threaten the credibility of the station as

a whole depends on the proportion of the audience that will already agree with the action proposed. Thus a local station that wants to raise money for handicapped children is unlikely to create opposition among its listeners. Even if the newsroom originates the campaign, the standing of the normal bulletin material will probably remain unaffected. If, however, the station is advocating action on a more contentious issue – the publication of lists of convicted paedophiles, the introduction of random breath checks as a deterrent to drunken drivers, or mandatory blood sampling to detect carriers of the AIDS virus – then the station must expect opposition from those who prefer more liberty, some of whom will criticise any story on the subject which the station carries.

In general, campaigning is best kept away from the newsroom. The news editor should be allowed to pursue the professional reporting of daily factual truth without being involved in considerations of what other people – governments, councils, advertisers or radio managements – want reported, or unreported. This at least minimises the danger of one programme's editorial policy jeopardising news credibility. Voices associated with news almost always run some risk when they appear in another broadcasting context. Journalists who lend weight to a particular view, however worthy, easily damage their reputation as dispassionate observers. Thus it was a group of campaigning Zambian women, not the newsroom, that broadcast a programme on the need for clean water that persuaded the authorities to drill new boreholes.

A producer wanting to promote a cause must obviously seek the backing of management and be aware of the possible effect of any campaign on other programmes, especially news. Partiality of view in itself may become counter-productive to the very issue it is supposed to promote.

The news reporting function

The reporter out on the street and the sub-editor at the newsdesk are the people who make the decisions about news. Their concerns are accuracy, intelligibility, legality, impartiality and good taste.

Before looking at the key principles it is important to say something about one of the more difficult aspects of the job.

Most of the work is relatively straightforward; some of it is routine. Chronicling events and the reasons for them requires much rewriting of other people's copy received by a multitude of means. It entails hours spent on the phone checking sources, and days out on location recording interviews and filing reports. It is during these times away from the newsroom, when the reporter is working alone, that there is a need for some self-sufficiency – an apparent self-confidence, not always felt – to tackle the unknown and sometimes dangerous situation.

Accuracy, realism, and truth

A reporter's first duty is to get the facts right. Names, initials, titles, times, places, financial figures, percentages, the sequence of events – all must be accurate. Nothing should be broadcast without the facts being double-checked, not by hearsay or suggestion but by thorough reliability. 'Return to the primary source' is a useful maxim. If it is not possible to check the fact itself, at least attribute the source declaring it to be a fact. Under pressure from a tight deadline, it is tempting to allow the shortage of time to serve as an excuse for lack of verification. But such is the way of the slipshod to their ultimate discredit. Even in a competitive situation, the listener's right to be correctly informed stands above the broadcaster's desire to be first. The radio medium, after all, offers sufficient flexibility to allow opportunity for continuing intermittent follow-up. Indeed, it is ideally suited to the running story.

Sometimes accuracy by itself is not enough. With statistics, the story may be not in their telling but in their interpretation. For instance, according to the traffic accident figures the safest age group of motorbike users is the 'over 80s' – not one was injured last year! So a story concerning a 20 per cent increase in the radioactivity level of cows' milk over two years may be perfectly true, but is it significant? How has it varied at other times? Was the level two years ago unusually low? Were the measurements on precisely the same basis? And so on. Statistical claims always need care.

Accuracy is required too in the sounds that accompany a report. The reporter working in radio knows how atmosphere is conveyed by 'actuality' – the noise of a building site, the shouts of a demonstration. It is important in achieving impact and establishing credibility to use these sounds, but not to make them 'bigger' than they really are. How fair is it then to add atmospheric music to an interview recorded in an otherwise silent café? It may be typical of the café's music (and it is useful in covering the edits) – but is it honest and real? Is it right to add small-arms fire to a report from a battle area? Typically it is there, it was just that the guns happened to be silent when I was recording. In other words does the piece have to *be* reality, or to convey reality? The moment you edit you destroy real-time accuracy. It is a question of motive. The accurate reporter, as opposed to a merely sensationalist one, will need a great deal of judgement if the object is to excite and interest, but not mislead.

The basic structure for the news interviewer is first to get the facts, then to establish the reasons or cause lying behind them, and finally to arrive at their implication and likely resultant action. These three areas are simply past, present and future – 'What's happened? Why do you think this is? What will you do next?' At another level, a news story is to do with the personal motives for decision and action, and it is these which have to be exposed and, if need be, challenged with accurate facts or authoritative opinion quoted from elsewhere.

Intelligibility in the writing

Conveying immediate meaning with clarity and brevity is a task that requires refinement of thought and a facility with words. The first requisite is to understand the story so that it can be told without recourse to the scientific, commercial, legal, governmental or social gobbledegook which can often surround the official giving of information. A reporter determined to show a familiarity with such technicalities through their frequent application has little use as a communicator. News must be the translator of jargon not its disseminator.

In recognising where to start it's necessary to have an insight into how much the listener already knows and how ideas are expressed in everyday speech. In being understood, the reporter's second requirement is therefore a knowledge of the audience – it is unwise to deal only with colleagues and professional sources, for you could find yourself subconsciously broadcasting only to them. If the audience is distant rather than local, then from time to time travel among them, or at the very least set up whatever means of feedback is possible.

The third element in telling a story is that it should be logically expressed. This means that it should be chronological and sequential – cause comes before effect:

Not: 'Two thousand jobs are to be cut, the XYZ Bank announced today.'
But: 'The XYZ Bank today announced it was to cut two thousand jobs.'

The key to intelligibility, therefore, is in the reporter's own understanding of the story, of the listener and of the language of communication.

Putting these three aspects together the news writer's job is to tell the story, putting it in an understandable, logical sequence, answering for the listener questions such as:

- 'What has happened?'
- 'When and where did it happen?'
- 'Who was involved?'
- 'How did it happen and why?'

The first technique is to ensure that of these six basic questions, at least three are answered in the first sentence:

1   The Chancellor of the Exchequer told parliament this afternoon that
    he would be raising income tax by an average of 4 per cent in the
    autumn. (who, where, when, what, when)
2   Eight people were killed and over 60 seriously injured when two trains
    collided just outside Amritsar in northern India in the early hours of
    this morning. (what, where, when)

The second and subsequent sentences should continue to answer these
questions:

1   He said it would be applied only to the upper tax bands and would not
    affect the basic rate. In answer to an opposition question he said this
    was necessary to reduce the government deficit. (how, who, why)
2   The crowded overnight express from Delhi was derailed and over-
    turned by a local freight train as it left a siding near Amritsar sta-
    tion. Railway workers and police are still taking the injured from the
    wreckage and it is feared that the death toll may rise. (how, what)

A fault commonly heard on the air, which we looked at in Chapter 6, is
that of the misplaced participle.

> 'The Prime Minister will have to defend the agreement he signed, in the House
> of Commons.'

Without the comma this sounds as though we are talking about an agreement he signed in the House of Commons. But no, the story means:

> 'The Prime Minister will have to defend in the House of Commons the agree-
> ment he signed.'

This error is frequently made in references to time.

> 'He said there was no case to answer last July.'

What was actually meant was:

> 'He said last July there was no case to answer.'

And another, following a report of child abuse by a nanny.

> 'There was a demand to register all nannies with local councils.'

But what, you might ask, about nannies who are not with local councils?
This would be better as:

'There was a demand for all nannies to be registered with local councils.'

Two more examples from real life:

'Action has been taken against the hospitals which removed the organs of patients who have died, without the knowledge or permission of their relatives.'

'The government has agreed to allow a variety of GM maize to be grown.'

These ambiguities have to be removed. In the first it must be made clear that it is the removal of organs that was without the knowledge of relatives, and not the patients' death. In the second, it sounds as though lots of different maize crops would be allowed. A subsequent bulletin changed this to 'a single variety'.

Radio requires intelligibility to be immediate and unambiguous.

Being fair

The reporter does not select 'victims' and hound them, does not ignore those whose views are disliked, pursue vendettas, nor have favourites. He or she does not promote the policies of sectarian interests, resists the persuasions of those seeking free publicity, and above all is even-handed and fair in presenting different sides of an issue. Expressing no personal editorial opinion, but acting as the servant of the listener, the aim is to tell the news without making moral judgements about it. Broadcasting is a general dissemination and no view is likely to be universally accepted. 'Good news' of lower trade tariffs for importers is bad news for home manufacturers struggling against competition. 'Good news' of another sunny day is bad news for farmers anxiously waiting for rain. The key is a careful watch on the adjectives, both in value and in size. Superlatives may have impact but are they fair? News may report an industrial dispute but what right does the reporter have to describe it as 'a *serious* industrial dispute'? On what grounds may he refer to a company's '*poor* record', or a medical research team's '*dramatic* breakthrough'? Words such as 'major', 'crucial', and 'special' are too often used simply to convince people that the news is important. Much better to leave the qualifying adjectives to the actual participants and for the news writer to let the facts speak for themselves.

Reporters are occasionally concerned that they might not be able to be totally objective since they have received certain inbuilt values from their upbringing and education. While it may be true that broadcasting has more of its fair share of people from middle-class families and with a college education, a reporter should be aware of any personal motivations of background and experience, recognising that others might not share them. Certainly what must be watched is any conscious desire to persuade others to think the same way.

Unlike the junior newspaper journalist, whose every last adjective and comma can be checked before publication, the broadcast reporter is frequently alone in front of the microphone. To help guard against the temptation of unintended bias, reporters should not be recruited straight from school, but have as wide and varied a background as possible and preferably bring to the job some experience of work outside broadcasting. It is sensible to ensure that any significantly large ethnic group in the community is represented in the broadcasting staff. The matter of bias is something on which members of the station's board should keep a watchful ear.

Giving offence

In avoiding needless offence there must first be a professional care in the choice of words. People are particularly sensitive, and rightly so, about descriptions of themselves. The word 'immigrant' means someone who entered a country from elsewhere, yet it tends to be applied quite incorrectly to people whose parents or even grandparents were immigrants. Human labels pertaining to race, disability, religion or political affiliations must be used with special care and never as a social shorthand to convey anything other than their literal meaning. Examples are 'black', 'coloured', 'Muslim', 'guerrilla', 'southern', 'Jewish', 'communist', etc. – used loosely as adjectives they tend to be more dangerous than as specific nouns.

The matter of giving offence must be considered in the reporting of sexual and other crimes. News is not to be suppressed on moral or social grounds but the desire to shock must be subordinate to the need to inform. The journalist must find a form of words which, when spoken, will provide the facts without causing embarrassment, for example in homes where children are listening. With print, parents can divert their children from the unsavoury and squalid; in radio an immediate general care must be exercised at the studio end. A useful guideline is for the broadcaster to consider how actually to express the news to someone in the local supermarket, with other people gathered round.

More difficult is the assessment of what is good taste in the broadcasting of 'live' or recorded actuality. Reporting an angry demonstration or drunken crowd when tempers are frayed is likely to result in the broadcasting of 'bad' language. What should be permitted? Should it be edited out of the recording? To what extent should it be deliberately used to indicate the strength of feeling aroused? There are no set answers; the context of the event and the situation of the listener are both pertinent to what is acceptable. However, in using such material as news, the broadcaster must ensure that the motive is really to inform and not simply to sensationalise. It might be 'good copy', but does it genuinely help the listener to understand the subject? If so it might be valid but the listener retains the right to react as he or she feels appropriate to the broadcaster's decision.

Causing distress

News of an accident can cause undue distress. It is necessary only to mention the words 'air crash' to cause immediate anxiety among the friends and relatives of anyone who boarded a plane in the previous 24 hours. The broadcaster's responsibility is to contain the alarm to the smallest possible group by identifying the time and location of the accident, the airline, flight number, departure point and destination of the aircraft concerned. The item will go on to give details of the damage and the possibility of survivors, but by then the great majority of air travellers will be outside the scope of the story. In the case of accidents involving casualties, for example a bus crash, it is helpful for listeners to know to which hospital the injured have been taken or to have a telephone number where they can obtain further information. The names of those killed or injured should not normally be broadcast until it is known that the next of kin have been informed.

A small but not unimportant point in bulletin compilation is the need to watch for the unintended and possibly unfortunate association of individual items. It could appear altogether too callous to follow a murder item with a report on 'a new deal for butchers'. Common sense and an awareness of the listener's sensitivities will normally meet the requirements of good taste but it is precisely in a multi-source and time-constrained process, which news represents, that the niceties tend to be overlooked.

Civil disturbance and war reporting

Tragedy should be reported in a sombre manner – the broadcaster always remaining sensitive to how the listener will react. When reporting on a riot or commentating from a battle zone it is the reporter's task to report and, as far as possible, not to get involved. It is sensible therefore always to get local advice on conditions and, as far as possible, to stay outside a disturbance, rather than try to work from inside the mêlée. It is then possible to see and assess what is happening as the situation develops. Under these conditions the reporter should remain as inconspicuous as possible and not add to or inflame the situation by his or her presence.

The primary ambience in a crisis is likely to be one of confusion. Asking for an official view tends to produce either optimistic hopes or worst fears, so any comment of this kind should be accurately attributed, or at least referred to as 'unconfirmed'. Apart from the scene as it is 'now', analysis and interpretation of an event takes two forms – the pressures and causes which led up to it, and the implications and consequences likely to stem from it. Unless the reporter is very familiar with the situation, it is best to leave reasons and forecasts to a later stage, and probably to others. On the spot, there is no room for speculation: the story should be told simply on the basis of what the reporter sees and hears, or knows as fact – even then it may be necessary to withhold the names and identities of people involved in a kidnapping or police siege.

Prior to reporting from an actual war zone, the essential reading is UNESCO's *Handbook for Journalists*. Originally published in French by Reporters Sans Frontières, it covers all the experience learned in recent years, especially from Afghanistan and the Middle East, but applies to all situations of danger and political instability. It is linked with the *World Index of Press Freedom* listing the documents intended to protect human rights and professional ethics (see Chapter 5). In practical terms it covers everything from vaccinations to the avoidance of land mines and booby-traps, being taken hostage, through to post-assignment counselling. It is very detailed.

In actual hostilities an accredited war reporter will be required to wear a flak jacket or other protective clothing – the military do not like to be held responsible unnecessarily for their own civilian casualties. This raised a particular ethical point concerning the war in Iraq.

Reports were filed by hundreds of journalists travelling with the American and British forces. The word 'embedded' was used for this integration. The advantage was that this gave the reporters the protection and some of the communication facilities of the military, and afforded access to the daily briefings. The disadvantage was that they were only given the information and movement that the coalition forces wanted them to have. It was up to the base station to restore some kind of objectivity. Reporters who wanted to travel independently had a tough time of it; some were killed in the crossfire.

In any case, under these circumstances it is necessary to liaise closely with the officer in charge and to accept limits sometimes on what can be said. Facts may have to be with held in the interests of a specific operation – for example, the size and intent of troop movements. This is for fairly obvious tactical reasons and it is generally permissible to say that reporting restrictions are in force. One of the now most memorable reports to come out of the Falkland Islands conflict arose from just such a situation. Brian Hanrahan reporting from the deck of the British aircraft carrier *Hermes*:

At dawn our Sea Harriers took off, each carrying three 1,000-pound bombs. They wheeled in the sky before heading for the islands, at that stage just 90 miles away. Some of the planes went to create more havoc at Stanley, the others to a small airstrip called Goose Green near Darwin, 120 miles to the west.

There they found and bombed a number of grounded aircraft mixed in with decoys. At Stanley the planes went in low, in waves just seconds apart. They glimpsed the bomb craters left by the Vulcan and left behind them more fire and destruction. The pilots said there'd been smoke and dust everywhere punctuated by the flash of explosions. They faced a barrage of return fire, heavy but apparently ineffective. I'm not allowed to say how many planes joined the raid, but I counted them all out and I counted them all back. Their pilots were unhurt, cheerful and jubilant, giving thumbs-up signs. One plane had a single bullet-hole through the tail – it's already been repaired.

(Courtesy of BBC News)

Expressed in a cool unexcited tone, it is worth noting the shortness of the sentences and ordinariness of the words used. It is not necessary to use extravagant language to be memorable (see also the section on live commentary in Chapter 19).

Working in conditions of physical danger, a basic knowledge of first aid is invaluable. Several organisations equip their staff reporting from areas of potential risk with a medical pack containing essentials such as sterile syringes, needles and intravenous fluid. Psychological as well as bodily safety remains important and reporters faced with violence, and the mutilated dead and dying – whether it be the result of a distant war or a domestic train crash – can suffer trauma for some time afterwards as a result of their experiences. The sometimes harrowing effects of news work should not be underestimated and the opportunity always provided for suitable counselling.

The newsroom operation

Almost every broadcasting station ultimately stands or falls by the quality of its news and information service. Its ability to respond quickly and to report accurately the events of the day extends beyond just news bulletins. The newsroom is likely to represent the greatest area of 'input' to a station and, as such, it is the one source capable of contributing to the whole of the output. Unlike a newspaper that directs its energies towards one or two specific deadlines, a radio newsroom is involved in a continuous process. The main sources of news coverage can be listed as:

1 *Professional:* staff reporters and specialist correspondents, e.g. crime, local government; freelances and 'stringers'; computer, fax and wire services; news agencies; syndicating sources including other broadcasting stations; newspapers.
2 *Official:* government sources both national and local; emergency services such as police, fire and hospitals; military and service organisations; public transport authorities.

Figure 7.2   Newsroom and production office at BBC Radio Solent. Each desk has a DAW for downloading audio, mixing and editing for final packages

3   *Commercial:* business and commercial PR departments; entertainment interests.
4   *Public:* information from listeners, taxi drivers, etc.; voluntary organisations, societies and clubs.

There is a danger that in basing its news too much on press releases and handouts, a station is too easily manipulated by government and business interests. The output begins to sound like the voice of the establishment. An editor will become wary of material arriving by hand from an official source just before a major deadline. Of course, a handout provides 'good', one-sided information – that is its purpose. It needs to be evaluated and cross-examined and questions asked about implications as well as immediate effects. A newsroom must be more than just a processor of other people's stories. The same is true of lifting items from newspapers – always look for new angles, and follow up if a story has appeared elsewhere, develop it, don't run it as it is, take it further.

The heart of the newsroom is its diary. This may be in book form or held on computer. As much information as possible is gleaned in advance so that the known and likely stories can be covered with the resources available. The first editorial meeting of each day will review that day's prospects and decide the priorities. Reporters will be allocated to the

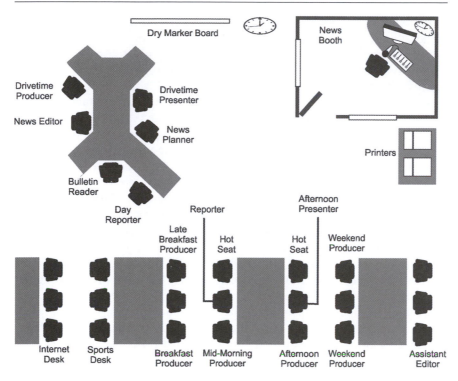

Figure 7.3   Typical local newsroom layout

monthly trade figures, the opening of a new airport or trunk road, the controversial public meeting, the government or industry press conference, the arrival of a visiting head of state, or the publication of an important report. The news editor's job is to integrate the work of the local reporters with the flow of news coming in from the other sources available, balancing the need for international, national or local news. But the news editor must also consider how to deal with the unforeseen – an explosion at the chemical works, a surprise announcement by a leading politician, a murder in the street. A newsroom, however, cannot wait for things to happen, it must pursue its own lines of enquiry, to investigate issues as well as report events.

Local newsrooms are sometimes tempted to select bulletin items in terms of geographical coverage – going for 20 stories representative of the area, rather than 10 items of more universal interest. This should be resolved in the form of a stated policy – that is, the extent to which a newsroom regards itself as serving several minorities as opposed to one audience of collective interest; The first is true of a newspaper, where each reader selects items of personal interest; the second is more appropriate to radio, where the selection is done by the station. Given considerably less 'space' the fewer stories have to be of interest to everyone.

The competent newsroom has to be organised in its copy flow. There should be one place which receives the input of letters, texts, tweets, emails, press releases and other written data. A reporter making 'routine' calls to the emergency services or other regular contacts collects the verbal information so that, after consulting the diary, the news editor can decide which stories to cover. A meeting in person or by phone then discusses the likely prospects, especially with specialist correspondents. One person will be allocated to write, or at least edit, the actual bulletins – a task not to be regarded as a committee job. If possible other writers are put on the shorter bulletins and summaries. Working from the same material, a two-minute summary needs a quite different approach. Omitting the last three sentences of each story in a five-minute bulletin will not do!

Reporters and freelances are allocated to the stories selected; each one is briefed on the implications and possible 'angles' of approach, together with suggestions for its treatment, and given a deadline for completion. Elsewhere in the room or nearby, material is edited, cues written, recordings are made of interviews over the phone, and previous copy 'subbed', updated or otherwise refreshed for further use. The mechanical detail will depend on the degree of sophistication enjoyed by the individual newsroom – the availability of a radio car or other mobile live inject equipment, OB and other 'remote' facilities, incoming lines, electronic data processes, off-air or closed circuit television, updated stories permanently available via computer screens in the studios, intercom to other parts of the building, etc.

In addition to all this the fully tri-media newsroom will not only be looking after the television side of things but may also have journalists to feed a variety of teletext, subtitle and website services.

A newsroom also requires systems for the speedy retrieval of information from a central file store. The computerised newsroom links together the diary information with 'today's' prospects, the current news programme running orders and bulletin scripts. Local is joined to national. Everyone has the same information updated at the same instant. The presenter in the studio has the latest news material constantly available, reading it from the screen. There may also be a physical file of newspaper cuttings or old scripts. For the short-term retention of local information, the urgent telephone number or message to a colleague on another shift, a chalk-board or similar device is simple and effective.

The important principle is that everyone should know exactly what they have to do to what timescale, and to whom they should turn in difficulty. The news editor or director in overall charge must be in possession of all the information necessary for him or her to control the output, and be clear on the extent to which the minute-by-minute operation is delegated to others. There is no place or time for confusion or conflict.

Smartphone

There are many different kinds of smartphones, tablets, cell phones and other portable digital devices which can be extremely useful when reporting away from the studio base. Audio quality varies, and often radio stations will recommend a particular device and network. They will also install additional equipment at base to enable the devices to be connected to the studio desk to provide cue and talkback. Some stations will allow the use of output from these devices straight on air, as the BBC does on some occasions. It must be remembered these consumer devices are using a public digital cell phone network, and therefore reception may vary due to location and time of day and could be subjected to signal variability or even break-up or loss. The reporter needs to hear the cue programme in order to go ahead, and as cell circuits are two-way the cue feed can be switched from the studio, often with the addition of studio talkback. All portable devices are improved by plugging in a professional mic, but if this is not available the basic quality can be improved by talking across the device to avoid distortion and 'pops'. Given the right app, such as Voice Memo, Viddio, or WavePad, the device can record and do multi-track editing. It can also do Internet research, make a phone call, or send a text back to the editor. It has the additional advantage of appearing 'normal', so in a situation of tension it doesn't draw attention to itself, thereby making the reporter's job safer. The cell phone – much improved by using an external mic with a windshield – offers great ease and flexibility of location reporting.

Style book

One of the editor's jobs is to maintain the collection of rules, guidelines, procedures and precedents that forms the basis of the newsroom's day-to-day policy. It is the result of the practical experience of a particular news operation and the wishes of an individual editor. The style book is not a static thing but is altered and added to as new situations arise. A large organisation will issue guidelines centrally which its local or affiliate stations can augment.

Sections on policy will clarify the law relating to defamation and contempt of court. It will define the procedure to be followed, for example,

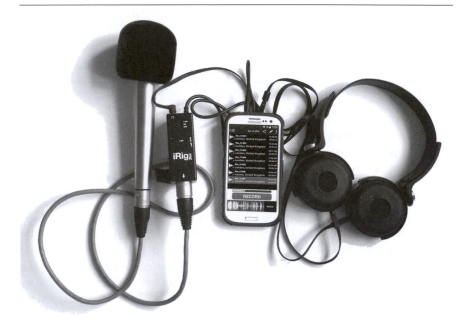

Figure 7.4   Reporter's basic smartphone rig

in the event of bomb warnings (whether or not they are hoax calls), private kidnapping, requests for a news black-out, the death of heads of state, observance of embargoes, national and local elections, the naming of sources and so on.

The book sets out the station's mission or purpose statement and the role of news within that. It will indicate the required format for bulletins, the sign-on and sign-off procedures, the headline style, correct pronunciation of known pitfalls, and the policy regarding corrections, apologies and the right or opportunity of reply. It will list on-station safety regulations, as well as provide advice on proper forms of address. Above all there will be countless corrections of previous errors of grammar and syntax – from the use of collective nouns to the use of the word 'newsflash'.

On arrival, the newcomer to a newsroom can expect to be given the style book and told to 'learn it'.

Radio car

The larger station will have a car, reserved for newsroom use, which the editor will send out with a reporter to cover a particular story. Some may be equipped with a telescopic mast capable of communicating back to the base station. Others will have a satellite dish (see p. 248). The principles

Figure 7.5 BBC Radio Norfolk radio car with mast partly extended. The reporter wears a hi-vis jacket

of using mobile facilities tend to be similar whatever their design, and the following forms a common basis for routine operation:

- ensure a proper priority procedure for every vehicle. Who controls and sanctions its use? Who decides if a booking is to be overridden to cover a more important story? Have all potential users been fully trained and do they know these procedures? Are they up to date with any modifications, or minor faults affecting the vehicle's operation?;
- remember that you are driving a highly distinctive vehicle. Whatever the hurry, be courteous, safe and legal;
- before leaving base, check that all necessary equipment is in the car, that the power supply batteries are in a good state of charge, and that someone is ready to listen out for you;

- switch on the two-way communications receiver in the car so that the base station can call you. Tune the car radio to receive station output;
- on arrival at the destination, take great care to avoid obstructions, such as overhead power and telephone lines, when using a telescopic mast. Set up the necessary link. Call the station and send level on the programme circuit;
- agree the cue to go ahead, duration of piece and handback. Check that you are hearing station output on your headphones;
- after the broadcast, replace all equipment tidily and lower the mast before driving off (cars with telescopic masts should have safety interlocks to prevent movement with the mast extended). Leave the communications receiver on until back at base in case the station wishes to call you;
- radio vehicles attract visitors; make sure vehicles are safely parked day and night.

Every car should be put 'on charge' when at base, and never garaged with the fuel tank less than half-full. It should carry a cell phone, area and local street maps, SatNav, a radio mic, a reel of cable for remote operation, spare headphones, batteries and a fuel can, extension leads, a pad of paper, a clipboard, a pencil, a cleaning cloth and a torch.

Equipment in the field

Distance from the station, of course, creates no problems in the use of the Internet, either for sending text via email or reports and interviews in digital form. This method of communication is used by the network of news correspondents worldwide. In addition, ISDN lines – the Integrated Services Digital Network – offer a relatively low-cost system using telephone circuits for high-quality voice transmission, where such circuits are provided. The reporter plugs a microphone or recorder playback into an ISDN 'black box' (Codec) which encodes the signal in digital form. The system works well for speech and may be 'stretched' for music.

Where phone lines are unreliable – providing a cell phone signal or reliable WiFi connection is available – a laptop or tablet and smartphone is the only kit required to get fully edited and packaged interviews and reports immediately on air from anywhere in the world.

However, while the high-tech solution may seem attractive, the reporter working away from base soon learns to become self-reliant when it comes to his or her technical equipment. It is often the low-tech which saves the day – the Swiss army knife or the collection of small screwdrivers. Everything has a back-up.

Figure 7.6   Interior of radio car showing transmission equipment

Experienced overseas reporters rely on such items as:

- battery recording machines – solid state recorders are robust and good for quick editing. A four-track recorder with two additional mic/line inputs capable of internal mixing;
- robust omni mics with built-in windshield;
- a small folding mic stand for use on a table;
- a lip mic to exclude untreated room acoustics or for use while travelling in a car;
- at least one long mic cable for putting a mic up-front for speeches or press conferences;
- assorted plugs and leads, including crocodile clips for connecting to phone lines;
- a smartphone with windshield and tablet with appropriate apps;
- waterproof insulating tape, gaffer/sticky tape;
- solar-powered or wind-up battery chargers can be useful for devices with integrated or rechargeable batteries. Cigarette lighter charging and power leads too, when in cars;
- headphones or an ear piece for monitoring and receiving the cue programme from a cell phone;

- an international double adaptor telephone plug. This is useful to be able to turn a single telephone socket, e.g. in a hotel room, into two for simultaneous connection of a telephone and a playback recorder;
- a radio – FM/MW/LW/SW with extra wire for an aerial;
- spare batteries, cards, leads, etc.

Such kit is invariably carried in a clearly labelled, tough foam-padded metal or high-impact plastic flight case – it is never left out of sight.

The news conference and press release

News conferences, company meetings, official statements and briefings of all kinds invariably set out to be favourable to those who hold them.

This is particularly so in the realm of politics. Ministerial aides, political advisers, press secretaries, public relations spokespersons, spin-doctors and the like abound. Their job is to create a positive appearance for all proposals and action – and to suppress the negative.

It is important therefore to listen carefully to what is being said – and what is not being said – and to ask key questions. Having been given facts or plans for action, the question is 'why?' It is necessary to quote accurately what is said – not difficult with a recorder – and to attribute the spokesperson or source. What is being said may or may not be totally true – what is true is that a named person is saying it.

Press releases, publicity handouts, notices and letters descend on the editor's desk in considerable quantity. Most end up on the spike, unused – although if it is not suitable for a bulletin, a well-organised newsroom will offer a story to another part of the output. From time to time editors are asked to define what should go into a press release and how it should be laid out. The following guidelines apply.

1 The editor is short of time and initially wants a summary of the story, not all the detail. Its purpose is to create interest and encourage further action.
2 At the top should be a headline which identifies the news story or event.
3 The main copy should be immediately intelligible, getting quickly to the heart of the matter and providing sufficient context to highlight the significance of the story – *why* it is of interest as well as *what* is of interest, e.g. 'this is the first 16-year-old ever to have won the award'. The writing style should be more conversational than formal; avoid legal or technical jargon.

4   Double-check all facts: names (first name and surname), personal qualifications, titles, occupations, ages, addresses, dates, times, places, sums of money, percentages, etc.

5   End with a contact name, office of origin, postal and email address, telephone numbers (including a home number), website, date and time of issue.

6   On paper or sent as a PDF document it should be typed, double-spaced with broad margins, on one side of a standard size such as A4, with a distinctive logo or heading – or on coloured paper to make it stand out. When sending copies to different addresses within a radio station, this should be made clear. An embargo should only be placed on a press release if the reason is obvious – for example, in the case of an advance copy of a speech, or where it is sensible to allow time to digest or analyse a complex issue before general publication. Radio is an immediate medium and editors have no inclination to wait simply in order to be simultaneous with their competitors.

A summary

To sum up the business of news. Good journalism is based on an oft-quoted set of values – it must be accurate and truthful, it stems from curiosity, observation and enquiry, and it must do more than react to events, it must report how those events affect people. In attempting to be impartial and objective it must actively seek out and test views. It has to make sense of what has happened and what is happening, for readers, listeners and viewers – that is, it has to be in the public interest – resisting the pressures of politicians, advertisers and others who may wish to cast the world in a light favourable to their own interests or cause.

The journalist might want to be an influence for good – a watchdog for society exposing corruption, driven by an interest in life, a desire to inform, and a love of the language. The motivation could be to become known as a good writer – accurate and ordered, with a distinctive style. Suffice it to say that any society founded on democratic freedom of choice requires a free flow of honest news. It is totally pointless to run a broadcast news service unless it is trusted and believed.

8

Interviewing

The aim of an interview is to provide, in the interviewee's own words, facts, reasons or opinions on a particular topic so that the listener can form a conclusion as to the validity of what he or she is saying.

The basic approach

It follows from the above definition that the opinions of the interviewer are irrelevant. An interview is not a confrontation, which it is the interviewer's object to win. A questioner whose attitude is one of battle, to whom the interviewee represents an opponent to be defeated, will almost inevitably alienate the audience. The listener quickly senses such hostility and, feeling that the balance of advantage already lies with the broadcaster, the probability is they will side with the underdog. For an interviewer to be unduly aggressive, therefore, is counter-productive. On the other hand, an interview is not a platform for the totally free expression of opinion. The assertions made are open to be challenged.

The interviewer should never get drawn into answering a question that the interviewee may put – an interview is not a discussion. We are not concerned here with what has been referred to as 'the personality interview' where the interviewer, often the host of a television 'chat show', acts as the grand inquisitor and asks guests their opinions on a whole range of topics. Within this present definition it is solely the interviewee who must come through and in the interviewer's vocabulary the word 'I' should be absent. Deference is not required but courtesy is; persistence is desirable, harassment not. The interviewer is not there to argue, to agree or disagree, nor to comment on the answers. The interviewer's job is simply to ask appropriate and searching questions, and this requires good preparation and astute listening.

The interview is essentially a spontaneous event. Any hint of its being rehearsed damages the interviewee's credibility to the extent of the listener believing the whole thing to be 'fixed'. For this reason, while the topic may be discussed generally beforehand, the actual questions should not be provided in advance. The interview must be what it appears to be – questions and answers for the benefit of the eavesdropping listener. The interviewer is acting on behalf of the listener, asking the questions that the listener would want to ask. More than this, it is an opportunity to provide not only what the listener wants to know, but also what the public may need to know. At least as far as the interviewing of political figures is concerned, the interview should represent a contribution towards a democratic society, i.e. the proper questioning of people who, because of the office they hold, are accountable to the electorate. It is a valuable element of broadcasting and care should be taken to ensure it is not damaged, least of all by casual abuse in the cult of personality on the part of broadcasters.

Types of interview

For the sake of simplicity three types of interview can be identified, although any one situation could involve all three categories to a greater or lesser extent. These are the informational, the interpretive and the emotional interviews.

Obviously, the purpose of the *informational interview* is to impart information to the listener. The sequence in which this is done becomes important if the details are to be clear. There may be considerable discussion beforehand to clarify what information is required and to allow time for the interviewee to recall or check any statistics. Topics for this kind of interview include: the action surrounding a military operation, the events and decisions made at a union meeting, or the proposals contained in the city's newly announced development plan.

The *interpretive interview* has the interviewer supplying the facts and asking the interviewee either to comment on them or to explain them. The aim is to expose the reasoning behind decisions and allow the listener to make a judgement on the implicit sense of values or priorities. Replies to questions will almost certainly contain statements in justification of a particular course of action, which should themselves also be questioned. The interviewer must be well briefed, alert and attentive to pick up and challenge the opinions expressed. Examples in this category would be a government minister on the reasons for an already published economic policy, why the local council has decided on a particular route for a new road, or views of the clergy on proposals to amend the divorce laws. The essential point is that the interviewer is not asking for the facts of the matter, since these will be generally known; rather, he or she is investigating the interviewee's reaction to the facts. The discussion beforehand may be quite brief, the interviewer outlining the purpose of the interview and the limits of the subject to be pursued. Since the content is reactive, it should on no account be rehearsed in its detail.

The aim of the *emotional interview* is to provide an insight into the interviewee's state of mind so that the listener may better understand what is involved in human terms. Specific examples would be the feelings of relatives of people trapped in the debris of an earthquake, the survivors of a terrorist incident, the euphoria surrounding the moment of supreme achievement for an athlete or successful entertainer, or the anger felt by people involved in an industrial dispute. It is the strength of feeling present rather than its rationality which is important and clearly the interviewer needs to be very sensitive in handling such situations. There is praise and acclaim for asking the right question at the right time in order to illuminate a matter of public interest, even when the event itself is tragic. But quick criticism follows for being too intrusive into private grief. It is in this respect that the manner of asking a question is as important as its content, possibly more so.

Another difficulty that faces the interviewer here is to reconcile the need to remain an impartial observer while not appearing indifferent to the suffering in the situation. The amount of time taken in preliminary conversation will vary considerably depending on the circumstances. Establishing the necessary relationship may be a lengthy process – but there is a right moment to begin recording and it is important for the interviewer to remain sensitive to this judgement. Such a situation allows little opportunity for retakes.

These different categories of interview are likely to come together in preparing material for a documentary or feature. First the facts, background information or sequence of events; then the interpretation, meaning or implication of the facts; finally their effect on people, a personal reaction to the issue. The *documentary interview* with, for example, a retired politician will take time but should be as absorbing for the interviewer as it will be for the listener. The process of recalling history should surprise; it should throw new light on events and people, and reveal the character of the person. Each interview is different but two principles remain for the interviewer – listen hard, and keep asking 'why?'

Related to the documentary interview, but not concerned necessarily with a single topic, is the style of interviewing that contributes to *oral history*. Every station, national, regional or local, should assume some responsibility for maintaining an archival record of its area. Not only does it make fascinating material for future programmes but it becomes of value in its own right as it marks the changes that affect every community. The elderly talking about their childhood or their parents' values, craftsmen describing their work, children on their expectations of growing up, the unemployed, music makers, shopkeepers, old soldiers. The list is endless. The result is an enriching library of accent, story, humour, nostalgia and idiosyncrasy. But to capture the voices of people unused to the microphone takes patience and a genuinely perceptive interest in others. It may take time to establish the necessary rapport and to put people at their ease. On the other hand, when talking to the elderly, it is often advisable to start recording as soon as the interviewee recounts any kind of personal memory. They might not understand whether or not the interview has begun and will not be able to repeat what they said with the same freshness. Preparation and research beforehand into personal backgrounds help to recall facts and incidents that the interviewee often regards as too insignificant or commonplace to mention. People generally like talking about themselves and it takes a quick-thinking flexibility to respond appropriately, to know when a conversation should be curtailed, and when moved on. The rewards, however, are considerable. A final point: such recordings need to be well documented – it is one thing to have them in the archives, but quite another to secure their speedy and accurate retrieval.

Securing the interviewee

An initial telephone call may secure consent to an interview, enabling the interviewer to arrive with a recorder later that day. Alternatively, arrange a studio interview several days in advance giving time for the interviewee to prepare. It has been known of course for interviewers to arrive equipped with a recorder and ask for an interview on the spot – or even to telephone and tell the interviewee that he or she is already on the air! This last technique is bad practice, and besides contravenes the basic right of an interviewee that no telephone conversation will be broadcast or recorded without consent. There can be no question of doing this without their knowledge (see Secret recording, p. 90). It should be a standard procedure for all broadcasters that in contacting a potential interviewee nothing is begun, recorded or transmitted without the interviewee being properly informed about the station's intention. Even so, to be rung up and asked for an interview there and then is asking a great deal. It is in no one's interest, least of all the listener's, to have an ill-thought-out interview with incomplete replies and factual errors.

What the interviewee should know

Since it is impossible to interview someone who does not want to be interviewed, it is reasonable to assume that the arrangement is mutually agreed. It may be acceptable for an interviewer to approach someone at a news event without any prior warning, but in a more formal setting it's best to contact a potential interviewee first to ask whether an interview might take place. The information that the interviewee needs at this point is:

1   What is it to be about? Not the exact questions but the general areas, and the limits of the subject.
2   Is it to be broadcast live or recorded?
3   How long is it to be? Is the broadcast a major programme or a short item? This sets the level at which the subject can be dealt with and helps to guard against the interviewee recording a long interview without being aware that it must be edited to another length.
4   What is the context? Is the interview part of a wider treatment of the subject with contributions from others or a single item in a news or magazine programme?
5   For what audience? A local station, network use, for syndication?
6   Where? At the studio or elsewhere?
7   When? How long is there for preparation?

No potential interviewee should feel rushed into undertaking an interview and certainly not without establishing the basic information outlined above. Sometimes a fee is paid but this is unlikely in community radio; it is worth making this clear.

Preparation before the interview

It is essential for the interviewer to know what he or she is trying to achieve. Is the interview to establish facts, or to discuss reasons? What are the main points that must be covered? Are there established arguments and counter-arguments to the case? Is there a story to be told? The interviewer must obviously know something of the subject and a briefing from the producer, combined with some personal research, is highly desirable. An essential is absolute certainty of any names, dates, figures or other facts used within the questions. It is embarrassing for the expert interviewee to correct even a trifling factual error in a question – it also represents a loss of control, for example:

> 'Why was it only 3 years ago that you began to introduce this new system?'
> 'Well actually it was 5 years ago now.'

It is important, although easily overlooked, to know exactly who you are talking to:

'As the chairman of the company, how do you view the future?'

'No, I'm the managing director . . . '

It makes no difference whatsoever to the validity of the question but a lack of basic care undermines the questioner's credibility in the eyes of the interviewee and, even more important, in the ears of the listener.

Having decided what has to be discovered, the interviewer must then structure the questions accordingly. Question technique is dealt with in a later section but it should be remembered that what is actually asked is not necessarily formulated in precise detail beforehand. Such a procedure could easily be inflexible and the interviewer may then feel obliged to ask the list of questions irrespective of the response by the interviewee. Preparation calls for the careful framing of alternative questions – with consideration of the possible responses so that the next line of enquiry can be worked out.

For example, you want to know why a government minister is advocating a reduction in government welfare payments, with possible hardship for many poorer people. If the question put is simply 'Why are you advocating . . . ?', the reply is likely to be a stock answer on the need to balance the books and reduce the country's deficit. Such a response is known by most people so the interview merely repeats the position, it does not carry the issue forward. To move ahead, the interviewer must anticipate and think laterally, to be in a position to put questions that *well-informed people* are asking – about other ways of achieving the same end – making welfare payments taxable, limiting the total that can be claimed, other ways of making up the shortfall, the cost to the country of additional help for the disabled, and so on.

To summarise, an interviewer's normal starting point will be:

1   to obtain sufficient briefing and background information on the subject and the interviewee;
2   to have a detailed knowledge of what the interview should achieve, and at what length;
3   to know what the key questions are;
4   by anticipating likely responses, to have ready a range of supplementary questions.

The pre-interview discussion

The next stage, after the preparatory work, is to discuss the interview with the interviewee. The first few minutes are crucial. Each party is sizing up the other and the interviewer must decide how to proceed.

There is no standard approach: each occasion demands its own. The interviewee might respond to the broadcaster's brisk professionalism or might better appreciate a more sympathetic attitude. He or she may need to feel important, or the opposite. The interviewee in a totally unfamiliar situation might be so nervous as to be unable to marshal their thoughts properly; their entire language structure and the speed of delivery might be affected. Under stress it might not even be possible to listen fully to your questions. The good interviewer will be aware of this and will work hard to enable the interviewee's thinking and personality to emerge. Whatever the circumstances, the interviewer has to get it right, and has only a little time in which to form the correct judgements.

The interviewer indicates the subject areas to be covered but is well advised to let the interviewee do most of the talking. This is an opportunity to confirm some of the facts, and it helps the interviewee to release some of the tensions while allowing the interviewer to anticipate any problems of language, coherence or volume.

It is wrong for the interviewer to get drawn into a discussion of the matter, particularly if there is a danger of revealing a personal attitude to the subject. Nor is it generally helpful to adopt a hostile manner or imply criticism. This might be appropriate during the interview, but even so it is not the interviewer's job to conduct a judicial enquiry, or to act as prosecuting counsel, judge and jury.

The interviewer's prime task at this stage is to clarify what the interview is about and to create the degree of rapport that will produce the appropriate information in a logical sequence at the right length. It is a process of obtaining the confidence of the interviewee while establishing a means of control. A complex subject needs to be simplified, and distilled for the purposes of, say, a three-minute interview – there must be no technical or specialist jargon, and the intellectual and emotional level must be right for the programme. Above all, the end result should be interesting.

It is common and useful practice to say beforehand what the first question will be, since in a 'live' situation it can help to prevent a total 'freeze' as the red light goes on. If the interview is to be recorded, such a question may serve as a 'dummy' to be edited out later. In any event it helps the interviewee to relax and to feel confident about starting. The interviewer should begin the actual interview with as little technical fuss as possible, the preliminary conversation proceeding into the interview with the minimum of discontinuity.

Question technique

An interview is a conversation with an aim. On the one hand, the interviewer knows what that aim is and knows something of the subject.

On the other, by taking the place of the listener, he or she is asking questions in an attempt to discover more. This balance of knowledge and ignorance can be described as 'informed naivety'.

The question type will provide answers of a corresponding type. In their simplest form they are:

1   Who?   asks for fact. Answer – a person.
2   When?   asks for fact. Answer – a time.
3   Where?   asks for fact. Answer – a place.
4   What?   asks for fact or an interpretation of fact.
            Answer – a sequence of events.
5   How?   asks for fact or an interpretation of fact.
            Answer – a sequence of events.
6   Which?   asks for a choice from a range of options.
7   Why?   asks for opinion or reason for a course of action.

These are the basic 'open' question types, of which there are many variations. For example:

'How do you feel about . . . ?'
'To what extent do you think that . . . ?'

The best of all questions, and incidentally the one asked least, is 'why?' Indeed, after an answer it may be unnecessary to ask anything other than 'why is that?' The 'why' question is the most revealing of the interviewee since it leads to an explanation of actions, judgements, motivation and values:

'Why did you decide to . . . ?'
'Why do you believe it necessary to . . . ?'

It is sometimes said that it is wrong to ask 'closed' questions based on the 'reversed verb':

'Are you . . . ?'
'Is it . . . ?'
'Will they . . . ?'
'Do you . . . ?'

What the interviewer is asking here is for either a confirmation or a denial; the answer to such a question is either yes or no. If this is really what the interviewer is after, then the question structure is a proper one. If, however, it is an attempt to introduce a new topic in the hope that the interviewee will continue to say something other than yes or no, it is an ill-defined question. As such it is likely to lead to the interviewer's loss of control, since it leaves the initiative completely with the interviewee. In this respect the reversed verb question is a poor substitute for a question that is specifically designed to point the interview in the desired direction. The reversed verb form should therefore only be used when a yes/no answer is what is required:

> 'Will there be a tax increase this year?'
> 'Are you running for office in the next election?'

Question 'width'

This introduces the concept of how much room for manoeuvre the interviewer is to give the interviewee. Clearly where a yes/no response is being sought, the interviewee is being tied down and there is little room for manoeuvre; the question is very narrow. On the other hand, it is possible to ask a question that is so enormously wide that the interviewee is confused as to what is being asked:

> 'You've just returned from a study tour of Africa, tell me about it.'

This is not, of course, a question at all; it is an order. Statements of this kind are made by inexperienced interviewers who think they are being helpful to a nervous interviewee. In fact the reverse is more likely, with the interviewee baffled as to where to start.

Another type of question, which again on the face of it seems helpful, is the 'either/or' question:

> 'Did you introduce this type of engine because there is a new market for it, or because you were working on it anyway?'

The trouble here is that the question 'width' is so narrow that in all probability the answer lies outside it, so leaving the interviewee little option but to say 'Well neither, it was partly . . . '. Things are seldom so clear cut as to fall exactly into one of two divisions. In any case it is not up to the interviewer to suggest answers; what the questioner wanted to know was:

> 'Why did you introduce this type of engine?'

A currently common but slightly vague question form is – 'What do you make of . . . ?' This is generally preceded by a synopsis of a situation, event or comment which ends with: 'What do you make of the report/ what she said/the minister's decision?', etc. Akin to 'What do you think of . . . ?', it's a very wide question leaving a lot of latitude for the response.

Devil's advocate

If an interviewee is to express their own point of view fully and to answer various critics, it will be necessary for those opposing views to be put. This provides the opportunity of confronting and demolishing the arguments to the satisfaction, or otherwise, of the listener. In putting such views the interviewer must be careful not to become associated with them, nor to be associated in the listener's mind with the principle of opposition. The role is to present propositions that are known to have been expressed elsewhere, or to voice the doubts and arguments that can reasonably be expected to exist in the listener's mind. In adopting the 'devil's advocate' approach, common forms of question are:

> 'On the other hand it has been said that . . . '
> 'Some people would argue that . . . '
> 'How do you react to people who say that . . . ?'
> 'What would you say to the argument that . . . ?'

The first two examples as they stand are not questions but statements, and if left as such will bring the interview dangerously close to being a discussion. The interviewer must ensure that the point is put as an objective question.

It has been said in this context that 'you can't play good tennis with a bad opponent'. The way in which broadcasters present counter-arguments needs care, but experienced interviewees welcome hard questions as a means of making their case more easily understood.

Multiple questions

A trap for the inexperienced interviewer, obsessed with the fear that the interviewee will be lacking in response, is to ask two or more questions at once:

> 'Why was it that the meeting broke up in disorder, and how will you prevent this happening in future? Was there a disruptive element?'

The interviewee presented with two questions may answer the first and then genuinely forget the second, or may exercise the apparent option to answer whichever one is easier. In either case there is a loss of control on the part of the interviewer as the initiative passes to the interviewee.

Questions should be kept short and simple. Long rambling circumlocutory questions will get answers in a similar vein; this is the way conversation works. The response tends to reflect the stimulus – this underlines the fact that the interviewer's initial approach will set the tone for the whole interview.

Beware the interviewer who has to clarify the question after asking it:

'How was it you embarked on such a course of events, I mean what made you decide to do this – after all at the time it wasn't the most obvious thing to do, was it?'

Confusion upon confusion, and yet this kind of muddle can be heard on the air. If the purpose of the question is not clear in the interviewer's mind, it is unlikely to be understood by the interviewee – the listener's confusion is liable to degenerate into indifference and subsequent total disinterest.

Leading questions

Lazy, inexperienced or malicious questioning can appear to cast the interviewee in an unfavourable light at the start:

'Why did you start your business with such shaky finances?'
'How do you justify such a high-handed action?'

It is not up to the interviewer to suggest that finances are shaky or that action is high-handed, unless this is a direct quote of what the interviewee has just said. Given the facts, the listener must be able to determine from what the interviewee says whether the finances were sufficient or whether the action was unnecessarily autocratic. Adjectives that imply value judgements must be a warning signal for interviewee and listener alike, that all is not quite what it appears to be. Here is an interviewer who has a point to make, and in this respect may not be properly representing the listener. The questions can still be put in a perfectly acceptable form:

'How much did you start your business with?' (fact)
'At the time did you regard this as enough?' (yes/no)
'How do you view this now?' (judgement)
'What would you say to people who might regard this action as high-handed?' (the 'devil's advocate' approach already referred to)

It is surprising how one is able to ask very direct, personally revealing, 'hard' questions in a perfectly acceptable way by maintaining at the same time a calmly pleasant composure. When a broadcaster is criticised for being over-aggressive, it is generally the manner rather than the content that is being questioned. Even persistence can be politely done:

'With respect, the question was *why* this had happened.'

In asking *why* something happened it is not uncommon to get, in effect, *how* it happened, particularly if the interviewee wishes to be evasive. Evasion a second or third time becomes obvious to the listener and there is no need for the interviewer to labour the point: it is already made.

Non-questions

Some interviewers delight in making statements instead of asking questions. Again, the danger is that the interview may become a discussion. For example, an answer might be followed by the statement: 'This wouldn't happen normally' instead of with the question 'Is this normal?' Again, the statement 'You don't appear to have taken that into account' instead of the question 'To what extent have you taken that into account?'

The fault lies in the question not being put in a positive way so that the interviewee can respond as he or she likes, perhaps by asking a question. The interviewer may then find it difficult to exercise control over both the subject matter and the timing.

Occasionally interviewers ask whether they can ask questions:

'Could I ask you if . . . ?'
'I wonder whether you could say why . . . ?'

This is unnecessary of course since in the acceptance of the interview there is an agreement to answer questions. There may occasionally be justification for such an approach when dealing with a particularly sensitive area where the interviewer feels the need to proceed gently. This phraseology can be used to indicate that the interviewer recognises the difficulty inherent in the question. Much more likely, however, it is used by accident when the questioner is uncertain as to the direction of the interview and is 'padding' in order to create some thinking time. Such a device is likely to give the listener the feeling that time is actually being wasted.

Non-answers

The *accidental evasion* of questions may be due to the interviewee genuinely misunderstanding the question, or the question may have been badly put; in either case the interview goes off on the wrong tack. When recording this is easily remedied, but if it happens on the air the listener may be unable to follow and lose interest, or regard the interviewee as stupid or the interviewer incompetent. One or other of the parties must bring the subject back to its proper logic.

The *deliberately evasive* technique often adopted by the non-answerer is to follow the interviewer's question with another:

'That certainly comes into it, but I think the real question is whether . . . '

If the new question genuinely progresses the subject, the listener will accept it. If not, the expectation is that the interviewer will put the question again. Rightly or wrongly the listener will invariably believe that someone who does not answer has something to hide and is therefore suspect.

There may be genuine reasons why '*I can't comment on that*' is an acceptable answer to a question. The facts might not yet be known with sufficient certainty, there may be a legal process pending, a need to honour a guarantee given to a third party, or the answer should properly come from another quarter. It might be that an interviewee legitimately wishes to protect commercially sensitive information – a factor that occurs in the sporting as well as the business world. Nevertheless, the interviewee must be seen to be honest and to say why an answer cannot be given:

> 'It would be wrong of me to anticipate the report . . . '
> 'I can't say yet until the enquiry is finished . . . '
> 'I'm sure you wouldn't expect me to give details, but . . . '

Even if the inability to give a particular answer has been discussed beforehand, an interviewee should still expect to have the question put if it is likely to be in the listener's mind.

Non-verbal communication

Throughout the interview the rapport established earlier must continue. This is chiefly done through eye contact and facial expression. Once the interviewer stops looking at the respondent, perhaps for a momentary glance at the recorder or a page of notes, there is a danger of losing the thread of the interview. At worst, the interviewee will look away, and then thoughts as well as eyes are liable to wander. The concentration must be maintained. The eyes of the interviewer will express interest in what is being said – the interviewer is never bored. It is possible to express surprise, puzzlement or encouragement by nodding one's head. In fact, it quickly becomes annoying to the listener to have these reactions in verbal form – 'ah yes', 'mm', 'I see'.

Eye contact is also the most frequent means of controlling the timing of the interview – of indicating that another question needs to be put. It may be necessary to make a gesture with the hand, but generally it is acceptable to butt in with a further question. Of course the interviewer must be courteous and positive to the point of knowing exactly what to say. Even the most talkative interviewee has to breathe and the signs of such small pauses should be noted beforehand so that the interviewer can use them effectively.

During the interview

The interviewer must be actively in control of four separate functions – the technical, the aims of the interview, the supplementary question and the timing.

The *technical* aspects must be constantly monitored. Is the background noise altering, so requiring a change to the microphone position? Is the position of the interviewee changing relative to the microphone, or have the voice levels altered? If the interview is being recorded, is the device continuing to function correctly and the level indicator giving a proper reading?

The *aims* of the interview must always be kept in mind. Is the subject matter being covered in terms of the key questions decided beforehand? Sometimes it is possible for the interviewer to make a positive decision and change course but in any event it is essential to keep track of where the interview is going.

The *supplementary question* – it is vital that the interviewer is not so preoccupied with the next question as to fail to listen to what the interviewee is saying. The ability to listen and to think quickly are essential attributes of the interviewer. This leads to the facility of being able to ask the appropriate follow-up question for clarification of a technicality or piece of jargon, or to question further the reason for a particular answer. Where an answer is being given in an unnecessarily academic or abstract way, the interviewer should ask to have it turned into a factual example.

The *timing* of the interview must be strictly adhered to. This is true whether the interview is to be of half an hour or two minutes. If a short news interview is needed, there is little point in recording 10 minutes with a view to reducing it to length later. There may be occasions when such a time-consuming process will be unavoidable, even desirable, but the preferred method must be to sharpen one's mind beforehand, rather than rely on editing afterwards. Thus the interviewer when recording keeps a mental clock running. It stops when it hears an answer which is known to be unusable but continues again on hearing an interesting response. This controls the flow of material so that the subject is covered as required in the time available. This sense of time is invaluable when it comes to doing a 'live' interview when, of course, timing is paramount. Such a discipline is basic to the broadcaster's skills.

One thing that must never happen when using a handheld mic, is for the interviewer to hand the mic to the interviewee. This leads to a loss of control of content, of timing and often of levels. Any attempt by the interviewee to take the mic has to be resisted.

Winding up

The word 'finally' should only be used once. It may usefully precede the last question as a signal to the interviewee that time is running out and that anything important left unsaid should now be included. Other signals of this nature are words such as:

'Briefly, why . . . '
'In a word, how . . . '
'At its simplest, what . . . '

It is a great help in getting an interviewee to accept the constraint of timing if the interviewer has remembered to say beforehand the anticipated duration.

Occasionally an interviewer is tempted to sum up. This should be resisted since it is extremely difficult to do without making some subjective evaluations. It should always be borne in mind that one of the broadcaster's greatest assets is an objective approach to facts and an impartial attitude to opinion. To go further is to forget the listener, or at least to underestimate the listener's own ability to form a conclusion. A properly structured interview shouldn't need a summary, much less should it be necessary to impose on the listener a view of what has been said.

If the interview has been in any sense chronological, a final question looking to the future will provide an obvious place to stop. A positive convention as an ending is simply to thank the interviewee for taking part:

'Mr Jones, thank you very much.'

However, an interviewer quickly develops an ear for a good out-cue and it is often sufficient to end with the words of the interviewee, particularly if they have made an amusing or strongly assertive point.

After the interview

The interviewer should feel that it has been an enlightening experience that has provided a contribution to the listener's understanding and appreciation of both the subject and the interviewee. If the interview has been recorded, it should be immediately checked by playing back the last 15 seconds or so. No more, otherwise the interviewee, if they are able to hear it, is sure to want to change something and one embarks on a lengthy process of explanation and reassurance. The editorial decision as to the content of the interview as well as the responsibility for its technical quality rests with the interviewer. If for any reason it is necessary to retake parts of a recording, it is generally wise to adopt an entirely fresh approach rather than attempt to recreate the original. Without making problems for the later editing, the questions should be differently phrased to avoid an unconscious effort to remember the previous answer. This amounts to having had a full rehearsal and will almost certainly provide a stale end product. The interviewee who is losing track of what is going into the final piece is also liable to remark ' . . . and as I've already explained . . . ' or ' . . . and as we were saying a moment ago . . . '. Such references to material which has been edited out will naturally mystify the listener, possibly losing concentration on what is currently being said.

If the interview has been recorded, the interviewee will probably want to know the transmission details. If the material is specific to an already scheduled programme, this information can be given with some confidence. If, however, it is a news piece intended for the next bulletin, it is best not to be too positive lest it be overtaken by a more important story and consequently held over for later use, or not used at all. Tell the interviewee when you hope to broadcast it but if possible avoid total commitment.

Thank the interviewee for their time and trouble and for taking part in the programme. If a journey to the studio is involved, it may be normal to offer travelling expenses or a fee according to station policy. Irrespective of how the interview has gone, professional courtesy at the closing stage is important. After all, you might want to talk again tomorrow.

Style

Evidence suggests that political interviewing reflects the conduct of government nationally. If government and opposition are engaged in hard-hitting debate, with individuals making not only party but personal points, then the media will assume this same style. If, however, opposition views are generally suppressed – or in some cases may not exist at all – then ministers will not expect, and may not allow, any challenge from the media.

Different cultures, different nations, have quite diverse views about authority, even when it has been elected by popular vote. In many places, broadcasters may not interview a government minister without first supplying the questions to be asked. In some it is the minister who provides the questions – a practice hardly in the public interest.

In the West, where authority of any kind is not held in particularly high regard, radio and television interviewing is frequently directly challenging, especially of publicly answerable figures. It should not, however, adopt a superior tone nor become a personal confrontation between interviewer and interviewee. The first sign of this is the interviewer making assertions in order to score points instead of asking questions. The producer must vigorously analyse interviewing style to cut short such tendencies that quickly attract public criticism as the media overreaching itself (see Chapter 5 on Ethics).

Interviewing 'cold'

One of the more challenging aspects of interviewing is in the long sequence programme, such as a breakfast show, where a number of interviewees have been lined up and brought to the studio. The producer/interviewer has little or no opportunity to prepare with each person beforehand. The situation can be improved either through the use of music to give two or three minutes' thinking time, or by having two presenters interviewing alternately. One essential is to have adequate research notes on each interviewee provided by a programme assistant.

If a presenter is meeting a guest for an immediate interview, the basic information that he or she needs is:

- topic title;
- the person's name;
- their position, role, job, status, etc.;
- the key issue at stake;
- this person's view of the issue – with actual quotes if possible;
- notes on possible questions or approaches, what other people with different views have said.

This kind of interviewing, jumping as it does from subject to subject in a quite unrelated way, calls for great flexibility of mind. Its danger is one of superficiality, where nothing is dealt with in depth or comes to a satisfactory conclusion. ' . . . I'm afraid we'll have to leave it there because of time.'

The situation is compounded when the interviewee is at the other end of a line and can't be seen. Judging the appropriate style is difficult, rapport between the interviewer and interviewee is slight, the result can feel distant and cold, and the need to press on with the running order can make a hurried ending seem rude as far as the listener is concerned – unaware of studio pressures. All these are factors that the producer of the programme should evaluate. The programme might be fast moving, but is it superficial? It might cover good topics but is it essentially lightweight? Is it brisk at the expense of being brusque?

Interviewing through a translator

News in particular may involve interviewing someone who does not speak your own language. When the interview is 'live', there is no option but to go through an interpreter with the somewhat laborious process of sequential translation – so keep the questions short and simple:

'When were the soldiers here?'
'What has happened to your home?'
'Why did they destroy the village?'
'What happened to your family?'

Depending of course on the circumstances, answers even in another language can communicate powerfully. The translation provides the content of what is said but the replies themselves will define the spirit and strength of feeling in a crisis situation. If the interview is recorded then simultaneous translation is possible through subsequent dubbing and editing. The first question is put, the reply begins and is faded down under the translation. The second question is followed by the translation of the second answer, and so on. It is good to have snatches of the interviewee's voice occasionally, especially in a long interview, as a reminder that we are using an interpreter. What is totally removed in the editing is the translation of the interviewer's questions. Naturally it is best to use a translator's voice similar to that of the interviewee, i.e. a man for a man, or a young girl for a young girl, etc. While it is not always possible to do this in the news context it should be carefully considered for a documentary or feature programme.

Location interviews

The businessman in his office, the star in her dressing room, the worker in the factory or out of doors – all are readily accessible with a Flashmic or other portable recorder and provide credible programming with atmosphere. Yet each may pose special problems of noise, interruption and right of access – difficulties also inherent in the vox pop (see Chapter 9).

To remain legal the producer must observe the rules regarding public and private places. Permission is usually required to interview inside a place of entertainment, business premises, a factory, shop, hospital or school. In this last case it is worth remembering that consent of parents or a guardian should normally be sought before interviewing anyone under the age of 16. Working in a non-public area also means that if a broadcaster is asked to leave it is best to do so – or else run the risk of a charge of harassment or trespass.

In any room other than a studio, the acoustic is likely to be poor, with too much reflected sound. It is possible to overcome this to an acceptable degree by avoiding hard, smooth surfaces such as windows, desktops, vinyl floors or plastered walls. A carpeted room with curtains and other furnishings is generally satisfactory, but in unfavourable conditions the

Location – try to avoid . . .

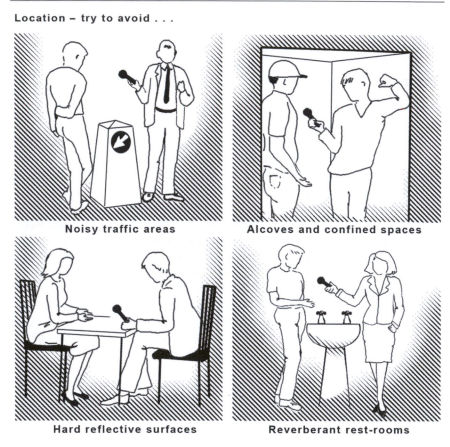

Noisy traffic areas   Alcoves and confined spaces

Hard reflective surfaces   Reverberant rest-rooms

Figure 8.1   Some interviewing situations to avoid. These would be noisy, acoustically poor, or at least asymmetrical or physically awkward

best course is to work closer to the microphone, while also reducing the level of input to the recorder.

The same applies to locations with a high level of background noise. Nevertheless, the machine shop or aeroplane cockpit need present no insuperable technical difficulty; again the answer is to work closer to the mic and reduce the record level. This will sufficiently discriminate between the foreground speech and the background noise. A greater problem arises where the sounds are intermittent – an aircraft passing overhead, a telephone ringing or clock striking. At worst these may be so overwhelming as to prevent the interview from being audible, but even if this is not the case, sudden noises are a distraction for the listener that a constant level of background is not. Background sounds which vary in volume and quality can also represent a considerable problem for later editing – a point

Location – try to find . . .

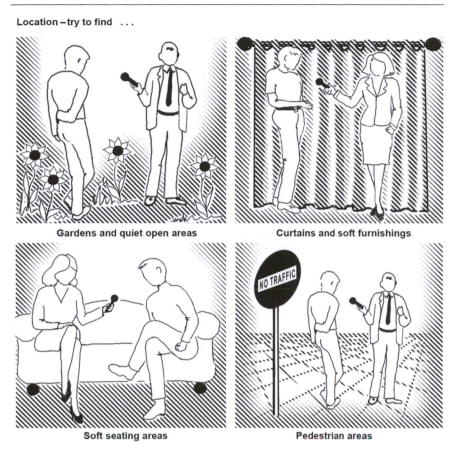

Gardens and quiet open areas   Curtains and soft furnishings

Soft seating areas   Pedestrian areas

Figure 8.2 Some good places for location interviews. These provide a low or at least a constant background noise, acoustically absorbent surroundings or a comfortable symmetry

interviewers should remember before they start. The greatest difficulty in this respect arises when an interview has been recorded against a background of music. It is then almost impossible to edit.

Only the very experienced should attempt interviewing using a stereo microphone – and it should then be fixed with a mic stand. Even small amounts of movement cause an apparent relocation of the environmental sounds with the consequent disorientation of the listener. Static stereo mics are good for recording location effects, but interviewing almost always calls for a handheld mono omni-directional mic, or two small clip-on personal mics. A handheld recorder with stereo mics is best used in its mono mode. A stereo recording can, of course, always be re-mixed on editing, adjusting the overlap between the two channels to provide the desired spatial relationship.

It is generally desirable for location interviews to have some acoustic effect or background noise and only practical experience will indicate how to achieve the proper balance with a particular type of microphone. When in doubt, priority should be given to the clarity of the speech.

As with the studio interview, the discussion beforehand is aimed at putting the interviewee at ease. When outside, using a portable recorder, part of this same process is to show how little equipment is involved. The recorder should be checked during this preliminary conversation. It is important to handle it in front of the interviewee and not spring the technicalities at the last moment. Before starting it is advisable to test the system by 'taking level', i.e. by recording some brief conversation to hear the relative volume of the two voices.

If a separate microphone is to be handheld, its cable should be looped around the hand, not tightly wrapped around it, so that no part of the cable leading to the recorder is in contact with the body of the microphone. This prevents any movement in the cable from making itself heard as vibration in the microphone. The mic should be out of the eye-line at a point where it can remain virtually stationary throughout. Only in conditions of high background noise should it be necessary to move the mic alternately towards interviewer and interviewee. Even so, this is preferable to the use of an automatic gain control (AGC) on the recorder, which affects the speaker's voice and background

Figure 8.3 When using a mic with a cable it is formed into two loops. The loop around the finger is kept away from the microphone case to prevent extraneous noises or mic rattles. A windshield should be used to prevent wind noise and voice 'pops'

noise together, i.e. it does not discriminate between them as microphone movement does. AGC should therefore be switched 'out'. A satisfactory playback of this trial recording is the final check before beginning the interview.

Some further rules in the use of portable recorders, including smartphones (see p. 103):

1 Where necessary check the ability of the machine to work in unusual conditions, e.g. bumpy vehicles, high humidity, electrical radiation or magnetic fields, or at low temperature; or its suitability for specialist functions, such as recording in a coal mine.
2 Always check it before leaving base – record and replay.
3 Always take spare batteries, a spare card and, where applicable, a charger. If the recorder has an internal cell make sure it's fully charged.
4 Always use a microphone windshield when recording out of doors.
5 When using a smartphone, speak into the side of the device not directly into it.
6 Use the best recording quality in the audio settings.
7 Take headphones if the recorder doesn't have its own speaker.

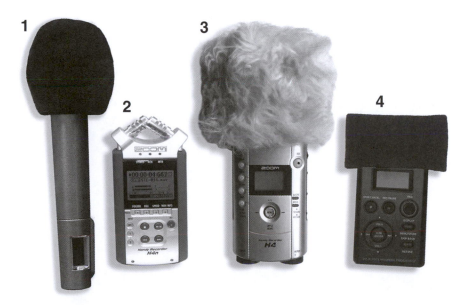

Figure 8.4 Handheld digital mics. 1. An HHB Flashmic. 2. The 4-channel Zoom H4n without a windshield to show the condenser stereo pair. It records on an internal SD card, and has two XLR sockets in the bottom. 3. The original 4-channel Zoom H4 with windshield. 4. A Marantz portable SD recorder

8   Do not leave equipment unattended, where it can be seen, even in a
     locked car.
9   Ensure that any rechargeable cells are put back on charge after use.

The triangle of trust

The whole business of interviewing is founded on trust. It is a three-way
structure involving the interviewer, the interviewee and the listener.

The interviewee trusts the interviewer to keep to the original statement
of intent regarding the subject areas and the context of the interview, and
also to maintain both the spirit and the content of the original in any subsequent editing. The interviewer trusts the interviewee to respond to questions in an honest attempt to illuminate the subject. The listener trusts the
interviewer to be acting fairly in the public interest without any secret
collusion between the interviewer and the interviewee. The interviewee
trusts the listener not to misrepresent what is being said and to understand
that within the limitations of time the answers are the truth of the matter.

This 'triangle of trust' is an important constituent not only of the media's
credibility but of society's self-respect as a whole. Should one side of
the triangle become damaged – for example, listeners no longer trusting
broadcasters, interviewees no longer trusting interviewers, or neither having sufficient regard for the listener – there is a danger that the process
will be regarded simply as a propaganda exercise. Under these conditions
it is no longer possible to distinguish between 'the truth as we see it' and
'what we think you ought to know'. Consequently, the underlying reason
for communication begins to disappear, thereby reducing broadcasting's
democratic contribution. The fabric of society is affected. This is to take
an extreme view, but every time a broadcaster misrepresents, every time
an interviewee lies or a listener disbelieves, we have lost something of
genuine value.

9

Vox pop

Vox populi is the voice of the people, or 'man in the street' interview. It's a different kind of interviewing from that discussed in the previous chapter, but the use of the opinions of 'ordinary' members of the public adds a useful dimension to the coverage of a topic that might otherwise be limited to a straight bulletin report or a studio discussion among officials or experts. The principle is for the broadcaster using a portable recorder to put one, possibly two, specific questions on a matter of public interest to people selected by chance, and to edit together their replies to form a distillation of the overall response. While the aim is to present a sample of public opinion, the broadcaster must never claim it to be statistically valid, or even properly representative. It can never be anything more than 'the opinions of some of the people we spoke to today'. This is because gathering material out on the streets for an afternoon magazine programme will almost certainly over-represent shoppers, tourists and the unemployed; and be low on businessmen, motorists, night shift workers and farmers! Since the interviewing is done at a specific time and generally at a single site, the sample is not really even random – it is merely unstructured and no one can tell what the views obtained actually represent. So no great claim should be made for the sample 'vox pop' on the basis of its being truly 'the voice of the people'.

It is easier to select a specific grouping appropriate to a particular topic – for example, early risers, commuters, children or lorry drivers. If the question is to do with an increase in petrol prices, one will find motorists, together with some fairly predictable comments, on any garage forecourt. Similarly, a question on medical care might be addressed to people coming out of a hospital. Incidentally, many apparently public places – such as hospitals, shopping malls, schools, and even railway stations – are in fact private property and the broadcaster should remember that he or she has no prescriptive right to work there without permission.

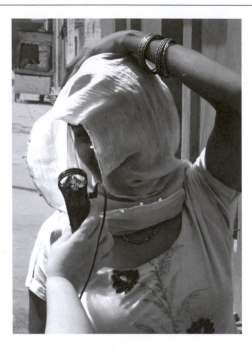

Figure 9.1   A vox pop in Northern India using a Zoom H1. A windshield is advised when recording outside (Photo: Kevin Keegan, FEBC)

As the question to which reaction is required becomes more specific, the group among which the interviews are carried out may be said to be more representative. Views on a particular industrial dispute can be canvassed among the pickets at the factory gate, opinions on a new show sought among the first-night audience. Nevertheless it is important in the presentation of vox pop material that the listener is told where and when it was gathered. There must be no weighting of the interview sample of which the listener is unaware. Thus the introductory sentence, 'We asked the strikers themselves what they thought' may mislead by being more comprehensive than the actual truth. A more accurate statement would be: 'We asked some of the strikers assembled at the factory gate this morning what they thought.' It is longer, but brevity is no virtue at the cost of accuracy.

Phrasing the question

Having decided to include a vox pop in the programme, the producer or the designated reporter must decide carefully the exact form of words to be used. The question is going to be addressed to someone with little preparation or 'warm-up' and so must be relatively simple and unambiguous. Since the object is to obtain opinions rather than a succession of 'yes/no' answers, the question form must be carefully constructed. Once decided, the same question is put each time, otherwise the answers cannot sensibly be edited together. A useful question form in this context is: 'What do you think of . . . ?' This will elicit an opinion which can if necessary be followed with the interviewer asking 'Why?' – this supplementary to disappear in the editing.

An example:

'What do you think of the proposal to raise the school-leaving age?'
'It sounds all right but who's going to pay for it?'
'I think it's a good idea, it'll keep the youngsters out of mischief.'
'It'll not do *me* any good, will it?'
'Bad in the short term, good in the long.'
('Why?')
'Well, it'll cause an enormous upheaval over teachers' jobs and classrooms and things like that, but it's bound to raise standards overall eventually.'
'I've not heard anything about it.'
'The cost! – and that means higher taxes all round.'
'I don't think it'll make much difference.'
('Why?')
'Because for those children who want to leave and get a job it'll be a waste and the brighter ones would have stayed on anyway.'
'I think it's a load of rubbish, there's too much education and not enough work.'

It is important that the question is phrased so that it contains the point to which reaction is required. In this example reference is made to the proposal to raise the school-leaving age, to which people can respond even if they had not heard about it. This is much better than asking 'What do you think of the government's new education policy?'

In addition to testing opinion, the vox pop can be used to canvass actual suggestions or collect facts, but where the initial response is likely to be short, a follow-up is essential. This question can be subsequently edited out. For example: 'Who is your favourite TV personality?' The answer is followed by asking: 'Why?'

Another example, this time for children:

'What's the best thing about school?'
And then: 'Why would you say this is?'

It is undoubtedly true that the more complex and varied the questioning, the more difficult will be the subsequent editing. The vox pop producer must remember that the process is not about conducting an opinion poll or assembling data, but making interesting radio which has to make sense in its limited context. The second example here may be useful in allowing the listener to compare his or her own schooldays with the views of current schoolchildren.

A characteristic of the vox pop is that in the final result, the interviewer's voice does not appear. The replies must be such that they can be joined together without further explanation to the listener, and hence the technique is distinguished from simply a succession of interviews. The conversations should not be so complex that the interviewee's contribution cannot stand on its own.

Interviewing children

Children's voices can be especially engaging in this context. However, it is too easy for an adult interviewer to be thought to be too dominating, even intimidating, in this respect. It is wise to put in place a few safeguards here. First, in Britain it is a legal requirement not to interview children under 16 without their parents' permission. Talking to a group is therefore best done within a family, or perhaps a school classroom, which obviously requires the permission of the teaching staff. This cannot be done on the spur of the moment but takes time to set up. Second, established Codes of Practice prohibit asking children about private family matters or for opinions on areas likely to be beyond their judgement. Third, as with any interviewee, level eye contact is needed so one sits down among the group. Fourth, unless the children know you, there will be some warm-up time when they can get used to you and the recorder. Fifth, with a mixed group of boys and girls it is sensible to have interviewers of both sexes present to do the questioning, and another adult to see that no pressure is imposed on the children. When children are brought into the studio, they should always be accompanied by their own independent chaperone.

Choosing the site

If vox pop questioning is to be carried out among a specific group, this may itself dictate the place – football fans at a match, shoppers at the market, holiday travellers at the airport, etc. If the material is to be gathered generally, the site or sites chosen will be limited by technical factors, so as to permit easy editing at a later stage. These are to do with a reasonably low but essentially constant level of background noise.

The listener expects to hear some background actuality and it would be undesirable to exclude it altogether. However, in essence, the broadcast

Figure 9.2   A reporter checking voice level using an iPad/iRig combination

is to consist of snatches of conversation in the form of remarks made off-the-cuff in a public place, and under these conditions immediate intelligibility is more difficult to obtain than in the studio. A side street will be quieter than a main road but a constant traffic background is preferable to intermittent noise. For this reason the interviewing site should not be near a bus stop, traffic lights or other road junction – the editing process becomes intrusively obvious if buses are made to disappear into thin air and lorries arrive from nowhere. Similarly, the site should be free from any sound which has a pattern of its own, such as music, public address announcements, or a chiming clock. Editing the speech so that it makes sense will be difficult enough without having to consider the effect of chopping up the background. A traffic-free pedestrian precinct or shopping mall is often suitable, but a producer should avoid always returning to the same place – one of the attractions of the vox pop, in the general form as well as in the individual item, is its variety.

The recorder

The recorder should be tested before leaving base, and on site a further check made to ensure an adequate speech level against the background noise. It's useful to record some 10 seconds of general atmosphere to provide spare background for editing the fades in and out. From here on, the recording level control should not be altered otherwise the level of the background noise will vary. In order to maintain this same background level, any AGC should be switched out. Different speech volumes are compensated for by the positioning of the microphone relative to the speaker, and of course the normal working distance will be considerably less than in a studio.

It may be possible to simplify the editing by recording only the replies. A machine with a rapid and unobtrusive means of starting is therefore a considerable asset. This may be done by holding the machine on 'pause' until required.

If using a smartphone don't forget to have spare batteries, a spare card, a windshield of some sort and headphones (for details see pp. 103, 131).

Putting the question

It is normal for the novice reporter to feel shy about his or her first vox pop but cases of assault on broadcasters are relatively rare. It might be helpful to remember that the passer-by is being asked to enter the situation without the benefit of any prior knowledge and is probably far more nervous. However, the initiative lies with the interviewer who needs to adopt a friendly and positive technique. So, explain quickly who you are and what you want, put the question and record the reaction.

First, the reporter should be obvious rather than secretive. Stand in the middle of the pavement holding the microphone for all to see. It is helpful for the microphone to carry an identifying badge so that the approaching pedestrian can already guess at the situation and, if necessary, take avoiding action. No one should be, or for that matter can be, interviewed against their will. Any potential but unwilling contributors should not be pursued or in any way harassed. In this sense, although the interviewer may receive the occasional rebuff, the contributors are only those who agree to stop and talk.

Seeing a prospective interviewee, the reporter approaches and says pleasantly 'Good morning. I'm from Radio XYZ.' At this the passer-by will either continue, protesting at being too busy, or will stop, being reassured by the truth of the statement since it confirms the station identification badge. The interviewee may possibly also be interested at the prospect of being on the radio. The reporter continues, 'Can I ask you what you think of the proposal to keep all traffic out of the city centre?', at which point the interviewer moves the microphone to within a foot or so of the contributor and switches on the recorder. In the chapter on interviewing, questions that began with 'Can I ask you' or 'Could you tell me' were generally disallowed on the grounds that they were superfluous; permission for the interview having already been granted, and that being unnecessary they were a waste of the listener's time. In the context of the vox pop, however, such a preamble is acceptable since it allows someone the courtesy of non-cooperation, and in any case the phrase will disappear in the editing.

The normal reaction of the 'man in the street' will vary from total ignorance of the subject, through embarrassed laughter and a collecting of thoughts, to a detailed or impassioned reply from someone who knows the subject well. All of this can be useful but there is likely to be a wastage rate of at least 50 per cent. If about 10 replies are to be used, then 20 should be recorded. If the final result is to be around two minutes – and a vox pop would seldom if ever exceed this – a total of four or five minutes of response should be recorded, by which time the interviewer will hope to have a diversity of views and some well-made argument.

Occasionally a group of people will gather round and begin a discussion. This might be useful, although inevitably some of it will be 'off-mic'. A developing conversation will be more difficult to edit and the 'one at a time' approach is to be preferred although, particularly with children, more revealing comment is often obtained if they are within a group talking among themselves.

Whatever the individual response, the interviewer remains friendly and courteous, obviously wanting to give a good impression of the radio station. It's important to avoid becoming sidetracked into a discussion of the subject itself, station policy or last night's programme. End by thanking each contributor for taking the trouble to stop and talk, remembering that it is they who have done you a favour.

The editing

Spontaneity, variety, insight and humour – these are the hallmarks of the good vox pop. Listening to the material back at the base studio, the first step is to remove anything that is not totally intelligible. This must be done immediately, before the editor's ear becomes attuned to the sound. It is a great temptation to include a prize remark, however imperfectly recorded, on the basis of its being intelligible after a few playings under studio conditions! The rejection of material that is not of first-class technical quality is the first prerequisite to preventing the finished result from becoming a confusing jumble. Using a digital recorder or smartphone with a suitable app, it might be possible to rearrange the recorded tracks and do some preliminary editing before getting back to the studio. This requires care, however, since it is easy, in a hurry, to erase wanted material. In fact at this stage it's best not to erase anything, but simply to mark

 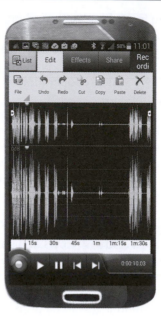

Figure 9.3 Recording with an Android smartphone: location recording on the left and on-screen editing on the right

the usable responses. Back at base, the material is downloaded and given its final edit.

The first piece of the finished vox pop needs to be a straightforward, clearly understood response to the question that will appear in the introductory cue. The subsequent comments are placed to contrast with each other, either in the opinions expressed or their style. Men's voices will alternate with those of women, the young with the old, the local accent with the 'foreign', the 'pros' with the 'antis'. The interviewer's voice is not used, except that occasionally it might be useful to be reminded of the question halfway through. What must be avoided, of course, is its continual repetition. Sometimes the answers themselves are similar, in which case sufficient should be used to indicate a consensus but not to become boringly repetitious. A problem can arise over the well-argued but lengthy reply that would be likely to distort the shape of the vox pop if used in its entirety. A permissible technique here is to use it in two or three sections, placing them separately within the final piece.

The editor needs a good comment to end on. Its nature will depend on the subject but it might be a view forcefully expressed, a humorous remark, or the kind of plain truth that often comes from a child. Good closing comments are not difficult to identify and the interviewer, after collecting the material, generally knows how the vox pop will end. The spare background noise can be used as required to separate replies, with a few seconds at the beginning and at the end as a 'fade up' and 'fade down' under speech – so much better than 'banging it in' and 'chopping it off' on the air.

It should go without saying that the finished vox pop will broadly reflect the public response found by the interviewer. It is possible, of course, for the editing to remove all views of a particular kind, so giving the impression that they do not exist. It may be that a producer would set out with the deliberate intention of demonstrating the overwhelming popularity of certain public attitudes – presumably those that the interviewer or reporter favours. Such manipulation, apart from betraying the trust that hopefully the listener has in the process, is ultimately self-defeating. Listeners do their own vox pops every day of their lives – they will know whether or not the radio station is biased in its reflection of public opinion. Probably more than the broadcaster, the listener knows reality when he or she hears it.

Used properly the vox pop represents another colour in the broadcaster's palette. It provides contrast with studio material and, in reflecting accurately what people are saying, it helps the listener to identify with the station and so enhances its credibility. The website illustrates the technique with an example of 'raw' and edited vox pop.

10

Cues and links

The information needed to accompany a recorded interview or other programme insert has two quite separate functions. The first is to provide studio staff with the appropriate technical data. The second is to introduce the item for the listener so that it makes sense in its context.

The general rules concerning cue material – to be interesting, to grab the listener and to be informative – apply equally to the links between items and to the introductory announcements given to whole programmes.

Information for the broadcaster

Before a recorded programme or item can reach the air, the producer, presenter or studio operational staff require certain information about it. Typically known as an 'audio label', this is generated by the computer system as the item is downloaded for editing and subsequent replay. In any given system this appears in a standard format and provides details of:

- the name of the piece, otherwise known as its slug or catchline;
- what the item is – interview, actuality or voice piece, etc.;
- the initials of the reporter/speaker;
- the 'in' and 'out' cues of the item as recorded;
- the duration, i.e. running time, of the item;
- the intended date of transmission.

When recording an insert, the reporter will generally introduce the piece with some background material about where it is being done, who is involved and the basics of the story. This is later edited off but it enables the producer or editor of the programme to write the introductory cue to the item, which the presenter will read. This is typed into the same file as the audio which, after editing, can be transferred to a 'hotbox' or transmission slot ready to be played on-air. Any item is also likely to have a 'back-announcement'. It's clearly important to have a means of preventing unedited material from appearing here, and it is up to the producer to see that only material ready for transmission arrives in the final running order.

When retrieving an item from the computer store, it is necessary to know its precise name or slug. To help in this, some systems give every item a number.

The studio presenter/host now has a clear indication of how to introduce the item. It might be necessary to alter words to suit a personal style, but the good cue writer will write in a way that fits the specific programme.

In the case of an interview, typically the first question is removed from the recording and transferred to the introduction so that the insert begins with the first answer. This style is a very common form of cue but it is only one of many and it should not be overused, particularly within a single programme. It is easy for cues to become mechanical – to 'write in a

Figure 10.1 News screen. The system provides information needed by the newsroom and the on-air studio. The column on the left shows the bulletin running order and which stories have audio inserts. Script and cues appear on the right. 'Messages' at the top are sent to a specific screen – like visual talkback. 'Flash Ticker' is information to the whole system

rut' – it is important to search for fresh approaches, and some are given in later examples.

Information for the listener

There is an art to writing good cue material. The piece of writing that introduces an item has three functions for the listener. It must be interesting, act as a 'signpost' and be informative.

The information must be *interesting*. The first sentence should contain some point to which the listener can relate. It should be written in response to likely questions: 'What is the purpose of this interview?' 'Why am I broadcasting this piece?' Having found the most interesting facet – the 'angle' most relevant to the listener – the writer starts from that. But more than this, it should be relevant to as many listeners as possible. It should not be written so as to exclude people. To take a local example:

Not: 'There's a big hole in the road at the corner of Campbell Street and the Broadway.'

But: 'They are digging up the road again. Everywhere you go traffic is diverted by a hole with workmen in it. What's going on? For example, at the corner of . . .'

The first intro will only interest people who drive down Campbell Street. The second is aimed at all road users, pedestrians too. To draw people in, write from the general to the particular. Another approach is to ask the occasional question as an attempt to involve the listener in a subconscious response.

A cue is a *signpost* and should make a promise about what is to follow. Having gained the listener's interest, it is then important to satisfy some expectations.

The introduction must be *informative*. One purpose of an introduction is to provide the context within which the item may be properly understood. There may have to be:

1   A summary of the events leading up to or surrounding the story.
2   An indication of why the particular interviewee was chosen.
3   Additional facts to help the listener's understanding. It might be necessary to clarify technical terms and jargon, or to explain any background noise or sounds that would otherwise distract the listener.
4   The name of the interviewer/reporter.

This last piece of information, generally the last words of the cue, can become a dreadful cliché: 'Our reporter, John Benson, has the details.'

This is a common introduction to reporter packages and wraps and if over-done becomes as boring as the 'and he asked her' introductions for interviews. Cue writing, therefore, needs a fertile imagination in order to avoid predictable repetition.

Unless the interview or voice piece is very short, say less than a minute, it will be necessary to repeat after an interview the information about the interviewee. There is a high probability that the listener is not wholly committed to the programme and heard the introduction only superficially, despite a compellingly interesting opening line. The listener's full attention frequently becomes engaged only during the interview itself. Having become fascinated or outraged, it is *afterwards* that you want to know the name of the interviewee and their qualification for speaking.

It's sometimes said that a 'back-announcement' like this slows the programme down, whereas introductions help to drive the programme forward. However, the argument in favour of a back-announcement puts listener information above programme pace, and helps to give the impression that the presenter has been listening. Without some reference to the interview, the presenter who simply continues with the next item can sound discourteous. Broadcasters should remember that, to the listener, pre-recorded items are people rather than computer playouts. They should therefore be referred to as if they had actually taken part in the programme.

The practice of omitting the back-announcement is probably an example of radio being influenced by television. In vision, it is possible to superimpose on an interview a caption giving the name and qualification of the interviewee. This can happen at any time throughout a piece and makes a verbal back-announcement unnecessary. The two channels of television information, sound and vision, can be used simultaneously for different purposes. This is not the case with radio where a statement of the interviewee's name is often the simplest and most logical way of 're-informing' the listener.

Two further examples will illustrate the functions of cue material – that is, to obtain the listener's interest, provide context, explain background noise, clarify technicalities and to 'back-announce'.

Example 1

```
ANNCT:      The strike at Abbots Electrical is over. Involving 45
            assembly workers and lasting nearly two weeks, the
            strike has meant a loss of production worth over half a
            million pounds.

            The dispute began when three men were sacked for what
            the management called 'persistent lateness affecting
            the productivity of other workers'.

            The Union objected, saying that the men were being
            'unduly victimised'.
```

```
                    Two of the men have since been reinstated. Is such
                    a stoppage worth it? On the now busy shop floor, our
                    reporter spoke to the Union representative, Joe
                    Frimley.

CUE IN:             (noise 3") No stoppage is ever . . .

CUE OUT:            . . . making up for lost time. (noise 2")
```

Duration 2' 08"

```
ANNCT:              Joe Frimley, the Union representative at Abbots
                    Electrical.
```

When a recording is made against background noise, it is useful to begin the piece with two or three seconds of the sound alone. The insert can be started before the cue and faded up under speech so that its words begin neatly after the introduction. Similarly at the end of the insert, the background noise is faded down behind the presenter's back-announcement. Such a technique is preferable to the jarring effect of 'cutting' on to noise.

In the interests of fairness and objectivity, such an item would invariably need to be followed and 'balanced' by a management view of the situation.

Example 2

```
ANNCT:              Space research and your kitchen sink. It seems an
                    unlikely combination but the same advanced technology
                    that put man on the moon has also helped with the
                    household chores.

                    For example, the non-stick frying pan uses a chemical
                    called polytetrafluoroethylene (Pron: poli-tetra-flooro-
                    éthileen). Fortunately it's called PTFE for short. Used
                    now for kitchen pans, this PTFE was developed for the
                    coating of hardware out in space.

                    Dr John Hewson of the National Research Council
                    explains.

CUE IN:             We've known about PTFE for some time . . .

CUE OUT:            . . . always looking to the future.
```

Duration 3' 17"

```
ANNCT:              Dr Hewson
```

In this case the cue had the job of explaining what PTFE is. It was referred to in the interview without clarification, so the listener has to be prepared for the term. The introductory cue is a great way of solving any problem of this sort in a pre-recorded item.

The last name in the cue is generally – but not always – the first voice on the insert. To cue the 'wrong' person is confusing. For example: 'Our reporter Bill West has been finding out how the building work is going.' The voice that follows is assumed to be that of Bill West; if it is actually the site manager it may be some time before the listener realises the fact.

In addition, therefore, to the several functions of cue material, the writer seeks both variety and a lack of ambiguity. Cues and links that are well thought-out will make a real difference in lifting a programme above the rest. It may take preparation time, but it will be time well spent.

Links

What do you say *between* items? One must get away from the 'that was, this is' approach. The last item may need an explanatory back-announcement and the real question is whether there is a logical progression between that and the next item. If not, because you are going into the weather forecast, then it's better not to try to contrive one. On the other hand, if there is a natural and easy way of moving from one scene to the next, it helps the programme flow. Do the items have anything in common? Consider the function of mortar in building a house. Does it keep the bricks apart or hold them together? It does both of course and so it is with the presenter. Rather than make the programme seem jerky and disconnected, presenters do well to make such transitions as smooth as possible – even by going into an ident or time check.

Some presenters do well to ad-lib, to do everything off the cuff, but it has to be said that for most of us the preparation of interesting, informative, humorous, provocative, friendly or insightful current remarks or comments in the links takes thinking about. This is where the style of the programme comes from. The links more than anything else give substance rather than waffle.

11

Newsreading and presentation

Presentation is radio's packaging. It hardly matters how good a programme's content, how well written or how excellent its interviews; it comes to nothing if it is poorly presented. It would be like taking a beautiful perfume and marketing it in a medicine bottle.

Good presentation stems from an understanding of the medium and a basic concern for the listener. The broadcaster at the microphone should genuinely care whether or not the listener can follow and understand what he or she is saying. If a newsreader or presenter is prone to the destructive effect of studio nerves, it is best to 'think outwards', away from yourself. This also helps to counter the complacency of over-familiarity, and is therefore more likely to communicate meaning. Since it's not possible to know the listener individually, adopt the relationship of an acquaintance rather than that of a friend.

The news presenter is friendly, respectful, informative, helpful – and personal. You know you have something to offer the listener, but this advantage is not used to exercise a knowledgeable superiority or to assume any special authority. The relationship is a horizontal one. We refer to 'putting something across', not down or up. In informing the listener we do not presume on the relationship but work at it, always taking the trouble to make what is being said interesting – and sound interesting – by ourselves being interested.

Of course, newsreading tends to be more formal than a music programme but there is room for a variety of approaches. Whatever the overall style of the station, governed by its basic attitude to the listener, it should be fairly consistent. While the sociologist may regard radio as a mass medium, the person at the microphone sees it as an individual communication – talking to some*one*. Thinking of the listener as one person it's better to say, 'If you're travelling south today, . . .' not 'Anyone travelling south . . .'. The presenter does not shout. If you are half a metre from the microphone and the listener is a metre from the radio, the total distance between you is one and a half metres. What is required is not volume but an ordinary clarity. Too much projection causes the listener psychologically to 'back off' – it distances the relationship. Conversely, by dropping his or her voice the presenter adopts the confidential or intimate style more appropriate to the closeness of late night listening.

The simplest way of getting the style, projection and speed right is to visualise the listener sitting in the studio a little way beyond the microphone. The presenter is not alone reading, but is talking with the listener. This small exercise in imagination is the key to good presentation.

The seven Ps

Here are the recognised basics of good presentation:

1   *Posture.* Is the sitting position comfortable, to allow good breathing and movement? Cramped or slouching posture doesn't make for an easy alertness.
2   *Projection.* Is the amount of vocal energy being used appropriate to the programme?
3   *Pace.* Is the delivery too fast? Too high a word rate can impair intelligibility or cause errors. Too slow is ponderous and boring.
4   *Pitch.* Is there sufficient rise and fall to make the overall sound interesting? Too monotonous a note can quickly become very tiring to listen to. However, animation in the voice should be used to convey natural meaning rather than achieve variety for its own sake. So even if the pitch varies, is it forming a predictable repetitive pattern?
5   *Pause.* Are suitable silences used intelligently to separate ideas and allow understanding to take place?
6   *Pronunciation.* Can the reader cope adequately with worldwide names and places? If a presenter is unfamiliar with people in the news, or musical terms in other languages, it may be helpful to learn the basics of phonetic guidelines.
7   *Personality.* The sum total of all that communicates from microphone to loudspeaker, how does the broadcaster come over? What is the visual image conjured up? Is it appropriate to the programme?

The rate of delivery depends on the style of the station and the material being broadcast. Inter-programme or a continuity announcements should be at the presenter's own conversational speed, for example newsreading at 160–200 words per minute, but slower for short-wave a transmission. Commentary should be at a rate to suit the action. If a reader is going too fast it may not help simply to slow down – this is likely to make the voice sound stilted and over-careful. What is required is to leave more pause *between* the sentences – that is when the understanding takes place. It is not so much the speed of the words that can confuse, but the lack of sufficient time to make sense of them.

Newsreading

The first demand placed on the newsreader is that he or she understands what is being read. You cannot be expected to communicate sensibly if you have not fully grasped the sense of it yourself. With the reservations expressed later about 'syndicated' material, there is little place for the newsreader who picks up a bulletin with 30 seconds to go and hopes to read it 'word perfect'. So, be better than punctual, be early. Neither is technically faultless reading the same thing as communicating sense. A newsreader should be well informed and have an excellent background knowledge of current affairs in order to cope when changes occur just before a bulletin. Take time to read it out loud beforehand – this provides an opportunity to understand the content and be aware of pitfalls. There may be problems of pronunciation over a visiting Chinese trade mission, a Middle East terrorist, or a statement by an African foreign minister. There may be a phrase that is awkward to say, an ambiguous construction, or a typing error. The pages should be verbally checked by the person who, in the listener's mind, is responsible for disseminating them. While a newsroom may like to give the impression that its material is the latest 'hot off the press' rush, it is seldom impossible for the reader to go through all the pages of a bulletin, on paper or off screen, as it is being put together. Thorough preparation should be the rule, with reading at sight reserved only for emergencies.

Of course in practice this is often a counsel of perfection. In a small station, where the newsreader may be working single-handed, the news can arrive within seconds of its deadline. It has to be read at sight. This is not the best practice and runs a considerable risk of error. It places on the syndicating news service, and on the sending keyboard operator, a high responsibility for total accuracy. The reason for poor broadcasting of on-screen material may lie with the station management for insufficient staffing, or with the news agency for less than professional standards. The fact of the matter is that in the event of a mistake on the air, from whatever cause, the listener blames the newsreader.

The person at the microphone, therefore, has the right to expect a certain level of service. This means a well-written, properly punctuated and set out bulletin, accurately typed, arriving a few minutes before it is needed. It's then possible to check if the lead story has changed and scan it quickly for any unfamiliar names. Pick out figures and dates to make sure they make sense. In the actual reading, your eyes are a little ahead of your speech, enabling you to take in *groups of words*, understanding them before passing on their sense to the listener.

The idea of syndicated news is excellent but it should not become the cause nor be made the excuse for poor microphone delivery.

In the studio the newsreader sits comfortably but not indulgently, feeling relaxed but not complacent, breathing normally and taking a couple of extra deep breaths before beginning.

Here are some other practicalities of script reading:

- Don't eat sweets or chocolate beforehand: sugar thickens the saliva.
- Always have a pen or pencil with you for marking alterations, corrections, emphasis, etc.
- If you wear them, make sure you have your glasses.
- Don't wear anything that could knock the table or rattle – bangles, cufflinks.
- Place a glass of water near at hand.
- Remove any staples or paper-clips from the script and separate the pages so that you can deal with each page individually.
- Make sure you have the whole script; check that the pages are in the right order, the right way up.
- Give yourself space, especially to put down the finished pages – don't bother to put them face down.
- Check the clock, cue light, headphones – for talkback and cue programme – and the mic-cut key if there is one.
- Check your voice level.
- Where timing is important, time the final minute of the script (180 words – perhaps 18 lines of typescript) and mark that place. You need to be at that point with a minute left to go and may have to drop items in order to achieve this.
- Once started, don't worry about your own performance; be concerned that you are really communicating to your imagined listener, 'just beyond the mic'.
- If reading from a screen connected to a local network, make sure it is secure and that a colleague on another terminal is not inputting to the news while you are broadcasting.

Pronunciation

A station should, as far as possible, be consistent over its use of a particular name. Problems arise when its output comprises several sources, e.g. syndicated material, a live audio news feed, a sustaining service. What should be avoided is one pronunciation in a nationally syndicated bulletin, followed a few minutes later by a different treatment in a locally read summary. The newsroom must listen to the whole of the station's news output, from whatever source, and advise the newsreader accordingly. Second, listeners are extremely sensitive to the incorrect pronunciation of names with which they are associated. The station that gets a local place name wrong loses credibility; one that mispronounces a personal name is regarded as either ignorant or rude. The difficulty is that listeners themselves might not agree on the correct form. Nevertheless, a station should make strenuous efforts to ensure a consistent treatment of place names within its area. A phonetic pronunciation list based on 'educated local knowledge' should be adopted as a matter of policy and a new broadcaster joining the staff, acquainted with it at the earliest possible time.

Alternatively, store correctly spoken pronunciations in audio form on a computer. It is then an easy matter for a presenter to bring a name up on screen and hear it being said.

Vocal stressing

An important aspect of conveying meaning, about which a script generally gives no clue at all, is that of stress – the degree of emphasis laid on a word. Take the phrase: 'What do you want me to do about that?'

With the stress on the 'you', it is a very direct question. On the 'me', it is more personal to the questioner; on the second 'do', it is a practical rather than a theoretical matter; on the 'that', it is different again. Its meaning changes with the emphasis. In reading news such subtleties can be crucial. For example, we may have in a story on Arab/Israeli affairs the following two statements:

> Mr Radim is visiting Washington where he is due to see the President this afternoon.
>
> Meanwhile the Israeli Foreign Minister is in Paris.
>
> (www.focalpress.com/cw/mcleish)

The name is fictional but the example real. Put the emphasis on the word 'Israeli', and Mr Radim is probably an Arab foreign minister. Put it on 'Foreign', and he becomes the Israeli Prime Minister. Try it out loud. Many sentences have a central 'pivot', or are counter-balanced about each other: 'While *this* is happening over *here*, *that* is taking place over *there*.' Many sentences contain a counter-balance of event, geography, person or time: 'Mr *Smith* said an election should take place *now*, *before* the issue came up. Mr *Jones* thought it should wait at least until *after* the matter had been debated.'

Listening to newsreaders it is possible to discern a widespread belief that there is a universal news style, where speed and urgency have priority over meaning, where the emphasis is either on every word or scattered in a random fashion, but always on the last word in every sentence. Does it stem from the journalist's need for clarity when dictating copy over the phone? The fact is that a single misplaced emphasis will cloud the meaning, possibly alter it.

The only way of achieving correct stressing is by fully understanding the implications as well as the 'face value' of the material. This must be a conscious awareness during the preparatory read-through. As has been rightly observed, 'take care of the sense and the sounds will take care of themselves'.

Inflection

The monotonous reader either has no vocal inflection at all, or the rise and fall in pitch becomes regular and repetitive. It is the predictability of the vocal pattern that becomes boring. A too typical sentence 'shape' starts at a low pitch, quickly rises to the top and gradually descends, arriving at the bottom again by the final full stop. Placed end to end, such sentences quickly establish a rhythm which, if it does not mesmerise, will confuse, because with their beginning and ending on the same 'note', the joins are scarcely perceptible. Meaning begins to evaporate as the structure disappears. However, in reality and without sounding artificial or contrived, sentences normally start on a higher pitch than the one on which the previous one ended – a new paragraph certainly should. There can often be a natural rise and fall within a sentence, particularly if it contains more than one phrase. Meaningful stressing rather than random patterning will help.

A newsreader is well advised occasionally to record some reading for personal analysis – is it too rhythmic, dull or aggressive? In the matter of inflection, try experimenting off-air, putting a greater rise and fall into the voice than usual to see whether the result is more acceptable. Very often when you may feel you are really 'hamming it up', the playback sounds perfectly normal and only a shade more lively. Even experienced readers can become stale and fall into the traps of mechanical reading, and a little non-obsessive self-analysis and experimentation is very healthy – alternatively, actively ask for the comments of others.

Quotation marks

Reading quotes is a minor art on its own. It is easy to sound as though the comment is that of the newsreader, although the writing should avoid this construction. Some examples:

While an early bulletin described his condition as 'comfortable', by this afternoon he was 'weaker'. (This should be rewritten to attribute both quotes.)

The opposition leader described the statement as 'a complete fabrication designed to mislead'.

He later argued that he had 'never seen' the witness.

(www.focalpress.com/cw/mcleish)

To make someone else's words stand out as separate from the newsreader's own, there is a small pause and a change in voice pitch and speed for the quote.

Alterations

Last-minute handwritten changes to the typed page should be made with as much clarity as possible. Crossings out should be done in blocks rather than on each individual word. Lines and arrows indicating a different order of the material need to be bold enough to follow quickly, and any new lines written clearly at the bottom of the page. To avoid confusion, a 'unity of change' should be the aim. It is amazing how often a reader will find their way skilfully through a maze of alterations only to stumble when concentration relaxes on the next perfectly clear page.

Corrections

But what happens when a mistake is made? Continue and ignore it or go back and correct it? When is an apology called for? It depends, of course, on the type of error. There is the verbal slip which it is quite unnecessary to do anything about, a misplaced emphasis, a wrong inflection, a word that comes out in an unintended way. The key question is 'could the listener have misconstrued my meaning?' If so, it must be put right. If there is a persistent error, or a refusal of a word to be pronounced at all, it is better to restart the whole sentence. Since 'I'm sorry I'll read that again' has become a cliché, something else might be preferred – 'I'm sorry, I'll repeat that', or 'Let me take that again'. It is whatever comes most naturally to the unflustered reader. To the broadcaster it can seem like the end of the world; it is not. Even if the listener has noticed it, what is needed is simply a correction with as little fuss as possible.

Lists and numbers

The reading of a list can create a problem. A table of sports results, stock market shares, fatstock prices or a shipping forecast – these can sound very dull. Again, the first job for the reader is to understand the material, to take an interest in it, so as to communicate it. Second, the inexperienced reader must listen to others, not to copy them, but to pick up the points in their style that seem right to use. There are particular inflections in reading this material which reinforce the information content. With football results, for example, the voice can indicate the result as it gives the score (www.focalpress.com/cw/mcleish). The same is true of racing results, which have a consistent format:

> 'Racing at Catterick – the three thirty.
> First, number 7, Phantom, 5 to 2 favourite,
> Second, number 9, Crystal Lad, 7 to 1
> Third, number 3, Handmaiden, 25 to 1
> Non-runner, number 1, Gold Digger.'

For obvious reasons care must be taken over pronunciation and prices. A highly backed horse might be quoted at '2 to 5'. This is generally given as '5 to 2 on'.

In passing, it is worth noting that sport has a good deal of its own jargon which looks the same on the printed page, for instance the figure '0'. The newsreader should know when this should be interpreted as 'nought', 'love', 'zero', 'oh' or 'nil' – the listener will certainly know what is correct. Unless it is automatic to the reader, it is well worth writing the appropriate word on the script. Whenever figures appear in a script, the reader should sort out the hundreds from the thousands and, if necessary, write the number on the page in words.

If it has been correctly written, a script consists of short unambiguous sentences or phrases, easily taken in by the eye and delivered with clear meaning, well within a single breath. The sense is contained not in the single words but in their grouping. To begin with, as children, we learn to read letter by letter, then word by word. The intelligent newsreader delivers scripted material phrase by phrase, taking in and giving out whole groups of words at a time, leaving little pauses between them to let their meaning sink in. The overall style is not one of 'reading' – it is much more akin to 'telling'.

In summary, the 'rules' of newsreading are:

1   Understand the content by preparation.
2   Visualise the listener by imagination.
3   Communicate the sense by telling.

Station style

Radio managers become paranoid over the matter of station style. They will regard any misdemeanour on-air as a personal affront, especially if they instituted the rule that should have been observed. It's nevertheless true that a consistent station sound aids identification. It calls for some discipline, particularly in relation to the frequently used phrases to do with time. Is it 3.25 or 25 past 3? Is it 3.40 or 20 minutes to 4? Is it 15.40? Dates: is it 'May the eleventh', 'the eleventh of May', or 'May eleven'? Frequencies, 1107 kilohertz, 271 metres medium wave, can be given in several ways: 'on a frequency of eleven-oh-seven', or 'two-seven-one metres medium wave', or 'two-seventy-one'. Temperatures: Celsius, centigrade or Fahrenheit? Or just '22 degrees'?

Some stations insist on a very strict form of identification, some prefer variety:

- radio Berkely
- Berkely radio – two-seven-one
- the county sound of Berkely
- the sound of Berkely county
- Berkely county radio, etc.

Idents can be by station name, frequency or wavelength, programme title, presenter name, or by some habitually used catchline:

- GZFM – where news comes first
- GZFM – the heartbeat of the county
- GZFM – serving the South
- GZFM – with the world's favourite music.

Learn the station policy and stick rigidly to it – even when sending in an audition tape, use the form you hear on-air.

A subtle facet of presentation is to remember that it is the presenter who joins the listener not the other way round. 'It's good to be with you' is a personal form of service, whereas 'Thank you for joining me' is more of an ego-trip for the presenter. The station should go to the bother of reaching out to its listeners, not expect them to come to it.

Continuity presentation

Presenting a sequence of programmes, giving them continuity, acting as the voice of the station, is very similar to being the host of a magazine programme responsible for linking different items. The job is to provide a continuous thread of interest even though there are contrasts of content and mood. The presenter makes the transition by picking up in the style of the programme that is finishing, so that by the time he or she has done the back-announcements and given incidental information, station identification and time check, everything is ready to introduce the next programme in perhaps quite a different manner. Naturally, to judge the mood correctly it's necessary to do some listening. It is no good coming into a studio with under a minute to go, hoping to find the right piece of paper so as to get into the next programme, without sounding detached from the whole proceedings. A station like this might as well be automated.

If there is time at programme breaks, trail an upcoming programme – not the next one, since you are going to announce that in a moment. The most usual style is to trail the 'programme after next'. But do so in a compelling and attractive way so as to retain the interest of the listener – perhaps by using an intriguing clip from the programme (see p. 159). If the trail is for something further ahead, then make this clear – 'Now looking ahead to tomorrow night . . .'.

Many programmes handover from one to the next, so that presenters may chat to each other at the break, or at least introduce what's coming. A frequent rule of presentation is 'never say goodbye'. It's an invitation for the listener to respond and switch off. At the end of a programme the presenter hands over to someone else – you (the station) never give the impression of going away – even for a commercial break.

Continuity presentation requires a sensitivity to the way a programme ends, to leave just the right pause, to continue with a smile in the voice or whatever is needed. Develop a precise sense of timing, the ability to talk rather than 'waffle', for exactly 15 seconds, or a minute and a half. A good presenter knows it is not enough just to get the programmes on the air, the primary concern is the person at the end of the system.

Errors and emergencies

What do you do when the computer fails to respond, the machine does not start or having given an introduction there is silence when the fader is opened? First, no oaths or exclamations! The microphone may still be 'live' and this is the time when one problem can lead to another. Second, look hard to see that there has not been a simple operational error. Are all the signal lights showing correctly? Is there an equipment fault that can be put right quickly? Is the item playing back on the computer? Is the right fader open? If by taking action the programme can continue with only a slight pause, five seconds or so, then no further announcement is necessary. If it takes longer to put right, 10 seconds or more, something should be said to keep the listener informed:

'I'm sorry about the delay, we seem to have lost that Report for a moment . . .'

Then, if action has assured it is possible to continue:

> 'We'll have it for you shortly . . .'

The presenter may assume personal or collective responsibility for the problem but what isn't right is to blame someone else:

> 'Sorry about that, the man through the glass window here pressed the wrong button!'

The same goes for the wrong item following a particular cue, or pages read in the wrong order. The professional does not become self-indulgent, saying how complicated the job is; you simply put it right, with everybody else's help, in a natural manner and with the minimum of bother. The job of presentation is always to expect the unexpected.

Sooner or later a more serious situation will occur which demands that the presenter 'fills' for a considerable time. Standby announcements of a public service type – an appeal for blood donors, safety in the home, code for drivers, procedure for contacting the police or hospital service. Also, the current weather forecast, programme trails and other promotional material can be used. These 'fills' should always be available to cover the odd 20 seconds, and changed once they have been used.

Standby music is an essential part of the emergency procedure. Something for every occasion – a break in the relay of a church service, the loss of Saturday football, an under-run of a children's programme. To avoid confusion the music chosen should not be identical to anything it replaces, simply of a sympathetic mood. Once it is on there is a breathing space to attempt to get the problem sorted out. The principle is to return to the original programme as quickly as possible. Very occasionally it may be necessary to abandon a fault-prone programme, and some stations keep a 'timeless talk' or 15-minute feature permanently on standby to cover such an eventuality. For longer-term schedule changes see p. 348.

Headphones

A vocal performer can sometimes become obsessed with the sound of his or her own voice. The warning signs include a tendency to listen to oneself continuously on headphones. The purpose of headphone monitoring is essentially to provide talkback communication, or an outside source or cue programme feed. Only if it is unavoidable should both ears be covered, otherwise presenters begin to live in a world of their own, out of touch with others in the studio. Remember, too, the possible danger to one's hearing of prolonged listening at too high a level. There's also a danger of getting into a rut if there is a great deal of routine work – the same announcements, station identifications, time checks and introductions. It becomes easy not to try very hard to find appropriate variations. Like the newsreader, all presenters should occasionally listen to themselves recorded off-air, checking a repetitive vocabulary, use of cliché or monotony of style.

Trails and promos

Part of a station's total presentation 'sound' is the way it sells itself. Promotional activity should not be left to chance but be carefully designed to accord with an overall sense of style. 'Selling' one's own programmes on the air is like marketing any other product, and this is developed in Chapter 17, but remember that the appeal can only be directed to those people who are already listening. The task is therefore to describe a future programme as so interesting and attractive that the listener is bound to tune in again. The qualities that people enjoy and which will attract them to a particular programme are:

- humour that appeals;
- originality that is intriguing;
- an interest that is relevant;
- a cleverness which can be appreciated;
- musical content;
- simplicity – a non-confusing message;
- a good sound quality.

If one or more of these attributes is presented in a style to which he or she can relate, the listener will almost certainly come back for more. The station is all the time attempting to develop a rapport with the listener, and the programme trailer is an opportunity to do just this. It is saying of a future programme, 'this is for *you*'.

Having obtained the listener's interest, a trail must provide some information on content – what the programme is trying to do, who is taking part and what form (quiz, discussion, phone-in, etc.) it will take. All this must be in line with the same list of attractive qualities. But this is far from easy – to be humorous *and* original, to be clever as well as simple. The final stage is to be sure that the listener is left with clear transmission details, the day and time of the broadcast. The information is best repeated:

'You can hear the show on this station tomorrow at six p.m. Just the thing for early evening – the "Kate Greenhouse Saga", on 251 – six o'clock tomorrow.'

Trails are often wrapped around with music which reflects something of the style of the programme, or at least the style of the programme in which the trail is inserted. It should start and finish clearly, rather than on a fade; this is achieved by prefading the end music to time and editing it to the opening music so that the join is covered by speech.

At its simplest, a trail lasting 30 seconds might look like this:

```
MUSIC:     Bright, faded on musical phrase, held under speech.    3"
SPEECH:    Obtains interest.                                     10"
           Provides information on content.
           (Music edit at low level under speech)                10"
           Gives transmission details.                            5"
MUSIC:     Fade up to end.                                         2"
```

There is little point in ordering people to switch on; the effect is better achieved by convincing listeners that they are really missing out and suffering deprivation if their radio isn't on. And, of course, if that is the station's promise then it must later be fulfilled. Trails should not be too mandatory, and above all they should be memorable.

The website illustrates station trails from Future Radio, Norwich. The music comes from a specialist 'mood music' library, licensed by MCPS-PRS. A useful source to get to know (see p. 386).

12

The discussion

The topic for a broadcast debate should be a matter in which there is genuine public interest or concern. The aim is for the listener to hear argument and counter-argument expressed in conversational form by people actually holding those views with conviction. The broadcaster can then remain independent.

Format

At its simplest, there will be two speakers representing opposing views together with an impartial chairman or host. The producer may of course decide that such an arrangement would not do justice to the subject, that it is not as clear-cut as the bi-directional discussion will allow and it might therefore be better to include a range of views – the 'multi-faceted' discussion.

In this respect the 'blindness' of radio imposes its own limitations and four or five speakers should be regarded as the maximum. Even then it is preferable that there is a mix of audibly different voices – male and female.

Under the heading of the discussion programme should also come what is often referred to as 'the chat show'. Here, a well-known radio personality introduces one or more guests and talks with them. It may incorrectly be described as an interview but since 'personalities' have views of their own which they are generally only too ready to express, the result is likely to be a discussion. The 'chairman plus one at a time' formula can be a satisfactory approach to a discussion, particularly with the more lightweight entertainment, the non-current affairs type of subjects. It works less well in the controversial, political, current affairs field since it is more difficult for the chairman to remain neutral as part of the discussion. In any case the danger for the broadcaster is that in order to draw out the guest contributor, it is necessary always to act as devil's advocate – the 'opposition' – and so

Figure 12.1 A studio discussion table with script, cue and information screens for the presenter, a mic and headphones for each contributor, a cue light and reverse talkback (Courtesy of BBC News)

become identified as 'anti-everything'. In such cases the more acceptable format is that of the interview. It's important for listeners and broadcasters to draw a clear distinction between an interview and a discussion. Having talked individually with each guest in turn, they can then talk together with the chairman as intermediary.

A very acceptable format is where questions are put – possibly from an audience – to a panel of speakers representing different political parties or specialist interests. Such 'Any Questions?' programmes will have a sharper edge by being topical and broadcast live.

Selection of participants

Do all participants start equal? A discussion tends to favour the articulate and well-organised. The chairman may have to create opportunities for others to make their case. It is possible, of course, to 'weight' a discussion so that it is favourable to a particular point of view, but since the listener must be able to come to a conclusion by hearing different views

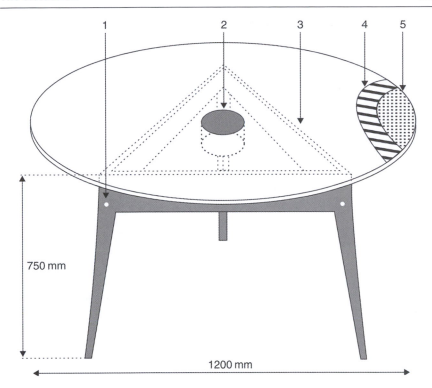

Figure 12.2 A talks discussion table for studio use. The table has three legs to reduce obstruction and stop it wobbling on an uneven floor. 1. Headphone jack carrying a programme feed with or without additional talkback. 2. Centre hole to take a mic, either on a special carrier or on a floor stand. 3. Acoustically transparent top consisting of a loose-weave cloth surface, a layer of padding (4) and a steel mesh base (5) (Courtesy of the BBC)

adequately expressed, the producer should look for balance – of ability as well as opinion. Often there are, on the one hand, the 'official spokesmen', and, on the other, the good broadcasters! Sometimes they combine in the same person but not everyone, in the circumstances of the broadcast debate, is quick thinking, articulate and convincing – however worthy they might be in other respects. In selecting the spokesperson for one political party it is virtually obligatory also to include their 'shadow' opposite number – whatever the quality of their radio performance.

There will obviously be times when it is necessary to choose the leader of the party, the council member, the chief executive of the company or the official PR person; but there will also be occasions when the choice is more open to the broadcaster, and in the multi-faceted discussion it is important to include as diverse a range of interests as is appropriate.

In general terms these can be summarised as: power holders and decision makers, legal representation and 'watch-dog' organisations, producers of goods and services, and consumers of goods and services. These categories will apply whether the issue is the flooding of a valley for a proposed hydro-electric scheme, a change in the abortion law or an increase in the price of food.

Listeners should also be regarded as participants and the topic should at least be one that involves them. If listeners are directly affected they can be invited to take part 'live' and 'now', or in a follow-up programme by phone, text, email or social media. In the event of a public meeting on the subject it may be that the broadcaster arranges to cover it with an outside broadcast.

The chairperson

Having selected the topic and the team, the programme will need someone to chair the discussion. The ideal is knowledgeable, firm, sensitive, quick thinking, neutral but challenging, and courteous. He or she will be interested in almost everything and will need a sense of humour – no mean task!

Once this paragon of human virtue, who also possesses a good radio voice and an acute sense of timing, has been obtained, there are several points that need attention before the broadcast. One of these is to decide what to call the chairperson – some people object to being called simply 'chair'. Here the term 'chairman' is used to include both men and women. The problem is solved by using the term 'host'.

Preparation

The subject must be researched and the essential background information gathered and checked. Appropriate reference material may be found in libraries, files of newspaper cuttings, on the Internet and in the radio station's own newsroom. The chairman must have the facts to hand and have a note of the views already expressed so as to have a complete understanding of the points of controversy. A basic 'plot' of the discussion is then prepared, outlining the main areas to be covered. This is in no sense a script; it is a reminder of the essential issues in case they should get side-tracked in the debate.

It is important that the speakers are properly briefed beforehand, making sure that they understand the purpose, range *and limitations* of the discussion. They should each know who is to take part and the duration of

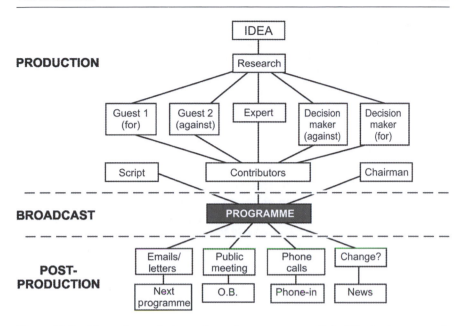

Figure 12.3 Stages in producing a discussion programme. Select the topic – research the information – choose the participants – coordinate the contributors – broadcast the programme – deal with the response – evaluate the possibility of follow-up

the programme. It is not necessary for them to meet before the broadcast but they should be given the opportunity to do their own preparation.

Advice to contributors

The producer wants a well-informed, lively discussion and might therefore have to help contributors to give of their best. Some advice to the newcomer may be needed. For example, suggest that a contributor needs to crystallise and hold on to the two or three most important points to put over – but not to come with a prepared list of things to say. Also, that the listener is likely to identify more readily with reasoned argument, based on a capacity to appreciate both sides of a case, than with dogma and bigotry. Accepting the existence of an opposite view and logically explaining why you believe it to be wrong is one of the best ways of sounding convincing on radio.

Whatever is said is enhanced by good illustration – a story – which underlines the point being made. Drawn from the contributor's own experience and conjuring up the appropriate pictures, the telling of an incident or conversation is a powerful aid to argument. It should be brief, factual, recent and relevant, and since such things seldom come to mind at the right time this needs thinking about in their own preparation. So also do facts – the specifics of the discussion topic. It is important not to appear too glib, but a contributor who is in possession of the facts is more likely to gain the respect of the listener.

Contributor nerves

It is not the slightest use telling anyone 'not to be nervous'. Nervousness is an emotional reaction to an unusual situation and as such it is inevitable. Indeed, it is desirable in that it causes the adrenalin to flow and improves concentration – with experience it is possible to use such 'red light' tensions constructively. On the other hand, if the contributor is too relaxed he or she may appear to be blasé about the subject and the listener may react against this approach. In practical terms participants should listen hard to the chairman and maintain eye contact – a great help to concentration.

Starting the programme

At the start of the broadcast the presenter/chairman introduces the subject, making it interesting and relevant to the listener. This is often done by putting a series of questions on the central issues, or by quoting remarks already made publicly.

The chairman should have everyone's name, and his or her designation, written down so as to be clearly visible – it is amazing how easy it is for one's mind to go blank, even when you know someone well. The introductions are then made, making sure that all their voices are heard as early as possible in the programme. During the discussion continue to establish the names, at least for the first two 'rounds' of conversation, and again at intervals throughout. It is essential that the start of the programme is factual in content and positive in presentation. Such an approach will be helpful to the less confident members of the team and will reassure the listener that the subject is in good hands. It also enables the participants to have something 'to bite on' immediately so that the discussion can begin without a lengthy warming-up period.

Speaker control

In the rather special conditions of a studio discussion, some people become highly talkative, believing that they have failed unless they have put their whole case in the first five minutes. On the other hand, there are the nervously diffident. It is not possible to make a hesitant speaker appear brilliant, yet someone with poor delivery might have significant points to make and need encouraging, while the verbose might actually have little to contribute. The chairman, listening to the *content*, must draw out the former and curb the latter. Even the most voluble have to breathe – a factor which repays close observation! The chairman's main task is to provide equal opportunity of expression for all participants. To do this may require suppression as well as encouragement, and such directions as are required should be communicated – with a hand gesture, non-verbally.

After having strongly expressed an opinion, that speaker should not be allowed to continue for too long before another view of the matter is introduced. The chairman can interrupt, and it's best done constructively: 'That's an important point, before we go on, how do others react to that? – Mrs Jones?' The chair must in these cases give a positive indication by voice, facial expressions and possibly hand signal as well of who is to speak. It is also necessary to prevent two voices from speaking at once, other than for a brief interjection, by a decisive and clear indication of 'who holds the floor'. It is not a disaster when there are two or more voices, indeed it may be a useful indicator of the strength of feeling. It has to be remembered, however, particularly when broadcasting monophonically, that when voices overlay each other, the listener is unlikely to make much sense of the actual content.

Subject control

The chairman has to obtain clarification of any technical jargon or specialist language a contributor might use. Acronyms and abbreviations, particularly of organisations, are generally far less well understood by the listener than people sitting round a studio table might think.

With one eye on the prepared 'plot' and the other on the clock, the chairman steers the subject through its essential areas. However, it's important to remain reasonably flexible and if one particular aspect is proving especially interesting, the chair may decide to depart from the original outline. Questions in the chairman's mind should be:

- Time gone – time to go.
- How long has this person had?
- Is it irrelevant?
- Is it boring?
- Is it incomprehensible?
- Next question.
- Who next?

Above all, the chairman must be able to spot and deal with red herrings and digressions. To do this it's important to know where the discussion should be going and have the appropriate question phrased so that it's possible to interrupt positively, constructively and courteously.

In a lengthy programme it may be useful to introduce a device that creates variety and helps the discussion to change direction. Examples are a letter, email or tweet, from a listener, or quote from an article read by the chairman, a pre-recorded interview, a piece of actuality, or a phone call or text received during the programme. If the chairman is to remain impartial such an insert should not be used to make a specific point but simply to raise questions on which the participants may then comment.

Technical control

The chairman has to watch for, and correct, alterations in the balance of voices that was obtained before the programme began. This could be due to a speaker moving back, turning 'off-mic' to someone at the side, or leaning in too close. Such changes can of course be overcome by giving each person their own lapel mic but there may still be wide variations in individual voice levels as the participants get annoyed, excited, discomfited or subdued. It's also necessary to be aware of any extraneous noise such as paper rustle, jingling bracelets or fingers tapping the table. Non-verbal signals should suffice to prevent them becoming too intrusive.

As an aid in judging the effect of any movement, changes in voice level or unwanted sounds, the chairman will often wear headphones. These should be on one ear only to avoid being isolated acoustically from the actual discussion. These headphones may also carry talkback from the

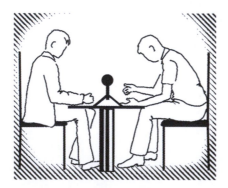

Figure 12.4 Voice level must be watched throughout. A person with a quiet voice will have to sit close to the table and the discussion chairman must prevent too much movement

producer, for example with additional ideas, questions, or a point of the discussion that might otherwise be overlooked. On occasion, everyone in the studio will require headphones. This is likely if the programme is to include phone calls from listeners, or when members of the discussion group are not physically present in the same studio but are talking over links between separate studios. In these circumstances, the talkback arrangements have to be such that the producer's editorial comments are confined only to the headphones worn by the chairman. To avoid embarrassment and confusion such a system must be carefully checked before the programme begins.

If the programme is in stereo, it's usual to place the chairman in the centre, perhaps with a neutral expert, and the participants panned half left and half right, to give some separation.

An important part of the technical control of the programme is its overall timing. The chairman must never forget the clock.

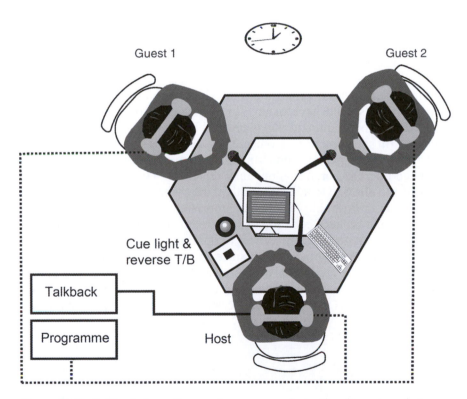

Figure 12.5 Talkback from the producer goes only to the discussion chairman. Other participants may need headphones carrying the programme in order to hear a remote contributor or phone call

Ending the programme

It is rarely desirable for the chairman to attempt a summing-up. If the discussion has gone well, the listener will already have recognised the main points being made and the arguments that support them. If a summary *is* required, it is often better to invite each speaker to have a 'last word'. Alternatively, the chairman may put a key question to the group which points the subject forward to the next step – 'Finally, what do you think should happen now?' This should be timed to allow for sufficient answering discussion.

Many a good programme is spoiled by an untidy ending. The chairman should avoid giving the impression that the programme simply ran out of time:

> 'Well, I'm afraid we'll have to stop there . . .'
> 'Once again the clock has beaten us . . .'
> 'What a pity there's no time to explore that last point . . .'

The programme should cover the material which it intended, in the time it was allowed. With a minute or less to go, the chairman should thank the contributors by name, giving any other credits due and referring to further programmes, public events or a helpline related to the subject.

After the broadcast comes the time when the participants think of the remarks they should have made. An opportunity for them to relax and 'unwind' is important, and this is preferably done as a group – assuming they are still speaking to each other. They are at this stage probably feeling vulnerable and exposed, wondering if they have done justice to the arguments and perhaps organisations they represent. They should be warmly thanked and allowed to talk informally if they wish. The provision of some refreshment or hospitality is often appropriate.

It is not the broadcaster's job to create confrontation and dissent where none exists. But genuine differences of opinion on matters of public interest offer absorbing broadcasting, since the listener may feel a personal involvement in the arguments expressed and in their outcome. The discussion programme is a contribution to the wider area of public debate and may be regarded as part of the broadcaster's positive role in a democratic society.

13

Phone-ins

Critics of the phone-in describe it as no more than a cheap way of filling airtime and undoubtedly it is sometimes used as such. But like anything else, the priority it is accorded and the production methods applied to it will decide whether it is simply transmitter fodder or whether it can be useful and interesting to the listener. *It's Your World* on the BBC's World Service had the potential for putting anyone anywhere in touch with a major international figure to question policy and discuss issues of the day.

Through public participation, the aim of a phone-in is to allow a democratic expression of view and to create the possibility of community action. An important question, therefore, is to what extent such a programme excludes those listeners who are without a landline or cell phone. Telephone ownership can vary widely between regions of the same country, and the cities are generally far better served than the rural areas. It is a salutary reminder that 50 per cent of the world's population has never used a phone. It's the same with 'send us an email', or 'see our website' – for many these are excluding suggestions.

So, it is possible to be over-glib with the invitation: 'If you want to take part in the programme just give us a ring on . . . '. Cannot someone take part simply by listening? Or to go further, if the aim is public participation, will the programme also accept texts, emails, letters, or people who actually arrive on the station's doorstep? The little group of people gathered round the door of an up-country studio in Haiti, while their messages and points of view were relayed inside, remains for me an abiding memory of a station doing its job.

It is especially gratifying to have someone, without a phone, go to the trouble of phoning from a public pay phone. To avoid losing the call when the money runs out, the station should always take the number and initiate such calls as are broadcast on a phone-back basis.

Technical facilities

When inviting listeners to phone the programme, it is best to have a special number rather than take the calls through the normal station telephone number, otherwise the programme can bring the general telephone traffic to a halt. The technical means of taking calls have almost infinite variation but the facilities should include:

1 off-air answering of calls
2 acceptance of several calls – say four or five simultaneously
3 holding a call until required, sending the caller a feed of cue programme
4 the ability to take two calls simultaneously on the air
5 origination of calls from the studio – phone-out
6 picking up a call by the answering position after its on-air use.

New equipment, such as phoneBOX solo, is coming on to the market all the time. Such devices will handle and route calls, apply EQ, dial out, record, edit and play back calls – all operated by a touch-screen or mouse.

Programme classification

The producer of a phone-in must decide the aim of the programme and design it so that it achieves a particular objective. If the lines are simply thrown open to listeners, the result can be a hopeless muddle. There are always cranks and exhibitionists ready to talk without saying anything, and there are the lonely with a real need to talk. Inexpert advice given in the studio will annoy those listeners who know more about the subject than the presenter, and could actually be harmful to the person putting the question. By adequate screening of the incoming calls in line with the programme policy, the producer limits the public participation to the central purpose of the show.
    Types of phone-in include:

1 the open line – general conversation with the studio host;
2 the specific subject – expert advice on a chosen topic;
3 consumer affairs – a series providing 'action' advice on detailed cases;
4 personal counselling – problems discussed for the individual rather than the audience.

The open line

This is a general programme where topics of a non-specific nature are discussed with the host in the studio. There need be no theme or continuity between the calls but often a discussion will develop on a matter of topical interest. The one-minute phone-in, or 'soap-box', works well when callers are allowed to talk on their own subject for one minute without interruption, providing they stay within the law.

Support staff

There are several variations on the basic format in which the host simply takes the calls as they come in. The first of these is that the lines are answered by a programme assistant or secretary who ensures that the caller is sensible and has something interesting to say. The assistant outlines the procedure – 'please make sure your radio set is not turned on in the background' (to avoid acoustic 'howl-round'); 'you'll hear the programme on this phone line and in a moment X (the host) will be talking to you'. The call is then held until the host wants to take it. Meanwhile the assistant has written down the details of the caller's name and the point he or she wishes

Figure 13.1 A local radio phone-in producer answers calls, phones callers back, selects the calls to put through to the host, deals with ISDN two-ways, and greets guests (Courtesy of BBC Radio Norfolk)

to make and this is passed to the host. The producer will decide whether or not to reject a call on editorial or other grounds, to take calls in a particular order, or to advise the host on how individual calls should be handled. If the staffing of the programme is limited to two, the producer should take the calls since this is where the first editorial judgement is made. The computer screen provides visual talkback, together with advance information about each call, between call-taker and host.

Choosing the calls

The person vetting the calls quickly develops an ear for the genuine problem, the interesting point of view, the practical or the humorous. Such people *converse.* They have something to say but can listen as well as speak; they tend to talk in short sentences and respond quickly to questions put to them. These things are soon discovered in the initial off-air conversation. Similarly, there are people whom one might prefer not to have on the programme:

- the 'regulars' who are always phoning in;
- the abusive, perverted, offensive or threatening caller;
- the over-talkative and uninterruptible;
- the boring, dull or slow;
- those who, for whatever reason, are very difficult to understand;
- the sycophantic, who only want to hear you say their name on the air.

Of course the programme must have the idiosyncratic as well as the 'normal', and perhaps anger as well as conciliation. But without being too rude there are many ways of ending a conversation, either on or off the air:

'Yes, thank you, we had someone making that point yesterday . . . '
'I can't promise to use your call, it depends how the programme goes . . . '
'There's no one here who can help you with that one – sorry . . . '
'I'm afraid we are not dealing with that today . . . '
'This is a very bad line, I can't hear you . . . '
'You are getting off the subject, we shall have to move on . . . '
'We have a lot of calls waiting . . . '

Or simply:

'Right, thank you for that. Goodbye.'

The good host will have the skill to turn someone away without turning them off, but in moments of desperation it is worth remembering that even the most loquacious person has to breathe.

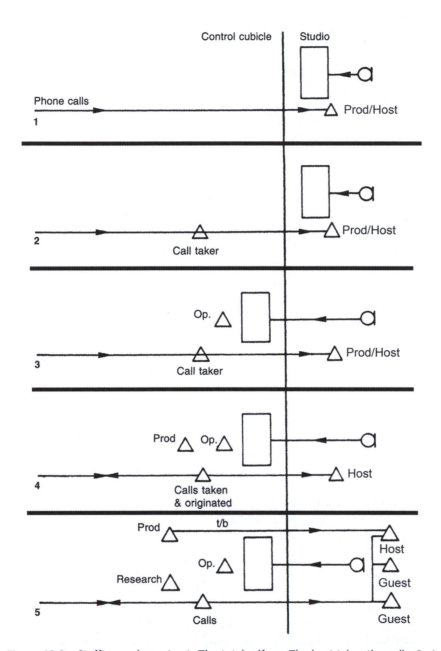

Figure 13.2 Staffing a phone-in. 1. The total self-op. The host takes the calls. 2. A 'call-taker' screens the calls and provides information to the presenter in advance of each call. 3. An operator controls all technical operation, e.g. levels, etc., while the producer/presenter concentrates on the programme content. 4. A separate producer in the control area makes programme decisions, e.g. to initiate 'phone-out' calls. 5. Guest 'experts' in the studio with research support available

The role of the host

The primary purpose of the programme is democratic – to let people have their say and express their views on matters that concern them. It is equiv-alent to the 'letters to the editor' column of a newspaper or the soap-box orator stand in the city square. The role of the host/presenter is not to take sides – although some radio stations may adopt a positive editorial policy – it is to stimulate conversation so that the matter is made interesting for the listener. The host should be well versed in the law of libel and defamation and be ready to terminate a caller who becomes obscene, overtly political, commercial or illegal in accordance with the programme policy.

Very often such a programme succeeds or fails by the personality of the person at the mic – quick thinking with a broad general knowledge, interested in people, well versed in current affairs, wise, witty, and by turn as the occasion demands, genial, sharp, gentle, possibly even rude. All this combined with a good characteristic radio voice; the presenter is a paragon of broadcasting.

Host style

In aural communication, information is carried on two distinct channels: *content* (what is said) and *style* (how it is said). They should both be under the speaker's control and, to be fully effective, one should reinforce the other. Due however to stress, it is not always easy, for example, to make a light point lightly. Without intention, it is possible to sound serious, even urgent, and the effect of making a light point in a serious way is to convey irony. The reverse – that of making light of something serious – can sound sarcastic. The problem therefore is how to appear natural in a tense situa-tion. It might be useful to ask yourself 'how should I come over?'

The following list may be helpful:

1   To be *sincere* – say what you really feel and avoid acting.
2   To be *friendly* – use an ordinary tone of voice and be capable of talk-ing with an audible smile. Avoid 'jargon' and specialist or technical language.
3   To appear *human* – use normal conversational language and avoid artificial 'airs and graces'. Admit when you do not know the answer.
4   To be *considerate* – demonstrate the capacity to understand views other than your own.
5   To be *helpful* – offer useful, constructive practical advice.
6   To appear *competent* – demonstrate an appreciation of the question and ensure accuracy of answers. Avoid 'waffle' and 'padding'.

This, of course, is no different from the ordinary personal contacts made hundreds of times each day without conscious thought. What is different for the broadcaster is that the stress in the situation can swamp the normal human qualities, leaving the 'colder' professional ones to dominate. Someone concerned to appear competent will all too easily sound efficient to the point of ruthlessness unless the warmer human characteristics are consciously allowed to surface.

The host has to be good at spontaneity – to think on his or her feet, and quickly sum up the caller's purpose. Probably the most valuable quality is credibility. Only when the listener believes the person speaking will they be prepared to take notice or act on what is being said. For this reason, style is initially more important than content.

The opposite of credibility may also be attractive, or at least entertaining. The 'shock jock' makes a point of being outrageous, and to some offensive. He or she will deliberately make fun of callers, belittling them, using them as the butt of humour with exaggerated comments and often risqué jokes. Frequently with a late night placing, callers tend to play the same game hoping to 'win' the conversation – they seldom do.

Reference material

The host may be faced with a caller seeking practical advice and it is important for the producer to know in advance how far the programme should go in this direction, otherwise it may assume expectations for the listener which cannot be fulfilled. Broadcasters are seldom recruited for their practical expertise outside the medium and there is no reason why they should be expected spontaneously to answer specialist questions. However, the availability in the studio of suitable reference material will enable the presenter to direct the caller to the appropriate source of advice or information. Reference sources may include a separate computer on the Internet, telephone directories, names and addresses of councillors, members of parliament or other elected representatives, government offices, public utilities, social services, health and education departments, welfare organisations and commercial PR people. This information is usually given on the air; but it is a matter of discretion. In certain cases it may be preferable for the host to hand the caller back to the programme assistant who will provide the appropriate information individually. If there is a great deal of factual material needed then a fourth person will be required to do the immediate Internet research.

Studio operation

At the basic level it is possible for the host alone to undertake the operation of the studio control desk. But as facilities are added, it becomes

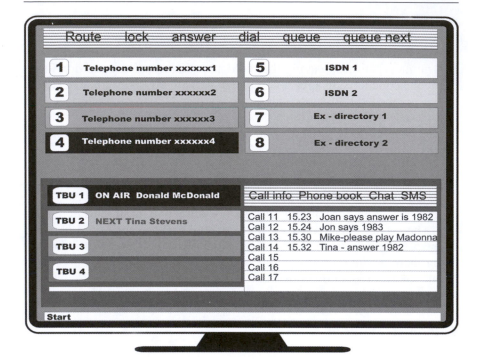

Figure 13.3 Phone-in screen. Incoming calls are taken by the producer on lines 1–4. Lines come from other studios and ex-directory lines are for 'phone-out'. The host is currently talking with Donald McDonald and knows that Tina Stevens is waiting with the answer '1982'

necessary to have a specialist panel operator, particularly where there is no automatic equipment to control the sound levels of the different sources. In this respect an automatic 'voice-over' unit for the host is particularly useful, so that when speaking, the level of the incoming call is decreased. It must, however, be used with care if he or she is to avoid sounding too dominating.

Additional telephone facilities

If the equipment allows, the host may be able to take two calls simultaneously, so setting up a discussion between callers as well as with the studio. The advice and cooperation of the telephone company may be required prior to the initiation of any phone programme. This is because there might well have to be safeguards taken to prevent the broadcasting function from interfering with the smooth running of the telephone service.

These may take the form of limitations imposed on the broadcaster in how the telephone might be used in programmes, or possibly the installation of special equipment either at the telephone exchange or at the radio station.

Use of 'delay'

The listening interest of a phone-in depends to an extent on the random nature of the topics discussed and the consequent possibility of the unexpected or outrageous. There is a vicarious pleasure to be obtained from a programme not wholly designed in advance. But it is up to the host to ensure that there is reasonable control. However, as an additional safeguard, it is possible to introduce a delay time between the live event and the transmission – indeed some radio stations and broadcasting authorities insist on it. Should any caller become libellous, abusive or obscene a delay device of, say, 10 seconds, enables that part of the programme to be deleted before it goes on the air. The programme is faded out and replaced by an ident or jingle (see Figure 13.4). With a good operator, this substitution can be made without it being apparent to the listener. Returning to the programme it is useful to have on hand news, music or other breaks to allow the host time to return to another call. If the caller is occasionally using words which the producer regards as offensive these can be faded out after a suitable warning. Overall, however, a radio station gets the calls it deserves and, given an adequate but not oppressive screening process, the calls will in general reflect the level of responsibility at which the programme itself is conducted.

The specific subject

Here, the subject of the phone-in is selected in advance so that the appropriate guest expert, or panel of experts, can be invited to take part. It may be that the subject lends itself to the giving of factual advice to individual questions, for example child care, motoring, medical problems, gardening, pets and animals, farming, antiques, holidays, cooking, financial issues or citizens' rights. Or the programme may be used as an opportunity to develop a public discussion of a more philosophical nature, for instance the state of the economy, political attitudes, education or religious belief.

A word of warning, however, about giving specific advice.

Doctors should not attempt diagnosis over the phone, much less prescribe drugs in individual cases. What tends to be most useful is general information about illness, the side-effects of drugs, what to expect when you go into hospital, the stages of pregnancy, children's diseases, reclaiming medical costs on social welfare, and so on.

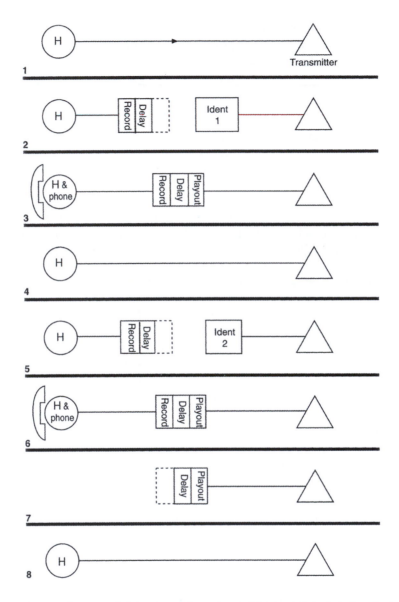

Figure 13.4 The use of delay in a phone-in. 1. The host P is fed directly to the output. 2. To introduce a delay, the host is recorded and held on a short-term (10 second) storage. The programme output is maintained for this duration by recorded ident 1. As ident 1 ends, the programme continues using the recorded playout. The transmission is now 10 seconds behind 'reality'. 4. If a caller says something that has to be cut, the delayed output is replaced by the host 'live'. 5. To reinstate the delay, the procedure is as in 2, but using a different ident. 6. The programme continues through the delay device. 7. The host brings the programme to an end, allowing the recorded playout to finish the transmission. 8. Normal live presentation

Of course there will be difficult calls and the producer should discuss a medical programme's policy with guests beforehand, rather than having to decide it live on-air. For example, how will you deal with cancer, AIDS and the terminally ill? What will you do with what appears to be a complaint against a hospital? At what point might you rule out a call as being too embarrassing, disgusting or voyeuristic? When is the caller just sending you up?

The legal phone-in too needs special care but can provide useful information on a wide range of subjects rooted in actual practice: employment law, motor insurance claims, marriage and divorce, wills, the rights of victims, disputes between landlord and tenant, land claims, house purchase, the responsibilities of children, arguments between neighbours – noise, nuisance, fences, trees, etc. A lawyer, like the doctor, should be wary of providing specific solutions over the phone – the caller is unlikely to give all the facts, especially those not supporting his or her case. There is also a real danger of being used as a second opinion in a case already in front of the courts, or where the caller is dissatisfied with advice already obtained. The professional guest will know when to respond with: 'I can't comment on your particular case without the details, but in general . . .'. The point of whether a radio service can be sued for giving wrong advice is of interest here. The answer will be 'no', so long as the programme does not try to imitate the private consultation. The producer serves the public best by providing advice of general applicability, illustrating how the law operates in specific areas.

'Early lines'

In order to obtain questions of the right type and quality, the phone lines to the programme may be opened some time before the start of the transmission – say half an hour. The calls are taken by a programme assistant who notes the necessary details and passes the information to the producer who can then select the most appropriate calls. For the broadcast these are originated by the studio on a phone-back basis.

The combination of 'early' lines and 'phone-back' gives the programme the following advantages:

1 The calls used are not random but are selected to develop the chosen theme at a level appropriate to the answering panel and the aim of the programme.
2 The order in which the calls are broadcast is under the control of the producer and so can represent a logical progression of the subject.

3   The studio expert, or panel, has advance warning of the questions and can prepare more substantial replies.
4   The phone-back principle helps to establish the credentials of the caller and serves as a deterrent to irresponsible calls. The programme itself may therefore be broadcast 'live' without the use of any delay device.
5   At the beginning of the programme there is no waiting for the first calls to come in; it can start with a call of strong general interest already established.
6   Poor or noisy lines can be redialled by the studio until a better quality line is obtained.

Consumer affairs

The consumer phone-in is related to the 'specific subject' but its range of content is so wide that any single panel or expert is unlikely to provide detailed advice in response to every enquiry. As the range of programme content increases, the type of advice given tends to become more general, dealing with matters of principle rather than the action to be taken in a specific case. For example, a caller complains that a car, item of clothing, holiday, electrical appliance, or service from a bank is unsatisfactory – what should they do about it? An expert on consumer legislation in the studio will be able to help distinguish between the responsibility of manu-facturer, retailer, or service provider, or advise whether the matter should be taken up with a particular complaints council, electricity authority or local government department. To provide a detailed answer specific to each case may require more information. What were the exact conditions of sale? Is there a guarantee period? Is there a service contract? Was the appliance being used correctly?

It's a useful area for discussion but again, expert knowledge of the law in retail and service matters is essential.

The need to be fair

Consumer affairs programmes, rightly tend to be on the side of the com-plainant, but it should never be forgotten that a large number of complaints disintegrate under scrutiny and it is possible that such fault as there is lies with the user. Championing 'the little man' is all very well, but radio stations have a responsibility to retailers and manufacturers too. Once involved in a specific case the programme must be fair, and be seen to be fair. Two further variations on the phone-in help to provide this balance:

1   *The phone-out.* A useful facility while taking a call is to be able to originate a second call and have them both on the air simultaneously. In response to a particular enquiry, the studio rings the appropriate head of sales, PR department or council/government official to obtain a detailed answer, or at least an undertaking that the matter will be looked into.

2   *The running story.* The responsibility to be fair often needs more information than the original caller can give or than is immediately available and an enquiry might need further investigation outside the programme. While the problem can be posed and discussed initially, it may be that the subject is one that has to be followed up later.

Linking programmes together

Unlike the 'specific subject' programme which is an individual 'one-off', programmes that deal with broad consumer affairs may run in series – weekly, daily or even morning and afternoon. A complex enquiry might run over several programmes and while it can be expensive of the station's resources, it can also be excellent for retaining and increasing the audience. For the long-term benefit of the community, and the radio station, the broadcaster must be sure that such an investigation is performing a genuine public service.

As with all phone-ins, consideration should be given to recording the broadcast output in its entirety as a check on what was said – indeed this may be a statutory requirement imposed on the broadcaster. People connected with a firm that was mentioned but who heard of the broadcast at second-hand may have been given an exaggerated account of what transpired. An ROT (recording off transmission) enables the station to provide a transcript of the programme and is a wise precaution against allegations of unfair treatment or threat of legal proceedings – always assuming of course that the station has been fair and responsible!

Personal counselling

With all phone-in programmes, the studio presenter is talking to the individual caller but has constantly to bear in mind the needs of the general listener. The material discussed has to be of interest to the very much wider audience who might never phone the station but who will identify with the points raised by those who do. This is the nature of broadcasting. However, once the broadcaster decides to tackle personal problems, sometimes at a deeply psychological and emotionally disturbing level, one cannot afford to be other than totally concerned for the welfare of the individual caller. For the duration, the responsibility to the one exceeds that to the many. Certainly, the host/presenter cannot terminate a conversation simply because it has ceased to be interesting or because it has become too difficult. Once a radio service says 'bring your problems here', it must be prepared to supply some answers.

This raises important questions for the broadcaster. Are we exploiting individual problems for public entertainment? Is the radio station simply providing opportunities for the aural equivalent of the voyeur? Or is there sufficient justification in the assumption that, without even considering the general audience, at least some people will identify with any given problem and so be helped by the discussion intended for the individual? It depends, of course, on how the programme is handled, the level of advice offered, and whether there is a genuine attempt at 'caring'.

Further, to what extent will the programme provide help outside its own airtime? The broadcaster cannot say 'only bring your depressive states to me between the transmission times of 9 to 11 at night'. Having offered help, what happens if the station gets a call of desperation during a music show? The radio station has become more than simply a means of putting out programmes, it has developed into a community focus to which people turn in times of personal trouble. The station must not undertake such a role lightly, and it should have sufficient contacts with community services that it can call on specialist help to take over a problem that it cannot deal with itself. But how can the programme director ensure that the advice given is responsible? Because of all the types of programme that a station puts out, this is the one where real damage could be done if the broadcaster gets it wrong. Discussing problems of loneliness, marriage and sex, or the despair of a would-be suicide, has to be taken seriously. It is important to get these callers to talk and to enlist their support for the advice given, and for this purpose a station needs its trained counsellors.

The presenter as listener

As with all phone-in programmes the host in the studio cannot see the caller. He or she is denied all the usual non-verbal indicators of communication – facial expression, gesture, etc. This becomes particularly important in a counselling programme when the caller's reaction to the advice given is crucial. The person offering the advice must therefore be a perceptive listener – a pause, a slight hesitation in what the caller is saying may be enough to indicate whether they are describing a symptom or a cause, or whether they've yet got to the real problem at all. For this reason many such programmes will have two people in the studio – the host, who will take the call initially and discuss the nature of the problem – and the specialist counsellor who has been listening carefully and who takes over the discussion at whatever stage it is felt necessary. Such a specialist could be an expert on marriage guidance, a psychiatrist, a minister of religion or a doctor.

Non-broadcasting effort

Personal counselling or advice programmes, also need off-air support – someone to talk further with the caller or to give names, addresses or phone numbers which are required to be kept confidential. The giving of a phone number over the air is always a signal for some people to call it, so blocking it as an effective source for the one person the programme is trying to help. Again, the broadcaster might need to be able to pass the problem to another agency for the appropriate follow-up.

The time of day for a broadcast of this type seems to be especially critical. It is particularly adult in its approach and is probably best at a time when it can be reasonably assumed that few children will be listening. This indicates a late evening slot – but not so late as to prevent the availability of unsuspected practical help arising from the audience itself.

Anonymity

Often a programme of this type, specialising in personal problems, allows callers to remain unidentified. Their name is not given over the air, the studio counsellor referring to them by first name only or by an agreed pseudonym. This convention preserves what most callers need – privacy. It is perhaps surprising that people will call a radio station for advice, rather than ask their family, friends or specialist, simply because they do not have to meet anyone. It can be done from a position of security, perhaps in familiar surroundings where they do not feel threatened.

People with a real problem seldom ring the station in order to parade it publicly – such exhibitionists should be weeded out in the off-air screening and helped in some other way. The genuine seeker of help calls the station because it's already known as a friend and as a source of unbiased personal and private advice. He or she knows that it's not necessary to act on the advice given unless they agree with it and want to. This is a function unique to radio broadcasting. It is perhaps a sad comment when a caller says, 'I've rung you because I can't talk to anyone about this', but it is in a sense a great compliment to the radio station to be regarded in this way. As such it needs to be accepted with responsibility and humility.

Phone-in checklist

The following list summarises what is needed for a phone-in programme.

If you are planning a genuinely massive national programme expecting an exceptional number of calls, it may be wise to talk to the telephone service to resolve any problems caused by the additional traffic. Do you want all the calls, even the unanswered ones, to be counted? Otherwise:

1   decide the aim and type of the programme;
2   decide the level of support staff required in the studio. This may involve a screening process, phone-back, immediate research, operational control and phone-out;
3   engage guest speakers;
4   assemble reference material;
5   decide if 'delay' is to be used;
6   arrange for ROT;
7   establish appropriate 'follow-up' links with other, outside, agencies.

14

Listener participation

Radio is not a good medium by itself for establishing a genuine two-way contact. Listeners may feel that the broadcaster comes into their home and they may even get the impression that they know an individual presenter. However, this is at best a substitute companionship rather than a genuine personal interaction. The broadcaster/listener relationship – or perhaps that of the station/listener – can be made more real through the broadcaster's ability to allow and encourage listeners to take part in the programme-making activity. This can go much further than the phone-in of the last chapter, for there are many ways of stimulating such involvement, extending its obligations as a public servant. But this should not be undertaken lightly for it can be expensive in time, money and effort. It therefore requires the backing of a management policy that understands and values additional forms of *personal* listener contact.

Letters, phone calls, emails, texts and tweets

Programme correspondence incoming to the station can be classified under three general headings – material intended for use on-air, that which requires a response, and things that can be forgotten. It is a matter of station policy whether or not an individual presenter is encouraged to become involved in replying to listeners. It is a time-consuming business and a hard-pressed station might not have the resources to do this. On the other hand, if a programme offers help, particularly for individual personal needs, then as with a phone-in, it must clearly honour its promise by meeting such requests.

On-air use includes music requests and dedications, competition replies, or correspondence addressed directly to a programme 'letter spot' i.e. to be read on-air. In general they either offer information or advice to the audience at large, or ask for help with a personal problem. Material like this can be a useful resource for programme makers – frequently raising questions of interest and making comments of substance, they can easily provide the sole content for a specialist programme. They can be dealt with in a variety of ways:

1   read by the presenter, who then responds;
2   read by male/female readers, the programme presenter then responding;
3   read by the presenter who then interviews an expert or introduces a discussion on the subject;
4   correspondence on similar or related topics grouped together for subsequent response;
5   a comment dramatised as a sketch to illustrate the point being made, followed by a response from the presenter or an interviewee, or by a discussion;
6   an issue raised by a listener read out by the presenter who, without replying, opens the discussion to the audience, inviting listeners to respond;
7   texts and emails are invited as the programme proceeds. These will come straight through to the screen of the producer or presenter. Depending on the nature of the content, some may be read out on-air in full – others might be edited or rejected;
8   listeners can be invited to follow the station's Facebook or Twitter accounts. These can provide effective two-way communications with the audience.

The producer of a 'letter' spot might consider the following in arriving at the most appropriate format:

• To maximise listener involvement, several pieces of correspondence can be dealt with in a single spot.
• A long letter/email might not be read in its entirety but extracts used to reflect accurately what the writer is saying.
• A matter with many questions is not allowed to monopolise a spot but is broken up and used in parts, perhaps over several spots.
• To give variety of pace and vocal interest a spot may use more than one of the response forms listed above.
• Avoid reading out full addresses on the air or giving any information likely to endanger the writer (see p. 204).

People contact a radio station either to get an issue aired publicly, or to get an authoritative reply. Such contact is not simply programme fodder, but deserves the same level of consideration that its sender gave it.

This becomes increasingly important for short-wave or long-distance broadcasters.

Programme follow-up

Mail items not intended for broadcast include requests for listener verification or QSL cards, scripts, programme information and offers, merchandise purchasing, and programme follow-up ranging from specific advice to pastoral and personal counselling. Much of this can be dealt with by means of standard replies – perhaps with some details to be filled in – or by fact-sheets, information booklets, website forms, and so on. It is the time-consuming one-off reply that poses the genuine problem.

Complaints in particular need special attention. It is tempting for a busy station to disregard them, and it has to be said that many are likely to seem unreasonable, resulting from an extreme or limited point of view. Nevertheless, if broadcasters are not to appear careless over alleged error and off-hand with the listener, complaints deserve a prompt but considered response at the appropriate level – programme presenter or producer, programme management or senior management – and if found to be of substance, a correction made. If the matter is serious and the station is genuinely in the wrong, an apology should be broadcast in the same slot as the offending transmission. The listening public deserves the highest standard of communication and, done well, a correction by letter or on-air can enhance a programme's reputation not only for the truth, but in its respect for the audience (see also pp. 354–355).

A larger category, we hope, comprises those items that query or seek clarification over something that has been said – or which wish to debate the issue further. This poses another question – is the broadcaster responsible for questions that a programme may have raised in a listener's mind? A small station might have to ignore such correspondence – it represents a drain on resources without providing any airtime. Besides, presenters might claim that the view of the listener even after a provocatively contentious programme is none of their business and that once the programme is over, the matter is finished. Furthermore, a philosophical, political or theological question raised could be outside their competence to answer.

On the other hand, a station may actively seek to develop a more personal relationship with its listeners through follow-up. This is often best done not by the programme presenters or producer individually but by a separate group – either a specialist department of the station staff, or volunteers closely associated with the programme. Such broadcasting support services are extremely useful for educational and religious programmes in particular, where individual listeners might need direct personal help – perhaps struggling with learning a language, coming to terms with illness or bereavement or coping with unemployment. Members of the replying group will need to be sensitive to any cultural differences that may exist between themselves and their correspondents. They will be selected as people experienced in a specialist field, but they will also rapidly develop their own expertise in this special form of one-to-one 'distance learning'. A computer database enables the station to keep track of the correspondence and perhaps to anticipate further enquiries. A final method is to put the listener directly in touch with a suitable college tutor, library, church, self-help group or other agency in his or her own vicinity. Follow-up can then be pursued on a personal basis.

Texting

Although some years ago sending text messages by smartphone may have been most popular with younger listeners, stations are now finding that texts have become the most common way for a wide range of listeners to contribute to programmes. From a production point of view texts are a most convenient way to communicate with the audience as, with a simple software application, the incoming texts can appear directly on a computer screen, like emails, for reference on-air. Texts are an ideal means of handling brief responses to programme items and a convenient way to accept answers to competitions.

Helpline

This is a useful item especially in a lengthy sequence – putting people with a specific need in touch with possible sources of help. For a small community station such links might vary from someone's search for a particular 'out of print' book or 'lost and found' information, to a noticeboard of job vacancies. Such requests are normally broadcast without charge but the station generally takes no responsibility for the outcome, a listener responding to an enquiry being put in touch with the originator off-air. In attempting to ensure the safety of this kind of transaction, it is essential that the caller gives their explicit willingness to be put in touch with a third party.

Other forms of help can be along the lines of a 'bought and sold' market or exchange, although again, the station must be sure to take no responsibility for the goods or services on offer. The station is merely supplying information and is in no sense a broker to the transaction.

Help- or action lines can also be opened *after* a particular programme, for example dealing with illness or disability, to provide further information or answer queries from listeners affected by the programme. Connecting callers with an established and recognised helping agency has often proved to be a crucial public service adjunct to the programme itself.

Visitors

Casual visitors to the radio station – adults and children – may be persuaded to become broadcasters and take part in a special spot, either live or by recording their comments via a dedicated and available facility, with subsequent editing. Very often they have a special story to tell – a childhood experience, war exploit, marriage encounter or holiday mishap. Like an on-station vox pop, the question to put is 'What's the most exciting/amusing/unusual/awful thing that's ever happened to you?'

Figure 14.1 A young listener is invited to read her own story (Courtesy of Radio Abracadabra)

Special involvement

Often, listener participation is sought for a particular purpose. Unlike unsolicited contact, it is expressly asked for and built into the format, and therefore becomes a particular station responsibility with a named producer in charge:

- run on-air courses with a correspondence element;
- establish a birthday club for children;
- set up a network of 'focus groups' to discuss and report back on programmes;
- invite programme contributions;
- raise money by asking for donations to a specific cause;
- ask for listeners' jokes or personal stories;
- run an unfinished drama/story for the listener to complete;
- offer a free telephone number to call for help with specific topics – finance, AIDS, unemployment, marriage/parenting problems;
- competitions, quizzes and games with prizes or forfeits, etc.

Travelling roadshow

A popular programme takes to the road, broadcasting each time from a different venue, generally 'live' in front of an audience. This could be one of the daily sequences such as the mid-morning or drivetime show, a music programme with its regular DJ, or a programme of the 'Any Questions' type. Events like this involve non-listeners as well as listeners and therefore with suitable questionnaires can be valuable in finding out why people do or don't listen to the station. Even a normally studio-based show may have the occasional OB, using a radio car or other mobile facility visiting its listeners by dropping in on a home discussion group or factory meeting.

Major events

The one-off major public event gives plenty of scope for a station to interact with its audience. These provide opportunities to find out about the likes, dislikes and needs of people, or simply to promote the station, extending its audience in size, reach or quality. This may take the form of a music festival or rock concert – sponsored by the station, perhaps with non-radio co-sponsors. This could be a commercial enterprise at an established venue, charging for admission and broadcasting the performance.

A national or local scriptwriting competition – for story, radio drama, poetry – could be held. Using eminent judges, it attracts public entry and may be sponsored, with considerable spin-off publicity and press coverage.

Sponsorship of a music or drama group has a double spin-off as it generates its own publicity while also creating output for broadcasting.

Another possibility is a radio conference with station management, producers and presenters hosting a question-and-answer session with the public, distributing and collecting questionnaires on listening habits, asking for and discussing ideas for future output, and selling merchandise. These events are probably not broadcast although the questions and answers could be recorded for transmission to demonstrate public service involvement.

A radio exhibition could be held, with the station illustrating its own history, to mark a special event, involving other appropriate organisations, sponsors, equipment manufacturers, and perhaps also providing an insight into the future. Held at its own premises, at a public venue or in a prestigious retail store, staff and presenters are on hand to give out publicity, answer questions, sell merchandise, and with a studio facility, produce inserts into station output.

Events such as these, and one will think of many more, involve a station well beyond its normal sphere of influence and resources. They require a good deal of planning, and may entail considerable up-front capital costs. This is why they need senior management intervention. Nevertheless for a station that wants to make its mark, to celebrate its anniversary, or to contribute to public life in a memorable way, it must look beyond the day-to-day schedule to the long term, discovering how best to raise its profile among the potential, as well as the actual, listenership.

15

Music programming

The filling of programme hours with recorded music is a universal characteristic of radio stations around the world. This is hardly surprising in view of the advantages for the broadcaster. Computer music stores, hard disks and CDs represent a readily available and inexhaustible supply of high-quality material of enormous variety that is relatively inexpensive, easy to use and enjoyable to listen to. Many stations allow little or no freedom for presenters in what they choose to play, but some leave it entirely up to the show's host. Most have all the music sources hidden away, to be brought up by the computer keyboard; others play CDs, even vinyl, in the studio. Suffice it to say that there is a wide range of operational practice. Before looking in detail at some of the possible formats and what makes for a successful programme, there are three important preliminaries to consider.

First, the matter of music copyright. Virtually every recording somewhere carries the words, 'all rights of the producer and of the owner of the recorded work reserved. Unauthorised public performance, broadcasting and copying of this recording prohibited.' This is to protect the separate rights of the composer, publisher, performers and recording company, who together enabled the recording to be made. The statement is generally backed by law – in Britain the Copyright, Designs and Patents Act (1988). It would obviously be unfair on the original artists if there were no legal sanctions against the copying of their work by someone else and its subsequent remarketing under another label. Similarly, broadcasters, who in part earn their living through the effort of recording artists and others, must ensure that the proper payments are made regarding the use of their work. As part of a 'blanket' agreement giving 'authorised broadcasting use', most radio stations are required to make some form of return to the societies representing the music publishers and record manufacturers, indicating what has been played. In Britain these are, respectively, the Performing Right Society (PRS for Music) and Phonographic Performance Ltd (PPL). In the United States it is ASCAP – The American Society of Composers,

Authors, and Publishers. It is the producer's responsibility to see that any such system is carefully followed.

Second, it must be said that in using recorded sound, broadcasters are apt to forget their obligation to 'live' music. Whatever the constraints on the individual station, some attempt should be made to encourage performers by providing opportunities for them to broadcast. Many recording artists owe their early encouragement to radio, and broadcasting must regard itself as part of the process that enables the first class to emerge. Having reached the top, performers should be given the fresh challenge that radio brings.

Third, top-flight material deserves the best handling. It is easy to regard a track simply as a 'thing' instead of a person. On the air someone's reputation could be at stake. Basic operational technique must be faultless – levels, accurate talk-overs, fades, etc. Most important is that music should be handled with respect to its phrasing.

Attitudes to music

Music, like speech, comes in sentences and paragraphs. It would be nonsensical to finish a voice piece other than at the end of a sentence, and similarly it is not good practice to fade a piece of music arbitrarily. A great deal of work has gone into its production and it should not be treated like water out of a tap, to be turned off and on at will – not unless the broadcaster is prepared to accept the degradation of music into simply a plastic filler material. The good operator, therefore, will develop an 'ear' for fade points. The 'talk-over' – an accurately timed announcement that exactly fits the non-vocal introduction of a song – provides a satisfying example of paying attention to such detail. Music handled *with respect to its phrasing* provides listening pleasure for everyone.

The presenter must accept the responsibility when music and speech are mixed through an automatic voice-over unit or 'ducker' so that any speech hurls the music into the background. It has its uses in particular types of high-speed DJ programmes, but to use music as a semi-fluid sealant universally applied seems to imply that the programme has cracks that have to be frantically filled!

The broadcaster's attitude to music is often typified by the level of care in the treatment of any CDs or records. They are worth looking after. This includes – for the stations that have them – an up-to-date library cataloguing system and proper arrangements for withdrawal and return; thus avoiding their being left lying around in the studio or production offices. Many broadcasters who play music professionally have domestic stereo equipment over which they take meticulous care – they would not dream of touching the playing surface of a disc with their fingers.

Metadata

Recordable CDs can store a wealth of material, from commercials and station jingles to individual presenter idents and programme inserts. Compilation CDs can be made to collect together the currently favoured music tracks, as well as being an excellent medium for station archives.

When tracks are converted to digital files, it is usual to 'normalise' the tracks. Normalisation ensures that the output levels of all tracks use the full amplitude range that is available, i.e. they all play back at roughly at the same volume level. In addition descriptive metadata tagging is usually added to the audio file. Metadata is the equivalent of liner or sleeve notes, embedded in every file. At the most basic level, the metadata consists of the album title, song title, artist, genre and timing. This can be expanded to include cue points and information regarding the length of the intro, how the track ends (fades Y or N), artwork, lyrics, complete credits – composer, sidemen in the band, technical notes, etc. The metadata can provide this additional information for the listener, as selected data can be sent directly to a digital radio system with text display, or to a streaming server. It can also be used to fulfil the station's obligations regarding music reporting.

Likes and dislikes

Programme makers, especially in the music area, should keep in touch with research on listener opinion. Nothing divides an audience as much as our likes and dislikes in music. Dislikes seem to be pretty universal and obvious. It's not difficult to see from the Ofcom research that people don't like 'crappy adverts', 'inane presenters', 'obviously repetitive playlists', and 'tunes they don't like'. This leads to a lot of 'zapping' or channel hopping. On the other hand, people appreciate 'genuine wit and passion for the music'. The research compared the use – especially by the younger generation – of radio with the iPad or tablet.

> 'On my tablet you always get to listen to the music you like – but it's kind of unpredictable at the same time.'

> 'Radio keeps up with new music, new artists and information.'

But radio is being challenged by digital and mobile media as the primary source of music and it can't be emphasised enough that music producers especially must use all possible means to stay close to their target audience to know the current flavours and sounds – what are the trade papers and popular blogs writing about? What is your competitor playing?

The programme areas now discussed in detail are: music formats, requests and dedications, guest programmes and the DJ show.

Clock format

Designing a music programme on a one-hour clockface has several advantages for both the presenter and the scheduler. It enables the producer/presenter to see the balance of the show between music and speech, types of music and the spread of commercials; it is a great help in maintaining consistency when another presenter has to take over; and it enables format changes to be made with the minimum of disruption. It is a suitable method to use regardless of the length of the programme. The clock provides a solid framework from which a presenter can depart if need be, and pick up again just as easily. It imposes a discipline but allows freedom.

Starting with the audience, the producer begins by asking questions. Who do I wish to attract? Is my programme to be for a particular age or demographic group? This will obviously affect the music chosen, and a reasonable rule of thumb assumes that the musical taste of many people was formed in their teens. For example, listeners in their forties are likely to appreciate the hits of 25 years ago. Of course there are many categories of music that have their own specialist format or may be used to contribute to a programme of wider appeal. Beware, however, of creating too wide a contrast – the result is likely to please no one. The basic categories, which contain their own sub-divisions, can be listed as follows:

Top 40	Classic hits (the last 25 years)
Progressive rock	Golden oldies
Black soul/funk	Jazz
Rap	Folk
Hip-hop	Country
Rhythm and blues	Latin American
Disco-beat	Middle of the road
House music	Light classical, orchestral – operetta
Adult contemporary	Classical, symphonic – opera, etc.

Having established the broad category, there are infinite combinations of tempo and sound with which to achieve the essential variety within any chosen consistency:

- sound – rock band, big/small, heavy metal, synthetic, swing;
- tempo – slow, medium, med/bright, up beat;
- vocal – male, female, duo, group, newcomer, star;

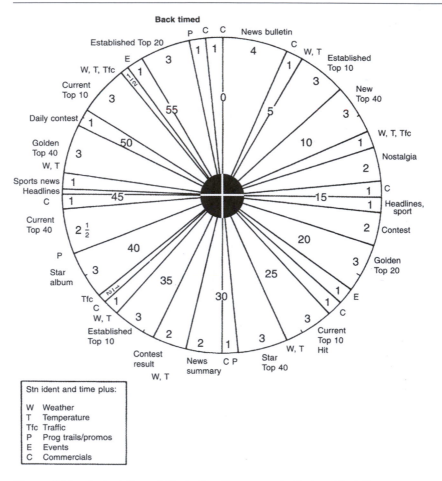

Figure 15.1 A clock format illustrating the concept of a breakfast time programme on the basis of: news on the hour, summary on the half-hour, headlines at quarters; eight minutes of advertising in the hour, employing news adjacency; music/speech ratio of about 60:40; generally MOR sound using current hits, golden or 'yesteryear' Top 40, star name and nostalgia tracks; speech includes current information, sports news, 'short' programme content and station daily competition

- era – forties, fifties, sixties, seventies, eighties, etc;
- ensemble – big band, string, military band, brass, orchestral, choral.

Lengthy formatted programmes need not be constructed to have a strong beginning and end; the skill lies in the host presenter providing the listener with a satisfying programme over whatever time-period the show is heard. The breakfast show and drivetime programmes are likely to be heard by many people over a relatively short time-span, perhaps 20–30 minutes. The afternoon output will probably be heard in longer durations. The producer sets out to meet the needs of the total audience – so the two key questions are: Does the plan contain the essential elements in the estimated normal period of listening? Does it contain the necessary variety over a longer time-span?

Station practice indicates the following guidelines:

- Come out of news bulletins on an up-tempo sound. Restore the music pace and listen carefully to any lyrics to avoid any unfortunate juxtaposition of the last news story and the first words.
- When programming new or unfamiliar music, place solid hits or known material on either side of it – 'hammocking'.
- Spread items of different types evenly up to the hour as well as on the downside.
- Mix items of different durations; avoid more than two items of similar length together.
- Occasionally break up the speech–music–speech sandwich by running tracks back to back, i.e. use the segue or segue plus talk-over.
- Place regular items at regular times.

Computerised selection

Many stations regulate the use of music in a totally systematic way, for example to prevent the same tracks being used in adjacent programmes. Entering titles as data, a computer is programmed to provide the right mix of material with music styles, performers, composers, etc. appearing at the desired frequency. The scheduling is then in the hands of one of the many types of computer software – SABLE, Selecta, Tune Tracker, MusicMaster, AirTrafficControl or other station system which decides the rotation of items, how many times a track will be played in a given period, and what type and length of music is appropriate for different times of day. The database keeps a history of the music played that not only provides the official music logs, but is the source for answering listener enquiries.

A computer system is used to generate a weekly playlist that is mandatory for presenters who are given no choice over their music. Less restricting is the playguide. This requires presenters to select a given percentage of their programme from material determined by the programme controller; the remainder they choose themselves. Station policy invariably lays down the extent of individual freedom in the matter of music choice.

Requests and dedications

There are now far fewer request shows. Where they do exist, it is all too easy to forget that the purpose is still to do with *broad*casting. There is a temptation to think only of those who have requested, rather than of the audience at large. Until the basic approach is clarified, it is impossible to answer the practical questions that face the producer. For example, to what extent should the same track be played in successive programmes because someone has asked for it again? Is there sufficient justification in reading out a lengthy list of names simply because a lot of people have asked for it? While the basis of the programme is clearly dependent on the initiative of the individual listener who requests a particular track, the broadcaster has a responsibility to all listeners, not least the great majority who do not write.

The programme aims may be summarised:

1   to entertain the general audience;
2   to give especial pleasure to those who have taken the trouble to send a request;
3   to foster goodwill by public involvement.

Programmes are given a good deal of individual character by the host who sets the guidelines. The idea may be only to have requests related to birthdays, weddings or anniversaries, or that each request has to be accompanied by a personal anecdote, joke or reminiscence to do with the music requested. References to other people's nostalgia can certainly add to the general entertainment value of the programme. The presenter should be consistent about this and it is important that style does not take over from content, for if the intention of the programme is to play music, it is this rather than speech that remains the central ingredient.

Dedications are often more frequent than requests, i.e. requests not related to a specific piece of music but which can be associated with any item already included in the programme. By encouraging 'open' requests of this type, the programme can carry more names, and listeners therefore have a greater chance of hearing themselves mentioned, which of course is what they like.

Having decided the character and format of the programme there are essential elements in its preparation.

Choosing music

With music items of two to four minutes each, there will be some eight or nine tracks in each half hour of non-advertising airtime. This allows for about a minute of introduction, signature tunes, etc. Given the volume of requests received, it soon becomes clear to what extent they can be met. The proportion that can be dealt with might be quite small and, assuming that their number exceeds the capacity to play them, a selection process is necessary.

The criteria of selection will include the host's desire to offer an attractive programme overall, with a variety of music consistent with the programme policy. It could be limited to the current Top 40, to pop music generally, or it may specifically deal with one area, for example gospel music. On the other hand, the choice might be much wider to include popular standards, light classical music or excerpts from symphonic works. The potential requester should know what kind of music a particular programme offers.

Another principle of selection is to choose requests that are likely to suit the presenter's remarks, i.e. those that will make for an interesting introduction, an important or unusual event, a joke or a particularly topical reference. It is frequently possible to combine requests using several different comments in order to introduce a single piece of music. The danger here is to become involved in several lists of names which may delight those who like hearing themselves referred to on the air, but which can become boring to the general listener. It is, however, a useful method of including a name check while avoiding a particular choice of music, either because to do so would be repetitious, or because it is out of keeping with the programme – or, of course, because the station does not have the named track.

Item order

After selecting the music, a decision has to be made about its sequence. This should not be a matter of chance, for there are positive guidelines in building an attractive programme. 'Start bright, finish strong' is an old music hall maxim and it applies here. A tuneful or rhythmic familiar up-tempo 'opener' with only a brief speech introduction will provide a good start with which the general listener can identify. A slower piece should follow and thereafter the music can be contrasted in a number of ways: vocal/non-vocal, female vocal/male vocal, group vocal/solo vocal, instrumental/orchestral, slow/fast, familiar/unfamiliar, and so on. It is sensible to scatter the very popular items throughout the programme and to limit the material that may be entirely new. Careful placing is required, with slow numbers which need to be followed by something brighter. The order in which music is played, of course, affects all music programmes whether based on recorded material or live.

Prefading to time

The last piece of music can be chosen to provide 'the big finish' and a suitable finale will often be a non-vocal item. This allows it to be faded in under the presenter's introduction. It may have been timed to end a little before the close of the programme, so leaving room for the final announcements and perhaps a signature tune. 'Prefading to time' (not to be confused with 'prefade' – audition or pre-hear) or 'back-timing' is the most common device for ensuring that music programmes run to time. A closing signature tune will almost certainly be prefaded to time.

In a similar way, items within a programme, particularly a long one, can be subject to this technique to provide fixed points and to prevent the overall timing drifting.

Preparing listeners' material

Programme hosts using material – written or electronic – from listeners, will know that they are not always intelligible and cannot be used in the studio without some form of preparation. Some retype the basic information to ensure clarity, others prefer to work directly from the originals. It is important that names and addresses are legible so that the presenter is not constantly stumbling, or sounding as though the problem of deciphering a correspondent's writing or typing is virtually insurmountable. In reality, this may often be the case, but a little preparation will avoid needlessly offending the people who have taken the trouble to write and

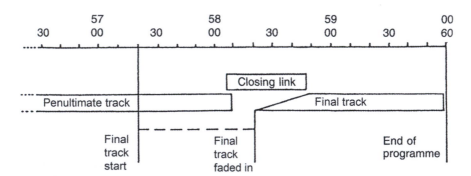

Figure 15.2 Prefading to time. This is the most usual method of ensuring that programmes run to time. In this example, the final track is 2′ 40″ long. It is therefore started at 57′ 20″ but not faded up until it is wanted as the closing link finishes. It runs for a further 1′ 10″ and brings the programme to an end on the hour

on whom the programme depends. In this respect, particular care has to be taken over personal information such as names, ages and addresses and it goes without saying that such things should be correctly pronounced. This also applies to streets, districts, hospitals, wards, schools and churches where an incorrect pronunciation – even though it might be caused by illegible handwriting – immediately labels the presenter as a 'foreigner' and so hinders listener identification. Such mispronunciations should be strenuously avoided, particularly for the community broadcasting station, and prior reference to telephone directories, maps and other local guides is advisable.

Reading through the requests beforehand enables the presenter to spot dates, such as anniversaries, which will have been passed by the time of the broadcast. This can then be the subject of a suitable apology coupled with a message of goodwill, or alternatively the item can be omitted. What should not happen is for the presenter to realise on-air during the introduction that the event which the writer anticipates has already happened. Not only does this sound unprofessional, but it gives the impression that the presenter does not really know what's happening, and worse, does not care. Anything that hinders the rapport between listener and presenter will detract from the programme. As has been noted previously, the trend is to emphasise this as a *personal* relationship.

In order to accommodate more requests and dedications, they can be grouped together, sharing a single piece of music. The music need not necessarily be precisely what was asked for, providing that it is of a similar type to that requested – for example, by the same artist.

Such preparation of the spoken material makes an important contribution to the host's familiarity with the programme content. While it may be the intention to appear to be chatting informally 'off the cuff', matters such as pronunciation, accuracy of information, relevance of content and timing need to be worked on in advance. The art is to do all this and still retain a fresh, 'live', ad-lib sound. The experienced presenter knows that the best spontaneity contains an element of planning.

Programme technique

On the air, each presenter must find his or her own style and be true to it. No one person or programme will appeal to everyone but a loyal following can be built up through maintaining a consistent approach. A number of general techniques are worth mentioning.

Never talk over vocals. Given adequate preparation, it is often pleasing to make an accurate talk-over of a non-vocal introduction, but talking over the singer's voice can be muddling, and to the listener may sound little different from interference from another station.

Avoid implied *criticism of the listener's choice* of music. The tracks might not coincide with your taste but they represent the broadcaster's intention to encourage public involvement. If a request is unsuitable, it is best left out.

Do not play less than one minute of anyone's request, or the sender will feel cheated. If the programme timing goes adrift it is generally better to drop a whole item than to compress. If the last item, or the closing signature tune, or both are 'prefaded to time' the programme duration will look after itself.

If by grouping requests together, there are several names and addresses, prevent the information from *sounding like a roll call*. Break up the list and intersperse the names with other remarks.

Develop the habit of talking *alternately* to the *general listener* and to the *individual listener* who asked for the piece. For example:

> Now here's a tweet from someone about to celebrate 50 years of married bliss – that's what it says here and she's Mrs Jane Smith of Highfield Road, Mapperley. Congratulations Mrs Smith on your half century, you'd like to dedicate this with all your love to your husband John. An example to the rest of us – congratulations to you too, John. The music is from one of the most successful shows of the sixties.

Avoid remarks which, combined with the address, could pose any kind of *risk to the individual*.

> ' . . . I'm asked not to play it before six o'clock because there's nobody at home till then . . . '

> ' . . . she says she's living on her own but doesn't hear too well . . . '

> ' . . . please play the record before Sunday the 18th because we'll all be away on holiday after that . . . '

> ' . . . he says his favourite hobby is collecting rare stamps.'

To reduce such risks, it is best to omit the house detail, referring only to the street and town. However, house numbers can generally be discovered by reference to voters' lists and other directories. Broadcasters must always be aware of the possible illegal use of personal information.

It's often sensible to keep notes of the programme detail for a week or so to provide any follow-up that may be needed.

Guest programmes

Here, the regular presenter invites a well-known celebrity to the programme and plays his or her choice of music. The attraction of hearing artists and performers talk about music is obvious but it is also of considerable interest to have the lives of others, such as politicians, sports-men and businessmen, illuminated in this way.

The production decisions generally centre on the ratio of music to speech. Is the programme really an excuse for playing a wide range of records or primarily a discussion with the guest but with musical punctua-tion? Certainly it is easy to irritate the listener by breaking off an interest-ing conversation for no better purpose than to have a musical interlude, but equally a tiny fragment only of a Beethoven symphony can be very unsatisfying. The presenter, through a combination of interview style and links into and out of the music, must ensure an overall cohesiveness to pre-vent the programme sounding 'bitty'. While it may be necessary to arrive at a roughly half and half formula, the answer could be to concentrate on the music where the guest has a real musical interest but to increase the speech content where this is not the case, using fewer but perhaps slightly longer music inserts.

Once again the resolution of such questions lies in the early identifica-tion of a programme aim appropriate to the target audience.

DJ shows

The radio disc jockey defies detailed categorisation. His or her task is to be unique, to find and establish a distinctive formula different from all other DJs. The music content may vary little between two competing pro-grammes, and in order to create a preference the attraction must lie in the way it is presented. To be successful, therefore, the DJ's personality and programme style must not only make contact with the individual listener, but in themselves be the essential reason for the listener's attention. The style may be elegant or earthy, raucous or restrained, but for any one pre-senter it should be consistent and the operational technique first class. See also the comments on the 'shock jock' (p. 177).

The same rules of item selection and order apply as those already iden-tified. The music should be sufficiently varied and balanced within its own terms of reference to maintain interest and form an attractive whole. Even a tightly formatted programme, such as a Top 40 show, will yield in all sorts of ways to imaginative treatment. One of the secrets is to be absolutely certain of the target audience. Top 40 for the 20–30 age group is radically different from Adult Contemporary for 25–40-year-olds. Different again is Nostalgia for the 40+. The personal approach to this type of broadcast-ing differs widely and can be looked at under three broad headings.

The low-profile DJ

Here the music is paramount and the presenter has little to say – the job is to be unobtrusive. The purpose of the programme may be to provide

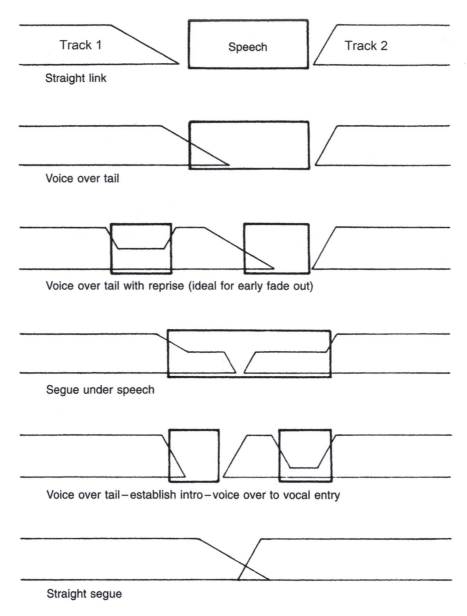

Figure 15.3 Six ways of going from one music item to the next, with or without a speech link. The use of several different methods helps to maintain programme variety and interest

Figure 15.4 Turntables used for a programme of 'oldie' vinyl discs (Courtesy of BBC Radio Derby)

background listening and all that is required is the occasional station identification or time check. In the case of a classical music programme the speech/music ratio should obviously be low. The listener is easily irritated by a host who tries to take over the show from Berlioz and Bach. The low-profile DJ has to be just as careful over what is said.

The specialist DJ

Experts in their own field of music can make excellent presenters. They spice their introduction with anecdotes about the artists and stories of happenings at recording sessions, as well as informed comment on performance comparisons and the music itself. Jazz, rock, opera and folk all lend themselves to this treatment. Often analytical in approach, the DJ's job is to bring alive the human interest inherent in all music. The listener should obviously enjoy the tracks but half the value of the programme is derived from hearing authoritative, possibly provocative, comment from someone who knows the field well.

The personality DJ

The most common of all DJ types, the role means doing more than just playing tracks with some spontaneous ad-libs in between. However popular the music, this simple form of presentation soon palls. The DJ must communicate personally, creating a sense of friendship with the audience. Therefore never embarrass the listener, either through incompetence or bad taste; the job is to entertain. To do this well, programme after programme, requires two kinds of preparation.

The first is in deciding what to say and when. This means listening to the music beforehand to decide the appropriate places for a response to the words of a song, a jokey remark or other comment, where to place a listener's comment, quiz question or phone call. The chat between records should be thought about in advance so that it doesn't sound pedestrian, becoming simply a repetitive patter. All broadcast talk needs some real substance containing interest and variety. This is not to rule out entirely the advantages of spontaneity and the 'fly by the seat of your pants' approach. The self-operating DJ, with or without a producer, is often capable of creating an entertaining programme, making it up as they go along. Undoubtedly though, such a broadcaster is even better given some preparation time.

Figure 15.5   DJ/presenter Makenda at Baraka FM, Mombasa (Courtesy of Roger Stoll)

The programme may contain identifications, weather and traffic information, commercials, time checks, trails for other programmes and news. It may contain so many speech items that it is better described as a sequence rather than a DJ show, and this is developed further in the next chapter. But no presenter should ever be at a loss as to what to do next. The prerequisites are to know in advance what you want to say, and be constantly replenishing your stock of anecdotes. Where possible these should be drawn from your own observation of the daily scene. Certainly for the local radio DJ, the more you can develop a personal rapport with the area, the more listeners will identify with you. The preparation of the programme's speech content will also include the timing of accurate talk-overs.

When a DJ is criticised for talking too much, what is often meant is that they are not interesting enough, i.e. there are too many words for too little content. It is possible to correct this by talking less, but similarly the criticism will disappear if the same amount of speech is used to carry less waffle and more substance. Much of what is said may be trivial, but it should still be significant for the listener through its relevance and point of connection. So talk about things to which the listener can relate. Develop that rapport by asking – what information, what entertainment, what companionship does my listener need? If you don't know, go and meet some of them. And remember that there is a key factor in establishing your credibility – it's called professional honesty.

The second kind of preparation for a DJ, and where appropriate the producer, is in actually making additional bits and pieces of programme material that will help to bring the show alive. Mostly pre-recorded, these may consist of snatches of conversation, sound effects, funny voices on echo, stings, chords of music, and so on. Presenters may even create extra 'people', or even animals – recall the DJ who always had a very realistic cat in the studio – playing the roles themselves on the air, talking 'live' to their own recording. Such characters and voices can develop their own personalities, appearing in successive programmes to become very much part of the show. Only the amount of time that is set aside for preparation and the DJ's own imagination set limits on what can be achieved in this way.

For the most part such inserts are very brief but they enliven a DJ's normal speech material, adding an element of unpredictability and increasing the programme's entertainment value.

Whether the programme is complex or simple, the personality DJ should, above all, be fun to listen to. But while the show may give the impression of a spontaneous happening, sustained success is seldom a matter of chance. It is more likely to be found in a carefully devised formula and a good deal of preparation and hard work.

16

Sequences and magazines

Of all programme types, it is the regular lengthy sequences and magazines that can so easily become boring or trivial by degenerating into a ragbag of items loosely strung together. To define the terms, a sequence or strip programme is the slot – generally between two and four hours – often daily, such as the breakfast or morning show, or the evening drivetime, etc., using music with a wide audience appeal, and with an emphasis on the presentation. A magazine, on the other hand, is usually designed with a specific audience in mind, and tightly structured with the emphasis on content. For both, the major problem for the producer is how best to balance the need for consistency with that of variety. Clearly there has to be a recognisable structure to the programme – after all, this is probably why the listener switched on in the first place – but there must also be fresh ideas and newness.

An obvious policy of marketing, which applies to radio no less than to any other product, is to build a regular audience by creating positive listener expectations, and then to fulfil or, better still, exceed them. The most potent reason for tuning in to a particular programme is that the listener liked it the last time. This time, therefore, the programme must be of a similar mould: not too much must be changed. It is equally obvious, however, that the programme must be *new* in the sense that it must have fresh and updated content and contain the element of surprise. The programme becomes boring when its content is too predictable, yet it fails if its structure is obscure. It is not enough simply to offer the advice 'keep the format consistent but vary the content'. Certainly this is important but there must be consistencies too in the intellectual level and emotional appeal of the material. From edition to edition there must be the same overall sense of style.

Since we have so far borrowed a number of terms from the world of print, it might be useful to draw the analogy more closely.

The newspapers and magazines we buy are largely determined by how we reacted to the previous issue. To a large extent purchases are a matter of habit and although some are bought on impulse, or by accident, changes in readership occur relatively slowly. Having adopted our favourite periodical, we do not care to have it tampered with in an unconsidered way. We develop a personal interest in the typography, page layout, length of feature article or use of pictures. We know exactly where to find the sports page, crossword or favourite cartoon. We take a paper that appeals to us as individuals; there is an emotional link and we can feel distinctly annoyed should a new editor change the typeface, or start moving things around when, from the fact that we bought it, it was all right as it was. In other words the consistency of a perceived structure is important since it leads to a reassurance of being able to find your way around, of being able to use the medium fully. Add a familiar style of language, words that are neither too difficult nor too puerile, sentences that avoid both the pompous and the servile, captions that illuminate not duplicate, and it is possible to create a trusting bond between the communicator and the reader, or in our case the listener. Different magazines will each decide their own style and market. It is possible for a single publisher, as it is for the manager of a radio station, to create a diverse output with a total range aimed at the aggregate market of the individual products.

For the individual producer, the first crucial decision is to set the emotional and intellectual 'width' of a programme and to recognise when there is a danger of straying outside it.

To maintain programme consistency several factors must remain constant. A number of these are now considered.

Programme title

This is the obvious signpost and it should both trigger memories of the previous edition and provide a clue to content for the uninitiated. Titles such as 'Farm', 'Today', 'Daytime', 'Sports Weekly', and 'Woman's Hour' are fairly self-explanatory. Whereas, 'Roundabout', 'Kaleidoscope', 'Miscellany', and 'Scrapbook' are less helpful, except that they do indicate a programme containing a number of different but not necessarily related items. With a title like 'Contact', or 'Horizon', there is little information on content, and a subtitle is often used to describe the subject area. Sequences, like DJ programmes, are often known by the name of the presenter – 'The Jack Richards Show' – but a magazine title should stem directly from the programme aims and the extent to which the target audience is limited to a specialist group.

Signature tune

The long sequence is designed to be listened to over any part at random – to dip in and out of. In this area of programming signature tunes are much less common than they were, but when they are used they can serve as a signpost intended to make the listener turn up the volume. A signature tune is best if it goes some way to convey the style of the programme – light-hearted, urgent, serious, or in some way evocative of the content. Fifteen seconds of the right music can be a useful way of quickly establishing the mood.

Magazine producers, however, should avoid the musical cliché. While 'Nature Notebook' may require a pastoral introduction, the religious magazine will often make strenuous efforts not to use opening music that is too churchy. If the aim is to attract an audience that already identifies with institutionalised religion, some kind of church music may be fine. If, on the other hand, the idea is to reach an audience that is wider than the churchgoing or sympathetic group, it will be better to avoid too strong a church connotation at the outset. After all, the religious magazine is by no means the same as the Christian magazine or the church programme.

Transmission time

Many stations construct their daily schedule with a series of sequences in fixed blocks of three or four hours. It is obviously important to have the right presenter and the right style of material for each time slot. The same principle holds for the more specialist magazine. Regular programmes must be at regular times and regular items within programmes given the same predictable placing in each programme. This rule has to be applied even more rigorously as the specialisation of the programme increases.

For example, a half-hour farming magazine may contain a regular three-minute item on local market prices. The listener who is committed to this item will tune in especially to hear it, perhaps without bothering with the rest of the programme. A long sequence having a wider brief, designed to appeal to the more general audience, is more likely to be on in the background and it is therefore possible to announce changes in timing. Even so, the listener at home in the mid-morning wants the serial instalment, item on current affairs, or recipe spot at the same time as yesterday, since it helps to orientate the day.

The presenter

Perhaps the most important single factor in creating a consistent style, the presenter regulates the tone of the programme by his or her individual approach to the listener. This can be outgoing and friendly, quietly companionable, informal or briskly businesslike, or knowledgeable and authoritative. It is a consistent combination of characteristics, perhaps with two presenters, that allows the listener to build a relationship with the programme based on 'liking' and 'trusting'. Networks and stations that frequently change their presenters, or programmes that 'rotate' their anchor people are simply not giving themselves a chance. Occasionally you hear the justification of such practice as the need 'to prevent people from becoming stale', or worse, 'to be fair to everyone working on the programme'. Most programme directors would recommend a six-month period as the minimum for a presenter on a weekly programme, and three months for a daily show. Less than this and they might hardly have registered with the listener at all.

In selecting a presenter for a specialist magazine, the producer could be faced with a choice of either a good broadcaster or an expert in the subject. Obviously the ideal is to find both in the same person, or through training to turn one into the other – the easier course is often to enable the latter to become the former. If this is not possible, an alternative is to use both. Given a strict choice, the person who knows and understands the material is generally preferable. Credibility is a key factor in whether or not a specialist programme is listened to, and expert knowledge is the foundation, even though it might not be perfectly expressed. In other words, if we have a doctor for the medical programme and a gardener for the gardening programme, should there not be children for the children's show, and a disabled person introducing a programme for the handicapped? In a magazine programme for the blind there might be a bit more paper shuffling 'off mic', and the Braille reading might not be as fluent as with a sighted reader, but the result is likely to have much more impact and be far more acceptable to the target audience.

Linking style

Having established the presenter, or presenters, and assuming he or she will write, or at least rewrite, much of the script, the linking material will have its own consistent style. The way in which items are introduced, the amount and type of humour used, the number of time checks, and the level at which the whole programme is pitched will remain constant. Although sounding as though it is spontaneous, what is said – even off the cuff – needs to be worked on in advance in order to contain any substance. The links enable the presenter to give additional information, personalised comment or humour. The 'link-person' is much more than a reader of item cues, and it is through the handling of the links more than anything else that the programme develops a cohesive sense of style. There is much to be said for sometimes introducing two items at a time – the one after next as well as the next one. This gives the listener something else to look forward to: 'We'll have that in a moment, but first, here's . . . '. It's often more interesting than simply wading through a list of items.

Information content

The more local a sequence becomes, the more specific and practical can be the information it gives. It can either be carried in the form of regular spots at known times or simply included in the chat. If a programme sets out with the intention of becoming known for its information content, the spots must be distinctive, yet standardised in terms of timing, duration, style, 'signposting', introductory ident or sound effect.

The types of useful information will naturally depend on the particular needs of the audience in the area covered by the station. The list is wide ranging and typical examples for inclusion in a daily programme are:

News reports	Time checks
Weather forecast	River conditions (for anglers)
Traffic information	Mobile library services
Sports results	Tide times
Job vacancies	Building planning applications
Tonight's TV	Changes to ferry times
Late-night chemists	Sea state – coastal waters
Road works for motorists	Shipping movements
Pollen count	Lighting-up times
What happened today x years ago	Obituaries of the famous
Auction sales/houses/cars	Police requests for help
Pavement works for the blind	What's on – films, entertainment
Rail/bus delays	Top 20 chart
Volunteer help needed	Coming events/parades/fairs
Airport information	Review of papers/journals
Local flying conditions	Club meetings
Mini music biography	Today's racing
Shopping prices, sales	Fatstock prices for farmers
Local market days/times	Storm warnings
Financial market trends	Blood donor sessions
Racing information	Holiday/travel deals
Supermarket offers	Station identification, etc.

Programme construction

The overall shape of the programme will remain reasonably constant. The proportion of music to speech should stay roughly the same between editions, and if the content normally comprises items of from three to five minutes' duration ending with a featurette of eight minutes, this structure should become the established pattern. This is not to say that a half-hour magazine could not spend 15 minutes on a single item given sufficient explanation by the presenter. But it is worth pointing out that by giving the whole or most of a programme over to one subject, it ceases to be a magazine and instead becomes a documentary or feature. There is an argument to be made in the case of a specialist programme for occasionally suspending the magazine format and running a one-off 'special' instead,

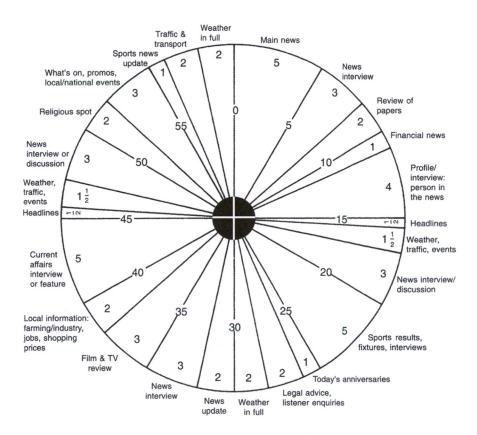

Figure 16.1   The clock format applied to an all-speech, news-based, non-commercial breakfast sequence. Typically programmed from 6.30 to 9.00 a.m. it requires a range of news, weather, sport, and information sources, plus five interviews in the hour – live, telephoned, or recorded. With some repeats this might total nine in the two and a half hours. Interviewees will be drawn mainly from politicians, people in the news, officials, 'experts' and media observers. Large organisations will have their own correspondents and local stringers. Presentation – preferably by two people – will include station idents, time, temperature, promos and incidental information

in which case this should, of course, be aimed at the same audience as the magazine it replaces. Another permissible variation in structure is where every item has a similar 'flavour' – as in a 'Christmas edition', or where a farming magazine is done entirely at an agricultural show.

Such exceptions are only possible where a standard practice has become established, for it is only by having 'norms' that one is able to introduce the variety of departing from them.

Sequences are best designed in the clock format described in the previous chapter.

Programme variety

Each programme should create fresh interest and contain surprise. First, the subject matter of the individual items should itself be relevant and new to the listener. Second, the treatment and order of the items need to highlight the differences between them and maintain a lively approach to the listener's ear. It is easy for a daily magazine, particularly the news magazine, to become nothing more than a succession of interviews. Each may be good enough in its own right, but heard in the context of other similar material, the total effect may be worthy but dull. In any case, long stretches of speech, especially by one voice, should be avoided. Different voices, locations, actuality, and the use of music bridges and stings can produce an overall effect of brightness and variety – not necessarily superficial, but something people actually look forward to.

Programme ideas

Producing a good programme is one thing, sustaining it day after day or week after week, perhaps for years, is quite another. How can anyone, with little if any staff assistance, set about the task of finding the items necessary to keep the programme going? First, a producer is never off duty but is always wondering if anything noticed, seen or heard will make an item. Record even the passing thought or brief impression, probably in a small notebook that is always carried. It is surprising how the act of writing down even a flimsy idea can help it to crystallise into something more substantial. Second, through a diary and other sources, note advance information on anniversaries and other future events. Third, cultivate a wide range of contacts. This means reading, or at least scanning as much as possible: newspapers, the Internet, children's comics, your 'tweets' and social media, trade journals, parish magazines, the small ads, poster hoardings – anything that experience has shown can be a source of ideas.

Producers need to get out of the studio and walk through the territory – easier in local radio than for a national broadcaster! Be a good listener, both to the media and in personal conversation. Be aware of people's problems and concerns, and of what makes them laugh. Encourage contributors to come up with ideas and, if there is no money to pay them, at least make sure they get the credit for something good. If you are too authoritarian or adopt a 'know it all' attitude, people will leave you alone. But by being available and open, people will come to you with ideas. Welcome your texts and emails. It is hard work but the magazine/sequence producer soon develops a flair for knowing which of a number of slender leads is likely to develop into something for the programme.

Having decided to include an item on a particular subject, the producer has several options on the treatment to employ. A little imagination will prevent a programme from sounding 'samey' – consider how to increase the variety of a programme by using the following radio forms.

Voice piece

A voice piece is a single voice giving information, as with a news bulletin, situation report, or events diary. This form can also be used to provide eye-witness commentary, or tell a story of the 'I was there' type of personal reminiscence.

A voice piece lacks the natural variety of the interview and must therefore have its own colour and vitality. In style it should be addressed directly to the listener – pictorial writing in the first person and colloquial delivery, as discussed in Chapter 6, can make compelling listening. But more than this, there must also be a *reason* for broadcasting such an item. There needs to be some special relevance – a news 'peg' on which to hang the story. The most obvious one lies in its immediacy to current events. Topicality, trends or ideas relevant to a particular interest group, innovation and novelty, or a further item in a useful or enjoyable series, are all factors to be used in promoting the value of items in a magazine.

Interview

The types of interview (see Chapter 8), include those that challenge reasons, discover facts or explore emotions. This sub-heading also includes the vox pop or 'man in the street' opinion interview. A further variation is the question and answer type of conversation done with another broadcaster – sometimes referred to as 'Q & A' or a 'two-way'. Using a specialist correspondent like this is often less formal, though still authoritative.

Discussion

Exploration of a topic in this form within a magazine programme will generally be of the bi-directional type, consisting of two people with opposing, or at least non-coincident, views. To attempt in a relatively brief item to present a range of views, as in the 'multi-faceted' discussion, will often lead to a superficial and unsatisfactory result. If a subject is big enough or important enough to be dealt with in this way, the producer should ask whether it shouldn't have a special programme.

Music

An important ingredient in achieving variety, music can be used in a number of ways:

1   as the major component in a sequence;
2   an item, concert performance or recording featured in its own right;
3   a new item reviewed;
4   music which follows naturally upon the previous item. For example, an interview with a choir about to make a concert debut, followed by an illustration of their work;
5   where there is a complete change of subject, music can act as a link – a brief music 'bridge' may be permissible. This is particularly useful to give 'thinking time' after a thoughtful or emotional item where a change of mood is required.

Music should be used as a positive asset to the programme and not merely to fill time between items. It can be used to supply humour or provide wry comment on the previous item, e.g. a song from *My Fair Lady* could be a legitimate illustration after a discussion about speech training. Its use, however, should not be contrived merely because its title has a superficial relevance to the item. It would be inappropriate, for example, to follow a piece about an expedition to the Himalayas with 'Climb Every Mountain' from the *Sound of Music*. For someone who looks no further than the track notes there appears to be a connection, but the discontinuity of context can lead to accusations of poor judgement.

One of the most difficult production points in the use of the medium is the successful combination of speech and music. Music is much more divisive of the audience since listeners generally have positive musical likes and dislikes. It is also very easy to create the wrong associations, especially among older people, and real care has to be taken over its selection.

Sound effects

Like music, effects or actuality noises can add enormously to what might otherwise be a succession of speech items. They stir the memory and paint pictures. An interview on the restoration of an old car would surely be accompanied by the sound of its engine, and an item on dental techniques by the whistle of a high-speed drill. The scene for a discussion on education could be set by some actuality of playground or classroom activity, and a voice piece on road accident figures would catch the attention with the squeal of brakes. These things take time and effort to prepare and, if overdone, the programme suffers as it will from any other cliché. But used occasionally, appropriately and with imagination, the programme will be lifted from the mundane to the memorable. You do not have to be a drama producer to remember that one of the strengths of the medium is the vividness of impression that can be conveyed by simple sounds.

Listener participation

Daily sequences in particular like to stimulate a degree of audience involvement, and the producer has several ways of achieving this as discussed in Chapter 14.

Regarding a spot for incoming letters and emails, it is important that programme presenters do not read these or reply to them in a patronising way, but respond as to a valued friend. Broadcasters are neither omniscient nor infallible and should always distinguish between a factual answer, best advice and a personal opinion. Some basic and essential reference sources – Internet, books, databases, available experts, etc. – will provide answers for questions like, why does the leaning tower of Pisa lean? Or, how do you remove bloodstains from cloth? But it is a quite different matter replying to questions about how to invest your money, a cure for cataracts, the causes of a rising crime rate or why a loving God allows suffering. Nevertheless, without being glib or superficial it is possible for such a question to spark-off an interesting and useful discussion, or for the presenter to give an informed and thought-out response with:

'On balance I would say that . . . ' or
'If you are asking for my own opinion I'd say . . . but let me know what you think.'

Research may have thrown up a suitable quote or piece of writing on the subject, which should, of course, always be attributed. Other phrases useful here are:

'Experts seem to agree that . . . '
'No one really knows, but . . . '

Competitions are a good method of soliciting response. These could be in reply to quizzes and on-air games with prizes offered, or simply for the fun of it.

A *phone-in* or *texting* spot in a live magazine helps the sense of immediacy and can provide feedback on a particular item. Placed at the end, it can allow listeners the opportunity of saying what they thought of the programme as a whole. Used in conjunction with a quiz competition it obviously allows for answers to be given and a result declared within the programme.

Listener participation elements need proper planning but part of their attraction lies in their live unpredictability. The confident producer or presenter will know when to stay with an unpremeditated turn of events and extend an item that is developing unexpectedly well. He or she will have also determined beforehand the most likely way of altering the running order to bring the programme back on schedule. In other words, in live broadcasting the unexpected always needs to be considered.

Features

A magazine will frequently include a place for a package of material dealing with a subject in greater depth than might be possible in a single interview. Often referred to as a featurette, the general form is either *person centred* – 'our guest this week is . . . '; or *place centred* – 'this week we visit . . . '; or *topic centred* – 'this week our subject is . . . '.

Even the topical news magazine, in which variety of item treatment is particularly difficult to apply, will be able to consider putting together a featurette comprising interview, voice piece, archive material, actuality and links, even possibly music. For instance, a report on the scrapping of an old ferry boat could be run together with the reminiscences of its former skipper, describing some perilous exploit, with the appropriate sound effects in the background. Crossfade to breaker's yard and the sound of lapping water. This would be far more interesting than a straight report.

The featurette is a good means of distilling a complex subject and presenting its essential components. The honest reporter will take the crux of an argument, possibly from different recorded interviews, and present them in the context of a linked package. They should then form a logical, accurate and understandable picture on which the listener can base an opinion.

Drama

The weekly or daily serial or book reading has an established place in many programmes. It displays several characteristics that the producer is attempting to embody – the same placing and introductory music, a consistent structure, familiar characters and a single sense of style. On the other hand, it needs a variety of new events, some fresh situations and people, and the occasional surprise. But drama can also be used in the one-off situation to make a specific point, for example in describing how a shopper should compare two supermarket products in order to arrive at a best buy. It can be far more effective than a talk by an official, however expert. Lively, colloquial, simple dialogue, using two or three voices with effects behind it to give it location – in a store or factory, on a bus or at the hospital – this can be an excellent device for explaining legislation affecting citizens' rights, a new medical technique, or for providing background on current affairs. Scriptwriters, however, will recognise in the listener an immediate rejection of any expository material which has about it the feel of propaganda. The most useful ingredients would appear to be: ordinary everyday humour, credible characters with whom the listener can identify, profound scepticism and demonstrable truth.

Programmes for children use drama to tell stories or explain a point in the educational sense. The use of this form may involve separate production effort, but nonetheless it can be effective, even limited to the dramatised reading of a book, poem or historical document.

Item order

Having established the programme structure, set the overall style and decided on the treatment of each individual item, the actual order of the items can detract from or enhance the final result.

In the case of a traditional circus or variety performance in a theatre, the best item – the top of the bill – is kept until last. It is safe to use this method of maintaining interest through the show since the audience is largely captive and it underwrites the belief that whatever the audience reaction, things can only get better. With radio, the audience is anything but captive and needs a strong item at the beginning to attract the listener to the start of the show, thereafter using a number of devices to hold interest through to the end. These are often referred to as 'hooks' – remarks or items designed to capture and retain the listener's attention. It is crucial to engage the audience early in the programme – an enticing, imaginatively written menu is one way of doing this – but then to continue to reach out, to appeal to the listener's sense of fun, curiosity, or need. Of course this applies to all programmes, but the longer strip programming is particularly susceptible to becoming bland and characterless. There should always be something coming up to look forward to.

The news magazine will probably start with its lead story and gradually work through to the less important. However, if this structure is rigidly applied the programme becomes less and less interesting – what has been called a 'tadpole' shape. News programmes are therefore likely to keep items of known interest until the end, e.g. sport, stock markets and weather, or at least end with a summary for listeners who missed the opening headlines. Throughout the programme as much use as possible will be made of phrases like 'more about that later'. Some broadcasters deliberately avoid putting items in descending order of importance in order to keep 'good stuff for the second half'; a dubious practice if the listener is to accept that the station's editorial judgement is in itself something worth having. A better approach is to follow a news bulletin with an expansion of the main stories in current affairs form.

The news magazine item order will be dictated very largely by the editor's judgement of the importance of the material, while in the tightly structured general magazine the format itself might leave little room for manoeuvre. In the more open sequence other considerations apply and it is worth noting once again the practice of the variety music bill. If there are two comedians they appear in separate halves of the show, something breathtakingly exciting is followed by something beautiful and charming, the uproariously funny is complemented by the serious or sad, the small by the visually spectacular. In other words, items are not allowed simply to stand on their own but through the contrast of their own juxtaposition and the skill of the compère they enhance each other so that the total effect is greater than the sum of the individual parts.

So it should be with the radio magazine. Two interviews involving men's voices are best separated. An urgently important item can be heightened by something of a lighter nature. A long item needs to be followed by a short one. Men's and women's voices, contributions by children or old people should be consciously used to provide contrast and variety. Tense, heated or other deeply felt situations need special care, for to follow with something too light or relaxed can give rise to accusations of trivialisation or superficiality. This is where the skill of the presenter counts, for it is his or her choice of words and tone of voice that must adequately cope with the change of emotional level.

Variations in item style combined with a range of item treatment create endless possibilities for the imaginative producer. A programme that is becoming dull can be given a 'lift' with a piece of music halfway through, some humour in the links, or an audience participation spot towards the end. For a magazine in danger of 'seizing up' because the items are too long, the effect of a brief snippet of information in another voice is almost magical. And all the time the presenter keeps us informed about what is happening in the programme, what we are listening to and where we are going next, and later. The successful magazine will run for years on the right mixture of consistency of style and unpredictability of content. It could be that apart from its presenter the only consistent characteristic is its unpredictability.

Examples

The following examples of the magazine format are not given as ideals but as working illustrations of the production principle. Commercial advertising has been omitted in order to show the programme structure more clearly but the commercial station can use its breaks to advantage, providing an even greater variety of content.

Example 1: Fortnightly half-hour industrial magazine

Structure	Running order	Actual timing
Standard opening (1' 15")	Signature tune	0 15
	Introduction	10
	Menu of content	15
	Follow-up to previous programmes	35
News (5 min)	News round-up	5 05
	Link	15
Item	Interview on lead news story	3 08
	Link - information	30
Item	Voice piece on new process	1 52
	Link	15
Item	Vox pop - workers' views of safety rules	1 15
	Link	20
Trades Council spot (2' 30")	Union affairs - spokesman	2 20
	Link	20
Participation spot (3 min)	Listeners' correspondence	2 45
	Link - introduction	20
Discussion	Three speakers join presenter for discussion of current issue - variable length item to allow programme to run to time	6 20
Financial news (3 min)	Market trends	3 00
Standard closing (50")	Coming events	30
	Expectations for next programme	10
	Signature tune	10
		29' 50"

In Example 1, the programme structure allows 1′ 15″ for the opening and 50″ for the closing. Other fixed items are a total of 8′ 00″ for news, 3′ 00″ for listeners' input and 2′ 30″ for the Union spot. About 2′ 00″ are taken for the links. This means that just over half the programme runs to a set format, leaving about 13′ 00″ for the two or three topical items at the front and the discussion towards the end.

So long as the subject is well chosen the discussion is useful for maintaining interest through the early part of the programme, since it preserves at least the possibility of controversy, interest and surprise. With a 'live' broadcast it is used as the buffer to keep things on time since the presenter can then bring it to an end and get into the 'Market trends' four minutes before the end of the programme. The signature tune is 'prefaded to time' to make the timing exact. With a recorded programme the discussion is easily dropped in favour of a featurette, which in this case might be a factory visit.

Example 2: Weekly 25-minute religious current affairs

Structure	Running order	Timing
Standard opening (15″)	Introduction	0 05
	Menu of content	0 10
Item	Interview – main topical interest	3 20
	Link	30
Item	Interview – woman missionary	2 05
	Link	10
Music (3 min)	Review of gospel record release	2 40
	Link – information	55
Item	Interview (or voice piece) – forthcoming convention	2 30
	Link	05
Featurette (7 min)	Personality – faith and work	7 10
	Link	15
News (5 min)	News round-up and What's On events	4 45
Standard closing (15″)	Closing credits	15
		24′ 55″

In Example 2, with no signature tune, the presenter quickly gets to the main item. Although in this edition all the items are interviews, they are kept different in character, and music is deliberately introduced at the midpoint. The opening and closing take half a minute and the other fixed spots are allocated another 15 minutes. If the programme is too long, adjustment can be made either by dropping some stories from the news or by shortening the music review. If it under-runs, a repeat of some of the record release can make a useful reprise at the end.

Example 3: Outline for a daily 2-hour afternoon sequence

Fixed times	Running order	Approximate durations
2.00	Sig. tune/intro-programme information/ sig. tune	1 min
	Music –	2
	Item – human interest interview	3
	Quiz competition (inc. yesterday's result)	3
	Music –	3
2.15	Listeners' comments and questions, emails, texts, letters	5
	Music –	3
	Voice piece – background to current affairs	2
	Music –	2
	Leisure spot-home improvements, gardening	3
	Humour on record	2
	Studio discussion – topical talking point	10
	Music –	3
	'Out and about' spot – visit to place of interest	5
	Music – (non-vocal)	2
3.00	News summary, sport and weather	2
	Phone-in spot	15
	Music –	3
	Film/theatre/TV review – coming events voice piece	4
	Music –	3
3.30	Special guest interview	10
	Music – (illustrative)	2
	Item – child care or medical interview	3
	Quiz result	2
	Music – (non-vocal)	2
3.50	Serial story – dramatised reading	9
	Closing sequence; items for tomorrow, production credits, sig. tune	1

In Example 3, the speech/music ratio is regulated to about 3:1. A general pattern has been adopted that items become longer as the programme proceeds. Each half of the programme contains a 'live' item which can be 'back-timed' to ensure the timekeeping of the fixed spots. These are, respectively, the studio discussion and the special guest interview. Nevertheless, the fixed spots are preceded by non-vocal music prefaded to time for absolute accuracy. Fifteen minutes is allowed for links. An alternative method of planning this running order is by the clock format illustrated on p. 215.

Production method

A regular magazine or sequence has to be organised on two distinct levels – the long term and the immediate. Long-term planning allows for local or national anniversaries, 'one-off editions', the booking of guests reflecting special events, or the running of related spots to form a series across several programmes. On the immediate timescale, detailed arrangements have to be finalised for the next programmes. Keeping a planning diary in one form or another, to which everyone has access, is a useful aid.

In the case of the fortnightly or weekly specialist magazine of the type represented in Examples 1 and 2, it is a fairly straightforward task for the producer to make all the necessary arrangements in association with the presenter and a small group of, possibly freelance, contributors acting as reporters or interviewers. The newsroom or sources of specialist information will also be asked for a specific commitment. The important thing is that everyone has a precise brief and deadline. It is essential too that while the producer will make the final editorial decisions, all the contributors feel able to suggest ideas. Good ideas are invariably 'honed up', polished and made better by the process of discussion. Suggestions are progressed further and more quickly when there is more than one mind at work on them. Almost all ideas benefit from the process of counter-suggestion, development and resolution. Lengthy programmes of this nature are best produced by team-work with clear leadership. This should be the pattern encouraged by the producer at the weekly planning meeting, by the end of which everyone should know what they have to do, by when, and with what resources.

The daily sequence illustrated by Example 3 is more complex and such a broadcast will require a larger production team. Typically, the main items, such as the serial story and special guest in the second half, will have been decided well in advance but the subject of the discussion in the first may well be left until a day or two beforehand in order to reflect a topical issue. A retrospective look at something recently covered, the further implications of yesterday's story, a 'whatever happened to . . . ?' spot – these are all part of the daily programme. Responsibility for collating listeners' comments, organising the quiz and producing the 'Out and about' item will be delegated to specific individuals, and the newsroom will be made responsible for the current affairs voice piece, the news, sport and weather package, and perhaps one of the other interviews. The presenter will write the links and probably select the music, according to station policy. The detail is regulated at a morning planning meeting, the final running order being decided against the format structure that serves as the guideline.

At less frequent intervals, say once a month, the opportunity should be taken to stand back to review the programme and take stock of the long-term options. In this way it might be possible to prevent the onset of the more prevalent disease of the regular programme – getting into a rut – while at the same time avoiding the disruptive restlessness that results from an obsession with change. When you have become so confident about the success of the programme that you fail to innovate, you have already started the process of staleness that leads to failure. A good approach is constantly to know what you are up against. Identify your competition and ask, 'What must I do to make my programme more distinguishable from the rest?'

Responding to emergency

Sooner or later all the careful planning has to be discarded in order to cope with emergency conditions in the audience area. Floods, snowfall, hurricane, earthquake, power failure, bush fire or major accident will cause a sequence to cancel everything except the primary reason for its existence – public service. It will provide information to isolated villages, link helpers with those in need, and act as a focus of community activity. 'Put something yellow in your window if you need a neighbour to call.' 'Here's how to attract the attention of the rescue helicopter which will be in your area at 10 o'clock. . . .' 'The following schools are closed. . . .' 'The army is opening an emergency fuel depot at. . . .' 'A spare generator is available from. . . .' 'Can anyone help an elderly woman at. . . .?' The practical work that radio does in these circumstances is immense; its morale value in keeping isolated, and sometimes frightened, people informed is incalculable. Broadcasters are glad to work for long hours when they know that their service is uniquely valued.

We saw on p. 60 an example of how a station coped with a night of coastal flooding – how it gave out information and also received reports from listeners via social media. People at the scene will not only send texts and tweets but also voice audio from their smartphones. Yes, the broadcaster must always be aware of the hoax call, but given some elementary checking this represents a valuable joint community service.

Flexibility of programming enables a station, local or national, to cover a story, draw attention to a predicament, or devote itself totally to audience needs as the situation demands. Radio services should have their contingency plans continually updated – at least reviewed annually – in order to be able to respond quickly to any situation. Programmers cannot afford to wait for an emergency before making decisions but must always be asking the question – 'what if. . .?'

17

Making commercials

The purpose of an advertisement is to sell things. It is not there simply to amuse people or to impress the chairman; it is designed to move goods off shelves, cars out of showrooms, and customers eagerly towards services. The radio advertiser must use a good deal of skill in motivating a target audience to a specific action. The effective advertisement will:

- interest
- inform
- involve
- motivate
- direct.

Many commercials are made by advertising agencies in conjunction with specialist production houses. They arrive at the station, mostly by email, or by downloading as a WAV. file with a password from the advertising agency's server. However, producers may be called upon to make their own local commercials for which the elements to be considered are:

- the target audience – for whom is this message primarily intended?;
- the product or service – what is the specific quality to be promoted?;
- the writing – what content and style will be appropriate?;
- the voice or voices – who will best reinforce the style?;
- the background – are music or sound effects needed?

The producer must also be familiar with the station copy policy and code of practice governing advertising. Regulatory standards form the essential background to commercial production.

Copy policy

'I've bought the time. I can say what I like.' Unfortunately not. A client does not have free rein with the station's airwaves but must comply with a set of rules, a copy of which should be readily available to all potential advertisers. In Britain the Communications Act 2003 makes the issuing of such rules the duty of Ofcom, which is also charged with enforcing them.

The principle is that all advertising should be 'legal, decent, honest and truthful' and that nothing should 'offend against good taste or public feeling'.

Ofcom's Code of Advertising Standards and Practice, and Programme Sponsorship sets out the rules, including specific prohibitions on advertising that:

- could be confused with programming;
- is on behalf of any political body;
- shows partiality in matters of current political or industrial controversy;
- unfairly attacks or discredits other products;
- includes sounds likely to create a safety hazard for drivers;
- exploits the superstitious or plays on fear;
- is on behalf of any body that practises or advocates illegal behaviour;
- makes claims that give a misleading impression.

No advertising is permitted within coverage of a religious service, a formal royal ceremony, or a schools' programme under 30 minutes. Programme presenters must not endorse a product in a presenter-read commercial. The Code lists those categories where central clearance of copy is required and includes detailed sections on financial advertising – investments, savings, insurance, etc. – alcohol, advertising for and by children, the advertising of medicines and treatments including contraceptives and pregnancy-testing services, charity advertising, appeals for disasters and advertising on behalf of religious bodies. It bans the advertising of cigarettes, guns and gun clubs, pornography, the occult, betting and gaming, escort agencies, products for the treatment of alcoholism, hypnosis and psychoanalysis. It also lists the current legislation relevant to broadcast advertising. The Code is essential reading for anyone involved in British commercial radio (see p. 386).

In the USA, radio advertising regulation is chiefly through the FCC Code and the self-adopted Code of the National Association of Broadcasters (see p. 385). This is similar to the Ofcom rules and has points to make about the use of the words 'safe' and 'harmless' related to pharmaceuticals, the presentation of marriage, and the sensitivity necessary in the use of material relating to race, colour or ethnic derivation. The advertising of hard liquor (distilled spirits) is prohibited. In addition to the National Code, network and local stations have their own policies that conform to State laws. Some of these are very detailed, defining terms such as 'like new', 'biggest', 'factory fresh' and 'guarantee'. For an Asian example of a Commercial Code, look at All India Radio (see p. 386).

Public Service Announcements made available to charitable and non-profit organisations without charge must conform to the same standards as the paid-for commercials and also require approval by the commercial department. Stations may define the standard of language acceptable – where payment is on wordage there is a temptation for advertisers to supply copy in 'telegraphese'. Copy policy is not a fixed and immutable thing. It goes out of date as standards and fashions vary. The commercial producer must therefore keep abreast of these regulations as well as being aware of changes in the law itself.

The target audience

Only on the very smallest station will the producer be required to sell airtime. The qualities of persistence, persuasion and patience belong to the sales and marketing team who will negotiate the price, the number of spots and the discount offered by the rate card. The placing of the spots will, of course, crucially affect the rate charged, and is something the producer also needs to know for two reasons: (1) because the transmission time must be appropriate to the intended listener, and (2) because it may affect the written copy. Clearly there is no point in broadcasting a message to children when they are in school, or to farmers when they are busy. Nothing is gained by exhorting people to 'buy something now' when the shops are shut. If you are selling holidays, who makes the purchasing decision for which type of vacation? What precise age buys what type of music? Which socio-economic group do you wish to attract for a particular drink?

To sell its airtime effectively, a commercial station needs to know about its audience. Through independent market research it has to show why it is better than its competitor at reaching a particular section of the public. What age groups does it attract? How much disposable income do they have, and what do they do with it? To what extent do they buy coffee, cars, furniture, holidays, magazines, mortgages or insurance? And what kind do they buy – expensive upmarket or 'cut-price'?

If you are the manufacturer of kitchen equipment you want to spend your advertising budget where it is most likely to pay off. So which is the best way of reaching newly married couples setting up home? A radio station with a well-researched audience profile is much more convincing than an amateur with a hunch.

The product or service 'premise'

In 30 seconds, it's not possible to say everything about anything. So identify one, or perhaps two, key features about the product which mark it out as especially attractive. Choose one of these – its usefulness, efficiency, simplicity, low cost, durability, availability, value for money, exclusivity, technical quality, newness, status, advanced design, excitement or beauty. There are other possibilities but a single memorable point about a product is far more effective than an attempt to describe the whole thing. In the case of food and drink the key phrases may be more subjective – easy to prepare, made in a moment, satisfying, long lasting, luscious, nutritious, economical and so on.

Now develop a short statement that connects the product's intention with a known and desirable effect. This becomes the 'critical premise' or consumer benefit, and often comprises a subject, an action verb and a result. Here are some examples:

- Clean breath helps personal relationships.
- Disinfecting your bathroom gives you a safer home.
- A slimming diet makes you more attractive.
- Serving rich coffee impresses your friends.
- Driving while drunk kills innocent people.
- Unsafe sex increases the risk of AIDS.
- Inoculation against disease can save your child's life.
- An energy breakfast promotes a successful day.

This is the writer's hypothesis and is the essential first building block for any radio spot. Before going further, test it out on other people – do they believe it? Does the action genuinely link the cause and the effect? Is the end result something that people in your target audience want? This brings us back to the point that effective advertising is grounded on thorough, relevant research. If the premise is shown to be true it is now necessary to connect the product firmly with it.

Whether the object of attention is a January sale, a fast food shop, a cosmetic cream or a life insurance policy, the client and producer/writer together must agree the primary distinctive feature to be sold – the USP, the Unique Selling Point.

It is important to consider the overall style or image to be projected. Is the impression required to be friendly, warm and domestic or is it unusual, lively and adventurous? If the idea is to convey reassurance, dependability and safety, this should be communicated in the writing, but also in the voicing and any music used. Every element must be consistent in combining to support the premise, and associate the product with it.

Writing copy

This is the heart of it, and it's worth remembering two things: (1) well-chosen, appropriate words cost no more than sloppy clichés; and (2) radio is a visual medium.

The American copywriter Robert Pritikin has pointed out the value of specifically writing for the eye as an aid to product recall. He wrote a now famous illustration of radio's ability to help the listener to visualise even something as intangible as a colour:

```
ANNCT:      The Fuller Paint Company invites you to stare with your
            ears at . . . yellow. Yellow is more than a colour.
            Yellow is a way of life. Ask any taxi driver about
            yellow. Or a banana salesman. Or a coward. They'll tell
            you about yellow. (Fx Phone rings)

            Oh, excuse me. Yello!! Yes, I'll take your order.
            Dandelions, a dozen; a pound of melted butter; lemon
            drops and a drop of lemon, and one canary that sings a
            yellow song. Anything else? Yello? Yello? Yello? Oh,
            disconnected. Well, she'll call back.

            If you want yellow that's yellow-yellow, remember
            to remember the Fuller Paint Company, a century of
            leadership in the chemistry of colour. For the Fuller
            colour center nearest you, check your phone directory.
            The yellow pages, of course.
```

(Robert C. Pritikin, writing in 'Monday Memo', *Broadcasting Magazine*, 18 March 1974, p. 22)

Here the listener will personalise the images in response to the ideas presented – and the key point about this paint is not its value or durability – it is its yellowness. Everything is geared to communicate its bright liveliness – even the shortness of the sentences. On reading the piece any good producer will be able to hear the appropriate voice for it and, if need be, the right music.

Creating something visual to produce a memorable image leading to product recall demands great imagination – especially for the more mundane. After all, as an on-station producer what would you write for a local windscreen repair service?

```
MAN'S VOICE:    You're not going to believe this.

                (Music under – orchestral strings, urgent 'thriller'
                theme)
```

It was about two in the morning and I was waiting
for the lights when a foreign looking woman jumped
into the car. 'Drive', she said. My foot hit the
floor. Five seconds later all hell let loose,
soldiers were everywhere, tracker dogs, helicopters
and armoured cars. I saw a rifle pointed at the
windscreen. She grabbed me and literally threw me
under the dashboard. There was a sharp crack and
the windscreen gave in. Moments later, I was alone
in the darkness – she'd gone, so had everyone else.
On the seat I saw a card. It read simply, 'silver
shield', they were with me in minutes. See – I said
you wouldn't believe me.

VOICE 2:   For the silver shield 24-hour windscreen service just
               dial one hundred and ask for Freephone Silver Shield –
               because you never know when you might need us.

               (Music: up to finish)

                                                    (50 sec)

(Courtesy of County Sound Radio)

In a few seconds of airtime the script must gain our interest, make the
key point about the product (in the above case, immediacy) and say clearly
what action the listener must take to obtain it. This is especially important
for station-produced marketing promotions aimed at potential buyers of
its own advertising spots. Here the voices wittily imitate two well-known
cricket commentators:

Fx: Cricket atmosphere (held under throughout)

VOICE 1:   Yes, hello everyone and welcome to Southern Sounds
               small-ads county classic. It's a marvellous day
               here and interestingly enough, it's 9-99 for 5
               transmissions. Quite an incredible offer – how did we
               arrive at that, John?

VOICE 2:   Well, it's 9-99 for 5 transmissions, once a night for
               a week. And that's the best small advertising offer in
               county radio in England since 1893.

VOICE 1:   Quite amazing, and all we have to do is call Alison
               small-ads, during playing hours on Brighton 4 triple
               2 double 8. (Fx: light applause) And here comes an
               advertiser now – running in with his cheque for 9-99
               (Fx: bat on ball, applause) – and it's superbly fielded
               by Alison small-ads – and it's on the air in a flash –
               very good effort that, I thought. And so with the offer
               still at 9-99 for 5 it's back to the studio.

                                                    (55 sec)

(Courtesy of Southern FM – station marketing promo)

Advertising based on familiar radio programmes obviously strikes a
chord with the listener, given the right placing. The next example is a

spoof on typical sports commentators – one of a series of ads parodying the style:

```
MAN 1:    So you want to be a football commentator, eh?

MAN 2:    Over the moon, Brian.

MAN 1:    Right, well you've gotta have all your football clichés -

MAN 2:    Ah ha.

MAN 1:    plenty of drama -

MAN 2:    I'm way behind you there.

MAN 1:    and put your emphasis on - all the wrong words.

MAN 2:    I'm sure I can, manage that.

MAN 1:    Right - I'll give you a bit of the old crowd.

Fx:       Football crowd noise (held under)

MAN 2:    (in football commentary style) And we go into the second
          half with the score standing at 1-nil. So the game really
          is perfectly balanced - and I'm not going to sit on the
          fence but this game could go either way. You could take
          the atmosphere and cut it up in a thousand pieces, dip it
          in custard and give it to the crowd. And there's Gray -
          Aston Villa's vibrant virtuoso who's decided to take his
          nerve by the horns and stamp his authority on this game
          in those Nike boots of his . . .

MAN 1:    (interrupting) Er - hang on, hang on. (Fx cut) What did
          you say?

MAN 2:    Er Gray - Andy Gray the footballer.

MAN 1:    No, no - you said Nike, I'm sure I heard you say Nike.

MAN 2:    Quite categorically, yes.

MAN 1:    Yes but you can't mention brand names - OK? I mean what
          do you think this is - a commercial or something?
```

                                                                    (60 sec)

(Courtesy of Grierson Cockman Craig & Druiff Ltd)

This approach, once the format, style and characters are established, is very effective in promoting the brand name. The time may come when, for a national product, the name need not be mentioned at all. The advantage is that the listener joins in the game and is almost certain to say the product name to himself – as in the famous 'Schhhh . . . you know who' campaign.

In the example below, Duracell batteries use only one voice – the same somewhat tired, 'ordinary', older man's voice which was used for a time in all their radio ads, plus their 'sound logo', used on TV as well:

```
VOICE:     As an ordinary HP-8 grade radio battery I have one great
           ambition – I want to be forgotten. If I'm remembered it
           means I'm dead. Admit it – that's the only time you ever
           remember your batteries. You never ask us how we are or
           take us out for a nice walk – only the walk to the bin.
           Well you can forget me for 145 hours of continuous radio
           noise. But there's a radio battery that can be forgotten
           for over 500 hours. Now that's what I call forgettable.

           You know the one – erm – oh well, wassisname.

ANNCR:     Wassisname (logo Fx: big door slam)

           No ordinary battery looks like it or lasts like it.

                                                            (50 sec)
```

(Courtesy of Dorland Advertising Ltd)

Voicing and treatment

Casting a commercial is a make or break business. Doing it well stems from having a clear idea of the overall impression required. Professional actors may be expensive but are much more likely than the station's office staff or the client firm's MD to provide what is required. They have greater flexibility, vocal range and, above all, are 'produceable'. Once they know what you want, trained performers will produce consistent results – and you will certainly need it done many times, if only to get the inflections and timing – with effects and music – absolutely right.

Advertising for a charity generally calls for a serious, uncluttered approach. Listen to the example for the Marie Curie Cancer Care organisation on the website. And here's another striking piece of no-nonsense writing without music or effects. The one voice has to be quiet, strong, tough and compassionate. It mustn't be associated with any single professional or social group. The speed of delivery is important.

```
MAN'S VOICE:   Our organisation currently has vacancies for people
               in this area.

               Successful applicants will get no company car, no
               luncheon vouchers, no holidays, no bonuses, no
               expense account, no business lunches, no glamour,
               and – no salary. If you're interested, and can spare
               a few hours a week to be a Samaritan, give us a
               ring for more information. We're in the phone book.
               But please don't phone unless you're serious –
               someone else may be trying to get through.

                                                            (30 sec)
```

(Courtesy of Saatchi & Saatchi Compton Ltd)

Notice that this doesn't give the telephone number to ring – 'we're in the book'. People don't remember numbers, especially when so many devices dial automatically. The exception may be when your number is a word that spells out your product. A florist for example with the number 01223569377, can say 'ring us on 0122FLOWERS'.

In production, vocal inflection, emphasis, pace and projection are infinitely variable. Even though you start with a clear idea of how a piece should sound, you might want to try it a number of ways. It's generally worth asking your speaker what sounds right or comfortable to him or her. Should it be confidential and relaxed or more excited? What emotional content is appropriate? Should you 'shout', literally, to call attention, for example, to a new store opening?

Here's a serious charity example that does include an element of humour – designed to develop some rapport with the listener.

```
POSH LADY (speaking politely):     Oi Cancer, you malignant little
                                   neoplasm.

                                   You're nothing but an ignorant
                                   carcinoma.

                                   We're wise to your tricks you
                                   worthless haematobium.

                                   We're going to rip your sarcoma off
                                   and shove it so far up your flobula,
                                   your lympho-proliferative system
                                   won't know its clonorchis from its
                                   vasculum.

                                   Cancer, we're coming to get you.

ORDINARY VOICE:                    Run, walk, dance, enter at
                                   raceforlife.org

                                   Cancer Research UK's Race for Life
                                   in partnership with Tesco.
```

(Courtesy of Mother London; www.focalpress.com/cw/mcleish)

There are times, however, when the end result is not the product of creative artifice and technique, but comes straight from reality. Here is the very real voice of genuine anguish – not an actress with words written for her – but the halting voice, careful and controlled, of a mother who had been interviewed after her son was killed in a road accident:

```
WOMAN'S VOICE:     They were crossing the road again after getting
                   off the bus – and this crazy car came from nowhere
                   and just took Simon – and he was killed instantly
                   apparently.

                   And I just knew when she said – I'm sorry,
                   will you come and sit down? (voice breaking) I
                   remember it so vividly.
```

```
                        When the police came and told us that he had been -
                        erm - charged - they just told us what an awful
                        state he was in, and - erm - couldn't sleep, and
                        was having nightmares, and he had a little boy of
                        his own, aged five.

                        The best way I can describe it is that when
                        I thought about what had happened, I was - I
                        preferred to be me than to be him. Because I
                        didn't think I could live myself with the idea
                        that I'd killed a small child.

MAN'S VOICE:            Drinking and driving wrecks lives.

                                                           (60 sec)

(Courtesy of COI/Dept for Transport, DMB&B)
```

The serious public service commercial such as this in particular needs the right voice – one that registers immediately to have impact. The presentation must reinforce the content. Here's a particularly good example of that, where the actual sound of the commercial *is* the message.

```
WOMAN:      You're four times more likely to have a crash when you're
            on a mobile phone.

            (simultaneously)

WOMAN:      It's hard to concentrate on two things at the same time.

            (Pause)

WOMAN:      You're four times more likely to have a crash when you're
            on a mobile phone.

            (simultaneously)

WOMAN:      It's hard to concentrate on two things at the same time.

            (Pause)

WOMAN:      You're four times more likely to have a crash when you're
            on a mobile phone.

            It's hard to concentrate on two things at the same time.

            It's illegal to use a handheld mobile phone when you're
            driving.

            Think - switch it off before you drive.

                                                           (30 sec)

(Courtesy of AMV BBDO/Dept for Transport; www.focalpress.com/cw/mcleish)
```

Many countries run advertising about the problems of AIDS and sexually transmitted disease – sensitive to the cultural context. Here's an award-winning British example from the Health Education Authority.

Music:	<u>Jingle Bells – non-vocal (establish and hold under)</u>
WOMAN 1 (friendly and brisk):	If you're wondering what to give your loved one this Christmas, the following gifts are always worth considering – Chlamydia, genital warts, or even gonorrhoea. At this time of year these little surprises are as common as ever, and although you may know of the need to protect against HIV, these other infections can be very serious despite sometimes showing no obvious signs. Using a condom of course can help prevent their spread. So whatever you do give your partner for Christmas, make sure it's properly wrapped.
Music:	(fade out)
	WOMAN 2 (formal and authoritative): For more information about HIV or other sexually transmitted infections, call the National AIDS helpline on 0800 567 123.

(40 sec)

(Courtesy of BMP DDB Ltd/Health Education Authority;
www.focalpress.com/cw/mcleish)

The producer has a range of techniques to alter a voice: filters, digital effects, graphic equaliser and 'presence' control to change the tonal quality or to give a more incisive cutting edge, compression to restrict the dynamic range and keep the levels up, multi-tracking and variable-speed recording to increase the apparent number of voices, echo for 'space', phasing effects for mystery, and so on. Beware of gimmicks, however. If one is tempted to use technical tricks to 'make it more interesting', check the writing again – are the words really doing their job? One of the most effective ads ever made by the authors was also the simplest. For a station in the tropics, its purpose was to sell a new iced lolly on a stick:

VOICE 1 (calling off):	Hey, where are you going?
VOICE 2 (calling, closer):	To get a Blums ice block.
ANNCR:	Available in the best stores.

(5 sec)

Run fairly frequently, because of its low cost, the phrases – or versions of them – could soon be heard all over town. An ad has to be right for its own culture. No matter how clever or complicated an advertisement is, it is never good on its own – only in relation to what happens after it is broadcast.

Music and effects

The main role of music is to assist in establishing mood. The biggest trap is to use a track from the library simply because of its title. The label may *say* 'Sunrise Serenade' but does it *sound* like an early morning promise of a new day, or is it cold, menacing, or just nondescript? Music in the context of the radio commercial must do what you want – immediately. If in doubt, play your choice to a colleague and ask 'what does this remind you of?' On your own you can convince yourself of anything.

The right music will almost certainly not be the right length. If you want the music to finish at the end, rather than to be faded, some judicious editing will be needed under the speech.

This was certainly the case in a witty ad using Mozart to promote holidays in Jamaica – in a British context a sensible guess at the tastes of the socio-economic target group likely to be interested in this type of holiday:

Fx:	Orchestra tuning up. Concert hall atmosphere
ANNCR (quietly):	And conducting Mozart's symphony number 40 in G minor – Arturo Barbizelli – looking tanned and fit from his recent holiday in Jamaica.
Fx:	Applause, quietens
MUSIC:	Mozart symphony No. 40 (4 bars) then accompanied by steel drums, calypso style (4 bars)
	(Music under)
ANNCR:	He was only there for a fortnight!
	(Music up, alternating between classical and calypso style. Music under)
ANNCR:	Find out about Jamaica – the island that warms you through and through.
JAMAICAN VOICE:	Ring Jamaica Tourist Board on 01-493-9007.
	(Music up and faded out)

(50 sec)

(Courtesy of Young and Rubicam Ltd)

As for effects, it would be hard to imagine a more dramatic use of sounds than this:

WOMAN:	Each gunshot you're about to hear represents a life lost to gun crime in the past year. (Actuality effects – gunshots – 20 sec)
WOMAN:	From March 31st to April 30th there'll be a national gun amnesty.
	It's your chance to hand in any gun to your local police station – no questions asked.
	Get guns off the streets.

(40 sec)

(Crown Copyright: The Home Office. Prodn. COI; www.focalpress.com/ cw/mcleish)

Some stations were reluctant to run an ad as graphic as that, but in the event the campaign, run alongside press and website advertising, brought in over 43,000 guns and a million rounds of ammunition.

If the budget will run to specially commissioned music, however simple, you can clearly make it do what you want. Even a small station ought to be able to offer clients the possibility of tailor-made music – perhaps using a local musician with a bank of synthesisers. It can make all the difference – as with this firm of solicitors singing their own song. A comic, and memorable, idea which certainly gives the impression of its being a lively and 'unstuffy' firm.

```
(Music – upbeat piano accompaniment)

VOICE 1:    I'm Underhill

VOICE 2:    I'm Wilcock

VOICE 3:    and I'm Taylor

ALL:        Pleased to meet you

VOICE 1:    We're solicitors

ALL:        and jolly proud of it

VOICE 2:    in Wolverhampton

ALL:        Waterloo Road –

VOICE 3:    If you've a problem

ALL:        then we can help you as only a solicitor can

VOICE 1:    so that's Underhill

VOICE 2:    and Wilcock

VOICE 3:    and Taylor

ALL:        solicitors in Wolverhampton's Waterloo Road

                                                              (20 sec)

(Courtesy of Beacon FM – station-produced commercial)
```

A word about adding new words to a well-known song. Parody is allowed so long as it is fair and not discriminatory. All published music used in advertisements should be cleared in the same way that the station deals with its other music. For ads and promos, performing rights and copyright societies will normally make special arrangements rather than insisting on details of individual use. Producers working in this area will find the various libraries of non-copyright music invaluable.

On a technical point, make sure that the music/speech mix is checked for audibility in both the mono and stereo versions. The point made earlier is crucial – it may not be satisfactory to broadcast the same mix, in terms of relative level, on stereo FM as on mono medium wave.

Sound effects, like music, have to make their point immediately and unambiguously. They are best used sparingly, unless the impression required is one of chaos or 'busyness'. The right atmosphere effect to set the scene, manipulated and added to at appropriate points in the script – as in the 'cricket' example earlier – works well. Producers should not be misled by the title of an effects track or what it *actually* is. It is only what it *sounds* like that matters. And again, there is an armoury of techniques, from filters to time stretch and speed change, for altering the sound.

Stereo

Commercials that deliberately exploit the stereo effect are relatively rare. There is the British Airways example, which intersperses phrases from their music logo – the 'Flower Duet' from *Lakmé* by Delibes – using a sitar on the left and a piano on the right, eventually bringing them together under the speech line:

```
'Listen - the world is closer than you think.
British Airways - the world's favourite airline'
```
```
                                                (60 sec)
```
```
(Courtesy of M & C Saatchi Ltd; www.focalpress.com/cw/mcleish)
```

Even more pointed is this advertisement for the SEAT Arosa car. It doesn't make much sense on medium wave mono because of the two voices apparently speaking together, but it's clearly written and nicely produced for FM stereo.

```
Music         (slow alternating string chords. Hold under
              throughout)
MAN'S VOICE:
              (L + R)   Your brain works in a mysterious way.
              (R)                         Creative and emotive
                                          impulses, dreams and
                                          desires are controlled
                                          by the right side of
                                          the brain.
              (L)       Practical thoughts
                        and decisions are
                        made by the left side
                        of the brain.
              (L + R)   So which will help you decide on a new SEAT?
                        Switch your stereo to left or right speaker
                        for the practical or emotive argument now.
```

```
WOMEN'S VOICES:  (L + R)
                        The SEAT Arosa. Neat,   The SEAT Arosa. A
                        classy, with a big      sub-seven thousand
                        car feel and finish.    pound price, and
                        The SEAT Ibitha         an equipment list
                        Fresca, eager and       to shame some mid-
                        sporty, puts the        range five-door
                        emphasis on control,    hatchbacks. The Arosa
                        handling, and good      is remarkably good
                        old fashioned           value. The SEAT Ibitha
                        entertainment. And      Fresca, economical
                        the Cooper sports       motoring, plenty of
                        series, any more        room, a spacious boot,
                        aggressive and it       central locking, and
                        would need a muzzle.    stereo radio cassette.
                        Twice Formula 2 World   The SEAT Cooper sports
                        Rally Champions, low    series brings race
                        cost, and seriously     proven capabilities
                        hairy. You're going     to everyday driving.
                        to enjoy this. SEAT     Practical, safe,
                                                economical. SEAT
MAN:         (L + R)                            SEAT
MAN/WOMAN:   (L) (R)    Enjoy yourself          Enjoy yourself
MAN:         (L + R):   Call the SEAT hotline on 0500 22 22 22 for
                        details of your nearest dealer.

                                                      (60 secs)
```

(Courtesy of Chrysalis Creative/Galaxy 101; www.focalpress.com/cw/
mcleish)

Humour in advertising

We all like to laugh, and there is a perfectly logical connection between our liking an advertisement because it makes us laugh and liking the product that it promotes. The brand name endears itself to us by being associated with something that is witty and amusing. But the danger is twofold – if the joke is too good it may obscure or send up the product, and if it is not good enough it will not stand up to one hearing, let alone the repetition that radio gives. The answer lies in genuinely comic writing that does not rely on a single punch line, and in characterisation that may be overplayed but which is nevertheless credible. The good commercial has much in common with the successful cartoon drawing. Even so, exposure of such wit should be carefully regulated. It is probably best to create a series of vignettes in a given style, and intermix them across the spot times to give maximum variety. The listener will enjoy the new jokes as well as welcoming old favourites – further, the hearer will recall the product long before the ad gets to it.

Some of these commercials are full-scale dramas in their own right. As has been said in Chapter 21, if the station facilities and expertise cannot properly undertake this kind of work, it is better to succeed with something simpler than embark on the complexity of a major production such as this:

(American voices)	<u>Fx hiss of space circuit, bleep</u>
MAN 1 (on filter):	OK base, I'm on the ladder now. (<u>Music: majestic, low level</u>) This is one small leap for man . . .
MAN 2:	Hank, er, just hold on there will you. (<u>Music cuts</u>) That's 'one small step for man' there.
MAN 1 (filter):	Er, yes sir. (To self) One small step – OK base. (<u>Music begins</u>) This is one tiny (<u>Music cuts</u>) – oh rats.
MAN 2 (laconic):	Relax, Hank, just take your time now. (<u>Music begins</u>)
MAN 1 (filter):	This is one big, small, step here, for one man to take off a ladder . . . (<u>music cuts</u>)
MAN 2:	OK, Hank, you're getting warm. 'One giant leap for mankind'.
MAN 3 (filter):	Hank, I'm getting cramp on the ladder up here, will you hurry it up please. (<u>Music begins</u>)
MAN 1 (filter):	This is one small leap for a giant – (<u>Music cuts</u>)
MAN 2 (interrupting):	One giant leap for mankind. (<u>Music begins</u>)
MAN 1 (filter):	This kind man is a small giant. (<u>Music cuts</u>)
MAN 2 (testily):	Giant leap. (<u>Music begins</u>)
MAN 1 (filter):	This leap year is gonna be the best (<u>Music cuts</u>) – aah
MAN 2:	Chuck, will you unload the Heineken bay and refresh that man's speech please?
Fx (close):	<u>Pouring liquid into a glass</u>
MAN 1 (filter):	OK, I'm ready – run the music. (<u>Music begins again</u>)
	This is one small step for man – one giant leap for mankind.
	<u>Fx applause up</u>
	<u>Music: men's voices sing slowly in the style of 'Space Odyssey'</u>
	'Heineken refreshes the parts other beers cannot reach'.

```
MAN 1 (filter; triumphant):   The Blue Tit has landed.
MAN 2:                        The Eagle, Hank, the Eagle.
```

(85 sec)

(Courtesy of Lowe Howard-Spink Marschalk)

That's a classic, but good humour is a proven way of developing listener rapport with your product.

Testing

Finally, before launching your ad you must test it. There are professional organisations that will advise you on testing but if you are simply making a one-off in-house commercial, play it to your family, your friends, your local club – see how they react. Does it make them smile? If so, are they laughing with it, or at it? Are they sympathetic to the essential message – in fact, what was the core message they got out of it? Is that what you wanted? Ask their opinion, find out their genuine feelings – does it motivate and direct them towards what you are trying to sell? If they think it second rate, you're wasting your time and you must change it. You may think it clever and bright – just the thing for the product – but if someone hearing it for the first time doesn't understand it, or thinks it's puerile or too complicated, then it's unlikely to work, despite the creative hours you put into it. It's a hard lesson to learn.

A radio commercial is trying to sell a real product or service: the advertisement itself must therefore have reality, however outrageous its style. In the end, people must believe it.

18

Outside broadcasts (remotes)

As has been noted elsewhere, there is a tendency for broadcasters to shut themselves away in studios, being enormously busy making programmes which do not originate from a direct contact with the audience. The outside broadcast (OB), or 'remote', represents more than a desire to include in the schedule coverage of outside events in which there is public interest. It is a positive duty for the broadcasters to escape from the confines of their buildings into the world that is both the source and the target for all their enterprise. The concert, festival, civic ceremony, church service, exhibition, school fête, sporting event, public meeting, conference or demonstration – these demand the broadcaster's attention. It is not only good for radio to reflect what is going on, it is necessary for the station's credibility to be involved in its own community, to debate the issues that matter. Radio must not only go to where people are, it must come from the interests and activities of many people. If its sources are too few, it is in danger of appearing detached, sectional, elitist or out of touch. Thus the OB is essential to broadcasting's health.

Planning

The producer in charge, together with the appropriate engineering staff, must first decide how much coverage is required of a specific event. The programme requirement must be established and the technical means of achieving it costed. Is it to be 'live' or recorded on-site? What duration is expected? Once there is a definite plan, the resources can be allocated – people, facilities, money and time.

It is also at this first stage that discussions must take place with the event organiser to establish the right to broadcast. It may be necessary to negotiate any fees payable, or conditions or limitations which the promoter or sponsor may wish to impose.

Visiting the site

A reconnaissance, or 'recce', is essential, but it may take considerable imagination to anticipate what the actual conditions will be like 'on the day'. Where possible include an engineer, who will assess the venue in a very different way from the producer or presenter. There are a number of questions that must be answered:

1 Where and of what type are the mains electricity supply points? Is the supply correctly earthed, safe and reliable? Do I need my own battery or power generator? What is the procedure if a fuse blows or a breaker trips out? Much modern digital equipment is very sensitive to mains fluctuations. A portable uninterruptible power supply or mains conditioner can avoid many a nasty splat or click being broadcast.

2 Where is the best vantage point to see the most action? Will there have to be more than one?

3 Will the sound mixing be done in a building, or in a radio OB vehicle outside?

4 What on-site communications are required, e.g. reverse talkback needing headsets?

5 How many microphones and what type will be needed?

6 If radio mics, is anyone else using the same or adjacent frequencies? Will a special events licence be needed for extra radio mic or radio talkback channels?

7 How long are the cable runs?

8 Will a public address system be in use? If so, where are the speakers? How is the volume controlled and by whom?

9 What else will be present on the day? For example, flags which obscure the view, vehicles or generators which might cause electrical interference, background music, other broadcasters.

10 What are the potential hazards and safety requirements?

11 What is the earliest the engineering staff can gain access to the venue for rigging?

Communications to base

If the programme is 'live', how is the signal to be sent to the controlling studio? What radio links are required? Is the site within available radio car range? Are ISDN landlines available? They may be expensive but

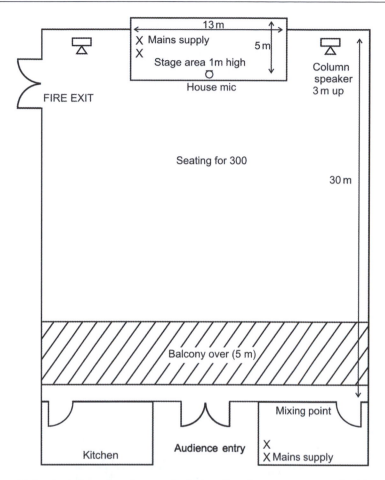

Figure 18.1 A sketch plan drawn up during the site visit is an invaluable aid to further planning

will additional programme or control circuits need to be ordered from the telecommunications department? If so, will the quality be good enough for music – or will the programme circuit have to be 'equalised'? Will WiFi work? Is a satellite dish the best solution? These questions need to be discussed at an early stage because, among other things, the answers will have a direct bearing on the cost of the programme.

Sooner or later an OB will be required from a hall, or from the middle of a field, where no lines exist. Such conditions are best realised well in advance so that either the telephone authority can be asked to make the appropriate connections, and if necessary build a suitable route, or the broadcaster must supply the necessary connections. A decision has to be made as to whether only a one-way programme circuit is required – the broadcaster at the site must then be able to obtain the cue to go ahead by listening off-air – or whether a second two-way telephone or control line is also needed. This additional facility is obviously preferred and perhaps a cell phone will suffice. Certainly for an OB of any length, or where a number of broadcasts are made from the same site, an ordinary communications link becomes essential. The same applies if radio links are used – is there to be a bi-directional control channel in addition to the programme circuit from OB to base? The specialist producer will certainly know about the mobile phone and ISDN options (described earlier on p. 106).

The advantage of a satellite dish over a car with a radio mast is its much greater range, reliability and freedom from interference. Remember though, the further south or north of the equator, the lower the dish will be pointing. In the UK, for example, the beam from the dish is only around 28 degrees above the horizontal. In a built-up area, relatively low buildings can soon block the signal. The satellite is very rarely overhead – as often imagined. In such locations great care must be exercised to ensure a workable parking place for the satellite uplink.

Figure 18.2   A radio car with a satellite dish will provide a signal to base in areas where other methods may be difficult or impossible (Courtesy of G. Jackson, BBC)

People

By this stage it should be clear how many people will be involved at the OB site. Anything more than a simple radio car or mobile phone job might require a number of skills – producer, engineer, floor manager, commentator, technical operator, secretary, driver, caterer, etc. A large event with the public present, such as an exhibition, might require the services of security staff or a publicity specialist. The list grows with the complexity of the programme, as does the cost.

The exact number of people is an important piece of anticipation – getting it right depends on being able to visualise whether, for example, there is a script writing and typing requirement on-site. It will also depend on whether the working day is to be so long as to warrant the employment of duplicate staff working in shifts.

Hazard assessment

Away from the familiarity of the home studio, OBs are full of uncertainties. During the reconnaissance visit there must be a careful assessment of the potential hazards and, for the safety of staff, performers and members of the public, plans made to minimise them. This applies not only to the broadcast but also to the rigging and derigging before and afterwards.

What measures are needed for crowd control – especially children? Are there any dangers from water, fire, heights, vehicles, traffic, animals, aircraft, etc.? Are hard hats, high visibility jackets, or any protective clothing needed? What contingency plans exist for dealing with an emergency? Should the police or other authorities be informed? No activity of this nature is entirely without risk, but responsible steps must be taken to anticipate and avoid possible accidents. See the model Hazard Assessment Form at: www.focalpress.com/cw/mcleish.

Equipment

This is best organised on a category basis by the individuals most closely involved:

1   *Engineering:* microphones including radio mics, cables, leads and connectors, audio mixers, computer kit and software, smartphone, tablet, TV video link, CD decks, recorders with windshields and spare cards, amplifiers, loudspeakers, headphones, editing facilities, digital Fx unit, radio, spare batteries of all kinds, power cables and distribution

boards, isolating transformers/'direct inject' (DI) boxes, circuit breakers, spare fuses, heavy-duty sticky tape, tool kit, fire extinguisher.

2 *Programme:* CDs, scripts, stopwatches, research notes.

3 *Administrative:* tables, chairs, paper, laptop, printer and ink cartridges, sticky labels, pens and pencils, marker pens, torches, clipboards, money, string, publicity and sign-writing materials, sticky tape, badges, cell phone, accurate large-faced clock.

4 *Personal:* food and drink, hi-vis jackets, hard hats, special clothing, first aid kit, sleeping bags, umbrellas, etc.

5 *Transport:* vehicles, fuel can, sat nav, maps, mallet, posts, rope, shovel.

There are always things that get forgotten, but if they are really important, this only happens once.

Safety

In a situation where crowds of people are present and their attention is inevitably drawn to the spectacle they have come to see, broadcasters have a special responsibility to ensure that their own operation does not present any hazards. They are, of course, affected by, and must observe, any by-laws or other regulations that apply to the OB site. Equipment must not obstruct gangways or obscure fire exit notices or the fire equipment itself. Everyone working on the site should have a fire drill briefing and know the location of the fire assembly point.

Cables across pavements or passageways must either be covered by a ramp, or should be lifted clear of any possibility of causing an obstruction.

Microphones suspended over an audience must be securely fixed, not just with sticky tape which can become loosened with a temperature change, but secured in such a way as to prevent any possibility of their being untied by inquisitive or malicious fingers. Safety chains should doubly secure any equipment rigged overhead.

Members of the public are generally curious of the broadcasting operation and all equipment must be completely stable, for instance microphone stands or loudspeakers should not be able to fall over. Nor, of course, must straying hands be able to touch mains electricity connections. Some fencing off might be necessary.

This will almost certainly be the case where a ground-level public address system is in use. For a DJ show where the broadcaster is providing an on-site loudspeaker output, the sound intensity close to the speakers may be sufficient to cause temporary, and in some cases permanent, damage to ears. To prevent this, some form of barrier three metres or so from the speaker is generally needed. A better alternative is to raise the speakers, fixing them securely well above head height.

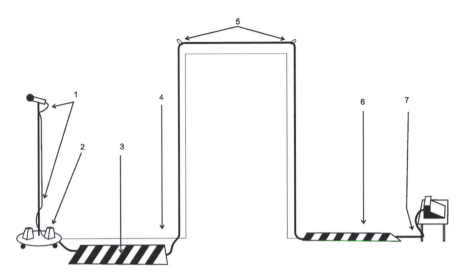

Figure 18.3 *Microphones and cables must be safely rigged. 1. Cable taped to mic stand. 2. Stand weighted. 3. Mat covers at walkway. 4. Cable laid without loops or knots. 5. Cable slung over doorways. 6. Cable taped to floor. 7. Cable made fast at mixer end*

Rainfall can cause disaster. Ensure that power cables in particular are sheltered from rain – even a waste-bin liner will suffice. Where there is a danger of water 'pooling' in dips, make sure such cabling is raised well above the ground. Finally, test and retest any overcurrent, or earth-leakage protection devices. A break in the programme may be undesirable, but serious injury or even death must, of course, be avoided.

Accommodation

In further discussion with the event organiser there has to be agreement on the exact location of the broadcasting personnel and equipment. There may be special regulations governing car parking or access to the site, in which case the appropriate passes and security badges need to be obtained.

In order to rig equipment it will be necessary to gain access well before the event – are any keys required? Who will be there – and what security exists to safeguard equipment once it is rigged? At this stage the producer must also be clear as to the whereabouts of lavatories, fire exits, catering facilities, lifts and any special features of the site, e.g. steps, small doorways, awkward passages, non-opening windows, or unusual acoustics.

When a public building is being used, the audio mixer frequently remains outside in the OB vehicle or is in a room from which the action cannot be seen. Under these conditions it is extremely useful to have a TV camera feeding a monitor close to the mixing desk so that the event can be followed – and anticipated – by the producer and engineer. The security and siting of cameras should be agreed with the event organiser in advance.

Programme research

Further discussion with the event organiser will establish the detailed timetable and list of participants. With an open-air event such as a parade or sports meeting, it is important to discover any alternative arrangements in case of rain. As much information as possible about who is taking part, the history of the event, how many people have attended on previous occasions, the order of what's happening and so on, is useful preparatory material for the broadcast itself. Additional research using libraries, press cuttings, the Internet, etc. may be necessary at this stage for the commentators (see Chapter 19).

The producer is then in a position to draw up a running order and to tell everyone his or her precise role both on and off the air. The running order

Figure 18.4   Omega 52 fader digital desk inside an OB vehicle. The operator follows the event on the TV monitors (Courtesy of Cloudbass Ltd)

should give as much relevant information as possible, including who is doing what, when, together with details of cues and timings, where these are known.

Liaison with the base studio

Particularly in the case of a 'live' OB, staff at the base studio need to be kept informed of what is planned and what alternatives exist. They should have copies of the running order and be involved in a discussion of any special fill-up material or other instructions in case of a technical failure. Arrangements should be made for a 'live' OB to be recorded at the base studio so that a possible feature can be made for a highlight or follow-up programme.

Publicity

The producer should ensure that everything has been done to secure the appropriate advance publicity. This may be in the form of programme billing, printed posters, a press release, or simply on-air trails and announcements. It is a matter of common experience that broadcasters go to immense pains to cover an important public event, but overlook the necessity of

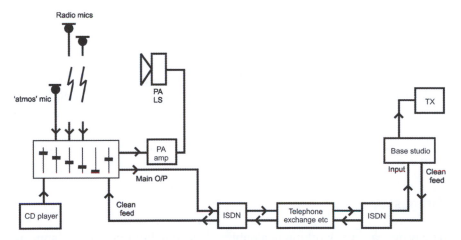

Figure 18.5 A live OB working with the base studio. Covering musicians and an audience, providing PA and playing in CDs, the OB provides programme via an ISDN line to the base studio for transmission. The base studio provides a clean feed back to the OB consisting only of material originating at base

telling people about it in advance. It is true that some promoters claim that broadcast coverage keeps people away from attending the event itself but, on the other hand, it is frequently the case that advance publicity will stimulate public interest and swell the crowds, many of whom will then follow the action via radio.

Conflicts of approach

Engineers are essential to the broadcasting process and programme people ought to know that they wouldn't get far without them – especially at an OB. While producers might well be criticised for taking too little interest in technical matters, it also has to be said that some engineers have a tendency to over-complicate things. They can use far too many mics, or have an undue liking for technical gimmicks, and they do not always fully explain to the producer the problems and possibilities of the situation. Furthermore, the mixing desk is often operated by a technician or specialist OB engineer and so the final mix – the programme sound – is not under the producer's direct control. So while an engineer at an OB may be aiming at studio speech quality and acoustically perfect sound, the producer might want much more the *event* – a sense of the occasion. Because of such differences of approach it is not surprising that arguments arise between producers and engineers. The problem is generally compounded by the stressful pressure of time.

A common difficulty with the broadcasting of live events involving music is the frequent lack of any proper rehearsal. It is often not possible to listen to the mix, discuss it, make adjustments and do it again. It is therefore essential that the producer discusses his or her objectives with the balance engineer well in advance and maintains close contact to resolve problems as they arise. Confronted by an engineer of many years' experience the young producer is likely to feel daunted by the older person and unable to question anything said or done. By all means use that experience, depend on that competence and learn from it – but also develop the personal skills to challenge precedent and make changes if you feel it right. After all, in most broadcasting organisations it is the producer who is in charge – and responsible for the end result.

Tidiness

The broadcaster is working in a public place and, both in terms of appearance and general behaviour, will contribute to the station's image. This is recognised by the more senior staff but might not always be appreciated by freelance contributors. A small but important aspect of public relations is the matter of leaving the OB site in a sensibly tidy state. It is clearly undesirable to leave any equipment behind but this should also apply to the accumulated rubbish of a working visit – scripts, notes, food tins, plastic bags, empty boxes, etc. To be practical, a good OB site will be required again and it is not in the broadcaster's interests to be remembered for the wrong reasons.

Gratuities

It is common sense to recognise that the broadcaster's presence at an OB site is likely to cause extra effort for those who normally work there. It will not be necessary to consider this point in every case, but a facility fee should be paid – certainly offered – where local assistance is provided beyond the normal level, and to any outsiders who supplied some special service, for example the use of a telephone, electricity or water, or the parking of vehicles. The amounts should obviously be related to the service provided – too much and one is open to a charge of profligate wastage, too little and one does more harm than good. It may, on the other hand, be sufficient to send a letter of thanks. These are the niceties of a well-judged operation.

The big disaster

Think the unthinkable. Nowhere on earth is immune from huge disaster, either natural or man-made. We have seen overwhelming floods destroy cities, communications put out of action by tsunami, fire, hurricane, earthquake, volcano, nuclear accident or terrorist bomb. What then is an appropriate response for the radio programme maker? Here is a practical course of action – set up an entirely new temporary station as an OB, based on the 'studio in a suitcase'.

Fitted into a standard suitcase, the studio comprises everything needed to get on the air – mic inputs to a mixer, MP3 digital player, on-air and production laptops, local and off-air monitoring, editing facility, USB, cell phone and Skype interfaces, recorders, headphones with built-in mics, cue lights and a spare guest mic. Connected to an FM transmitter it can broadcast up to 20 km.

This was used for 24/7 live broadcasting 72 hours after typhoon 'Haiyan' devastated huge areas of the Philippine Islands when all communications were down, affecting some 14 million people, with 26,000 injured, and a million homes damaged.

First Response Radio (FRR) arrived with all its equipment stowed in four standard flight cases – this studio, the dipole antenna and coaxial

Figure 18.6   A complete studio in a suitcase (Courtesy of Randall Concepts Ltd)

cables in a padded bag, the 600 watt FM transmitter, and a Honda genera-tor (empty tank). Frequencies had already been allocated for emergency use, and local people, who knew the territory, did the broadcasting. 'Where can I get food, clean water, or fuel?' 'What roads are open?' 'How can I trace relatives?' 'How can I build some shelter for the family?' 'How can I get medical help?' 'When will the power come back on?' Catchy songs, replies to texts, interviews and information helped to instil a sense of hope and bring some 'normality' back to daily life.

Setting up the radio isn't much use unless people can hear, so FRR gave out over 2,000 wind-up or solar powered radios to the villages around. Mobile loudspeaker trucks tuned to the station were placed at evacuation centres. The purpose was obviously to get expert information to as many people as possible as quickly as possible – to save lives. The radio helpers were faced with dealing with up to 2,000 messages a day. Radio in these circumstances is certainly not a one-way communication, it's a dialogue – people could express their needs, concerns and frustrations.

This is radio in the extreme. But, of course, the suitcase studio is very useful for ordinary OBs; it is used for staff training, or as a station start-up

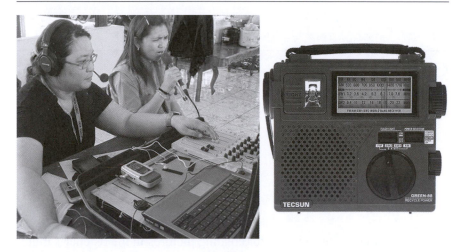

Figure 18.7 The First Response Radio team operating in Tacloban City, and the wind-up FM set distributed (Photo: Mike Adams)

before the actual station is ready, and in one African country it had to replace a whole station after all the studio equipment had been stolen.

This is the best possible example of an OB demonstrating real involvement in its own community, reflecting the interests, aspirations and pursuits of its own target audience.

19

Commentary

Radio has a marvellous facility for creating pictures in the listener's mind. It is more flexible than television in that it is possible to isolate a tiny detail without waiting for the camera to 'zoom in' and it can create a breadth of vision much larger than the dimensions of a glass screen. The listener does more than simply eavesdrop on an event; radio, more easily than television or video, can convey the impression of actual participation. The aim of the radio commentator is therefore to recreate in the listener's mind not simply a picture but a total impression of the occasion. This is done in three distinct ways:

1 The words used will be visually descriptive of the scene.
2 The speed and style of their delivery will underline the emotional mood of the event.
3 Additional 'effects' microphones will reinforce the sounds of the action, or the public reaction to it.

Attitude to the listener

When describing a scene, the commentator should have in mind 'a blind friend who couldn't be there'. It is important to remember the obvious fact that the listener cannot see. Without this it is easy to slip into the situation of simply chatting about the event to 'someone beside you'. The listener should be regarded as a friend because this implies a real concern to communicate accurately and fully. The commentator must use more than his or her eyes and convey information through all the senses, so as to heighten the feeling of participation by the listener. Thus, for example, temperature, the proximity of people and things, or the sense of smell are important factors in the overall impression. Smell is particularly evocative – the scent of newly mown grass, smoke from a fire, the aroma inside a fruit market or the timeless mustiness of an old building. Combine this with the appropriate style of delivery and the sounds of the place itself, and you are on the way to creating a powerful set of pictures.

Preparation

Some of the essential stages are described in Chapter 18, but the value of a pre-transmission site visit cannot be overemphasised. Not only must the commentator be certain of the field of vision and whether the sun is likely to be in their eyes, but it is important to spend time obtaining essential facts about the event itself. For example, in preparing for a ceremonial occasion, research:

1 the official programme of events with details of timing, etc.;
2 the background of the people taking part, their titles, medals and decorations, positions, relevant history, military uniforms, regalia or other clothing, personal anecdotes – for the unseen as well as the seen, for example organisers, bandmasters, security people, caretakers, etc.;
3 the history of where it's taking place, the buildings and streets, and their architectural detail;
4 the names of the flowers used for decoration, the trees, flags, badges, mottoes and symbols in the area. The names of any horses or make of vehicles being used;
5 the titles of music to be played or sung, and any special association it might have with the people and the place.

It adds immeasurably to the description of a scene to be able to mention the type of stone, carving, architectural history or usage relating to a building, or that 'around the platform are purple fuchsias and hydrangeas'. The point of such detail is to use it as contrast with the really significant elements of the event, so letting them gain in importance. Contrast makes for variety and for more interesting listening and mention of matters both great and small is essential, particularly for an extended piece. An eye for detail can also be the saving of a broadcast when there are unexpected moments to fill. There is no substitute for a commentator doing proper homework.

In addition to personal observation and enquiry, useful sources of information will be the Internet, reference section of libraries or museums, newspaper cuttings, back copies of the event programme, previous participants, specialist magazines and commercial or government press offices.

Having obtained all the factual information in advance, the commentator must assemble it in a form that can be used in the conditions prevailing at the OB. If perched precariously on top of a radio car in the rain, clutching a guardrail and holding a microphone, stopwatch and an umbrella, the last thing you want is a bundle of easily windblown papers! It might be easiest to have your script notes held on a tablet or smartphone. If you are using paper, use as few sheets as possible, held firmly on a clipboard. Cards can be useful since you can quickly change the order and they are silent to handle. The important thing is their order and logic. The information will often be chronological in nature, listing the background of the people taking part. This is particularly so where the participants appear in a predetermined sequence – a procession or parade, variety show, race meeting, athletics event, church service or civic ceremony. Further information on the event or the environment can be on separate pages so long as they can be referred to easily. If the event is non-sequential, for instance a football match or public meeting, the personal information might be more useful in alphabetical form, or better still memorised.

Figure 19.1   Rehearsing the commentary for a stone unveiling, the day before the event (Courtesy of BBC Radio Nottingham)

Working with the base studio

The commentator will need to know the precise handover details. This applies both from the studio to the commentator and at the end for the return to the studio. These details are best written down, for they easily slip the memory. There should also be clarity at both ends about the procedures to be followed in the event of any kind of circuit failure – the back-up music to be played, and who makes the decision to restore the programme. It might be necessary to devise some system of hand signals or other means of communication with technical staff, and on a large OB whether all commentators will get combined or individual 'talkback' on their headphones, etc. These matters are the 'safety nets' that enable the commentators to fulfil their role with a proper degree of confidence.

As with all outside broadcasts, the base studio should ensure that the commentary output is recorded. Not only will the commentator be professionally curious as to how it came over, but the material might be required for archive purposes. Even more important is that an event worthy of a 'live' OB will almost certainly merit a broadcast of edited highlights later in the day.

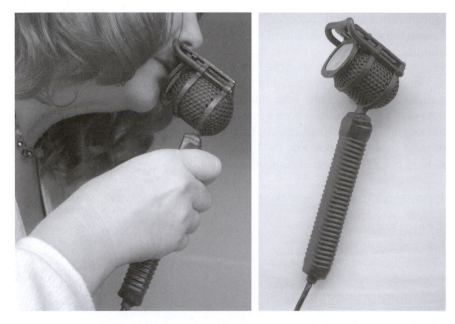

Figure 19.2 Lip mic in use, and showing handle containing the bass cut filter to compensate for the bass lift that results from working close to a ribbon microphone

Sport

First and foremost, the sports commentator must know his or her sport and have detailed knowledge of the particular event. What was the sequence that led up to this event? What is its significance in the overall contest? Who are the participants and what is their history? The possession of this background information is elementary, but what is not so obvious is how to use it. The tendency is to give it all out at the beginning in the form of an encyclopedic but fairly indigestible introduction. Certainly the basic facts must be provided at the outset, but a much better way of using background detail is as the game, race or tournament itself proceeds, at an appropriate moment or during a pause in the action. This way the commentator sounds much more a part of what is going on, instead of being a rather superior observer.

Traditionally, for technical reasons, the commentator has often had to operate from inside a soundproof commentary position, isolated from the immediate surroundings. It's easy to lose something of the atmosphere by creating too perfect an environment, and there is a strong argument in favour of the 'ringside seat' approach, provided that it is possible to use a noise-cancelling microphone and that communication facilities, such as headphone talkback, are secure.

Sports stadia seem to undergo more frequent changes to their layout than other buildings and unless a particular site is in almost weekly use for radio work, a special reconnaissance visit is strongly advised. It is easy to forego the site recce, assuming that the place will be the same as it was six months ago. However, unless there are strong reasons to the contrary, a visit and technical test should always be made.

Where the action is spread out over a large area, as with horse racing, golf, motor racing, a full-scale athletics competition or rowing event, more than one commentator is likely to be in action. Cueing, handovers, timing, liaison with official results – all these must be precisely arranged. The more complex an occasion, the more necessary is observance of the three golden rules for all broadcasts of this type:

1   meticulous production planning so that everyone knows what they are *likely* to have to do;
2   first-class communications for control;
3   only one person in charge.

Communicating mood

The key question is 'what is the overall impression here?' Is it one of joyful festivity or is there a more intense excitement? Is there an urgency to the occasion or is it relaxed? At the other end of the emotional scale there may be a sense of awe, a tragedy or a sadness that needs to be reflected in a sombre dignity – soldiers' bodies being brought back from a war zone. Look at people's faces – they will tell you. Whatever is happening, the commentator's sensitivity to its mood, and to that of the spectators, will control matters of style, use of words and speed of delivery. More than anything else this will carry the impressions of the event in the opening moments of the broadcast. The mood of the crowd should be closely observed – anticipatory, excited, happy, generous, relaxed, impressed, restive, sullen, tense, angry, solemn, sad. Such feelings should be conveyed in the voice of the commentator and their accurate assessment will indicate when to stop and let the sounds of the event speak for themselves.

Coordinating the images

It is all too easy to fall short of an overall picture but to end up instead with some accurately described but separate pieces of jigsaw. The great art, and challenge, of commentary is to fit them together, presenting them in a logically coordinated way that allows your 'blind friend' to place the information accurately in their mind's eye. The commentator must include not only the information relating to the scene, but also something about how this information should be integrated to build the appropriate framework of scale. Having provided the context, other items can then be related to it. Early on, it should be mentioned where the commentator's own position is relative to the scene; also giving details of distance, up and down, size (big and small), foreground and background, side to side – left and right, etc. Movement within a scene needs a smooth, logical transition if the listener is not to become hopelessly disorientated.

Content and style

The commentator begins with a 'scene-set', saying first of all where the broadcast is coming from, and why. This is best not given in advance by the continuity handover and duplication of this information must be avoided. The listener should be helped to identify with the location, particularly if it is likely to be familiar. The description continues from the general to the particular noting, as appropriate, the weather, the overall impression of lighting, the mood of the crowd, the colour content of the scene and what is about to happen. Perhaps two minutes or more should be allowed for this 'scene-setting', depending on the complexity of the event, during which time nothing much may be happening. By the time the action begins the listener should have a clear visual and emotional picture of the setting, its sense of scale and overall 'feel'. Even so,

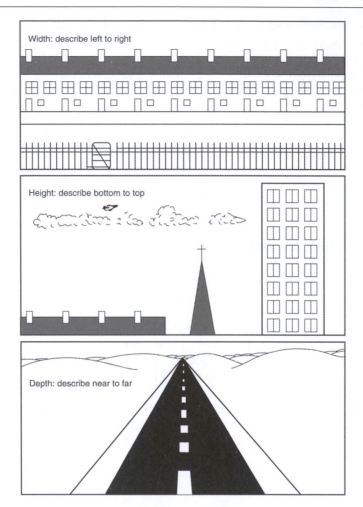

Figure 19.3 The dimensions of commentary. The listener needs three-dimensional information in which to place the action. Such orientation should not be confined to the scene-set but should be maintained throughout the commentary

the commentator must continually refer to the generalities of the scene as well as to the detail of the action. The two should be woven together.

Time taken for scene-setting does not, of course, apply in the case of news commentary where one is concerned first and foremost with what is happening. Arriving at the scene of a fire or demonstration, one deals first with the event and widens later to include the general environment. Even so, it is important to provide the detail with its context.

Many commentaries are greatly improved by the use of colour. Colour, whether gaudy or sombre, is easily recreated in the mind's eye and mention

Figure 19.4 London's Notting Hill Carnival, illustrating the need to convey width, height and depth

of purple robes, brilliant green plumage, dark grey leaden skies, the blue and gold of ceremonial, the flashes of red fire, the green and yellow banners, or the sparkling white surf – such specific references conjure up the reality much better than the shortcut of describing a scene simply as 'multi-coloured'.

In describing the action itself, the commentator should proceed at the same pace as the event, combining prepared fact with spontaneous vision. In the case of a planned sequence, as a particular person appears, or slightly in anticipation, reference is made to the appropriate background information, title, relevant history and so on. This is more easily said than done and requires a lot of practice – perhaps using a recorder to help perfect the technique.

News action

With little or no time for scene-setting it's important to answer the central questions, what's happening, where and why? Here, mood is especially important. Are people in a demonstration angry, dignified or relaxed? Read the banners and placards, describe any unrest or scuffles with police. How are the police reacting? Above all, get the event in proportion – how many people are there? How many stone-throwers? What is their age? Very often at such gatherings there are many inactive bystanders, and these are part of the scene too. At a large riot take a wide view, describing things from the edge not the centre of the action – keep out of the way of petrol bombs or water cannon.

At the scene of an accident, again it's important that you report what you see. This is not the time for analysis as to causes, blame or speculation on outcomes. Whether this is a two-minute piece for a bulletin, or a half-hour actuality programme, leave me, the listener, with an accurate picture of the event so that I can really appreciate what has happened.

Sports action

The description of sport, even more than that of the ceremonial occasion, needs a firm frame of reference. Most listeners will be very familiar with the layout of the event and can orient themselves to the action so long as it is presented to them the right way round. They need to know which team is playing from left to right; in cricket, which end of the pitch the bowling is coming from; in tennis, who is serving from which court; in horse racing or athletics, the commentator's position relative to the finishing line. It is not sufficient to give this information at the beginning only; it has to be used throughout a commentary, consciously associated with the description of the action.

As with the ceremonial commentator, the sports commentator is keeping up with the action but also noticing what is going on elsewhere, for instance the injured player, or a likely change in the weather. Furthermore, the experienced sports commentator can increase interest for the listener by highlighting an aspect of the event which is not at the front of the action. For example, the real significance of a motor race may be the duel going on between fourth and fifth place; the winner of a 10,000-metre race may already be decided but there can still be excitement in whether the athlete in second place will set a new European record or a personal best.

With slower games, such as cricket, the art is to use pauses in the action interestingly, not as gaps to fill but opportunities to add to the general picture or give further information. This is where another commentator or researcher can be useful in supplying appropriate information from the record books or with an analysis of the performance to date. Long stretches of commentary in any case require a change of voice, as much for the listener as for the commentators, and changeovers every 20–30 minutes are about the norm.

If for some reason the commentator cannot quite see a particular incident or is unsure what is happening, it is better to avoid committing oneself.

'I think that . . . ' is better and more positive than 'It looks as though . . . '. Similarly it is unwise for a commentator to speculate on what a referee is saying to a player in a disciplinary situation. Only what can be seen or known should be described – the red or yellow card. It is easy to make a serious mistake affecting an individual's reputation through the incorrect interpretation of what might appear obvious. And it must be regarded as quite exceptional to voice a positive disagreement with an umpire or referee's decision. After all they are closer to the action and may have seen something that the commentator, with a general view, missed. The reverse can also be true but in the heat of the moment it is sensible policy to give match officials the benefit of any doubt.

Scores and results should be given frequently for the benefit of listeners who have just switched on but in a variety of styles in order to avoid irritating those who have listened throughout. Commentators should remember that the absence of goals or points can be just as important as a positive scoreline.

Actuality and silence

It may be that during the event there are sounds to which the commentary should refer. The difficulty here is that the noisier the environment, the closer on-microphone will be the commentator so that the background will be relatively reduced. It is essential to check that these other sounds can be heard through separate microphones, otherwise references to 'the roar of the helicopters overhead', 'the colossal explosions going on around me' or 'the shouts of the crowd' will be quite lost on the listener. It is important in these circumstances for the commentator to stop talking and to let the event speak for itself.

There are times when it is virtually obligatory for the commentator to be silent – during the playing of a national anthem, the 'Last Post', the blessing at the end of a church service, or important words spoken during a ceremonial. Acute embarrassment on the part of the over-talkative commentator and considerable annoyance for the listener will result from being caught unawares in this way. A broadcaster unfamiliar with such things as military parades, church services or funerals must be certain to avoid such pitfalls by a thorough briefing beforehand.

The ending

Running to time is helped by having a stopwatch synchronised with the studio clock. This will provide for an accurately timed handback, but if open-ended, the cue back to the studio is simply given at the conclusion of the event.

It is all too easy after the excitement of what has been happening to create a sense of anti-climax. Even though the event is over and the crowds are filtering away, the commentary should maintain the spirit of the event itself, perhaps with a brief summary, or with a mention of the next similar occasion. Another technique is radio's equivalent of the television wide-angle shot. The commentator 'pulls back' from the detail of the scene, concluding as at the beginning with a general impression of the whole picture before ending with a positive and previously agreed form of words which indicates a return to the studio.

Many broadcasters prefer openings and closings to be scripted. Certainly if you have hit upon the neat, well-turned phrase, its inclusion in any final paragraph will contribute appropriately to the commentator's endeavour to sum up both the spirit and the action of the hour.

An example

One of the most notable commentators was the late Richard Dimbleby of the BBC. Of many, perhaps his most memorable piece of work was his description of the lying-in-state of King George VI at Westminster Hall in February 1952. The printed page can hardly do it justice; it is radio and should be heard to be fully appreciated. It is old now, but nevertheless it is still possible to see here the application of the commentator's 'rules'. A style of language, and delivery, that is appropriate to the occasion. A 'scene-set' that quickly establishes the listener both in terms of the place and the mood. 'Signposts' which indicate the part of the picture being described. Smooth transitions of movement that take you from one part of that picture to another. Researched information, short sentences or phrases, direct speech, colour and attention to detail, all used with masterly skill to place the listener at the scene.

> It is dark in New Palace Yard at Westminster tonight. As I look down from this old, leaded window I can see the ancient courtyard dappled with little pools of light where the lamps of London try to pierce the biting, wintry gloom and fail. And moving through the darkness of the night is an even darker stream of human beings, coming, almost noiselessly, from under a long, white canopy that crosses the pavement and ends at the great doors of Westminster Hall. They speak very little, these people, but their footsteps sound faintly as they cross the yard and go out through the gates, back into the night from which they came.

> They are passing, in their thousands, through the hall of history while history is being made. No one knows from where they come or where they go, but they are the people, and to watch them pass is to see the nation pass.

> It is very simple, this lying-in-state of a dead king, and of incomparable beauty. High above all light and shadow and rich in carving is the massive roof of

chestnut that Richard II put over the great hall. From that roof the light slants down in clear, straight beams, unclouded by any dust, and gathers in a pool at one place. There lies the coffin of the King.

The oak of Sandringham, hidden beneath the rich golden folds of the Standard; the slow flicker of the candles touches gently the gems of the Imperial Crown, even that ruby that King Henry wore at Agincourt. It touches the deep purple of the velvet cushion and the cool, white flowers of the only wreath that lies upon the flag. How moving can such simplicity be. How real the tears of those who pass and see it, and come out again, as they do at this moment in unbroken stream, to the cold, dark night and a little privacy for their thoughts.

(Richard Dimbleby, Broadcaster, pub. BBC 1966)

Coping with disaster

Sooner or later something will go wrong, sudden and quite unexpected. The VIP plane crashes, the football stadium catches fire, terrorists suddenly appear, a peaceful demonstration unexpectedly becomes violent or spectator stands collapse. The specialist war correspondent or experienced news reporter sent to cover a disaster knows how far to go in describing death and destruction. Sensitivity to the reactions of the listener in describing mutilated bodies or the bloody effect of shellfire is developed through experience and a constant reappraisal of news values. But the non-news commentator must also learn to cope with tragedy. From the crashing of the *Hindenburg* airship in 1937, the 1986 explosion of the *Challenger* space shuttle and the 2005 flooding of New Orleans, to the 9/11 destruction of the Twin Towers, commentators are required to react to the totally unforeseen, responding with an instant transition perhaps from national ceremony to fearful disaster. Certain kinds of events such as motor racing and airshows have an inherent capacity for accident, but when terrorists invade a peaceful Olympic Games village, commentators normally used to describing the excitement of the track are called on to cope with tensions of quite a different kind.

Here is an example of BBC sports commentator Peter Jones covering a football match at Hillsborough Stadium, Sheffield in 1989. The game had only just begun when more people crowding into the ground suddenly caused such a crush in the stands that supporters were climbing the fence and invading the pitch. The match was stopped and a few moments later:

At the moment there are unconfirmed reports, and I stress unconfirmed reports, of five dead and many seriously injured here at Hillsborough. Just to remind you what happened – after five minutes, at the end of the ground to our left where the Liverpool supporters were packed very tightly – and the report is that one of the gates in the iron fence burst open – supporters poured on the pitch. Police intervened and quite correctly the referee took the police advice and took both teams off. Since then we've had scenes of improvised stretchers with the advertising hoardings being torn up, spectators have helped, we've got medical teams, oxygen cylinders, a team of fire brigade officers as well to break down the fence at one end to make it easier for the ambulances to get through and we've got bodies lying everywhere on the pitch.

(Courtesy of BBC Sport)

Remembering that his commentary was being heard by the families and friends of people at the match, it was important here to describe the early casualty reports as 'unconfirmed' and also to avoid any attempt at identifying the cause of the situation, or worse, to apportion blame. Commentators do well to report only what they can personally see.

So what should the non-specialist do? Here are some guidelines:

- Keep going if you can. A sense of shock is understandable, but don't be so easily deterred by something unusual that you hand back to the studio. Even if your commentary is not broadcast 'live' it could be crucial for later news coverage.
- There's no need to be ashamed of your own emotions. You are a human being too and if you are horrified or frightened by what is happening, say so. Your own reaction will be part of conveying that to your listener. It's one thing to be professional, objective and dispassionate at a planned event, it is quite another to remain so during a sudden emergency.
- Don't put your own life, or the lives of others, in unnecessary danger. You may from the best of motives believe that 'the show must go on', but few organisations will thank you for the kind of heroics that result in your death. If you are in a building which is on fire, say so and leave. If the bullets are flying or riot gas is being used in a demonstration, take cover. You can then say what's happening and work out the best vantage point from which to continue.
- Don't dwell on individual anguish or grief. Keep a reasonably 'wide angle' and put what is happening in context. Remember the likelihood that people listening will have relatives or friends at the event.
- Let the sounds speak for themselves. Don't feel you have to keep talking; there is much value in letting your listener hear the actuality – gunfire, explosions, crowd noise, shouts and screams.
- Don't jump to conclusions as to causes and responsibility. Leave that to a later perspective. Stick with observable events; relay the facts as you see them.
- Above all, arrive at a station policy for this sort of coverage well before any such event takes place. Get the subject on the agenda in order to agree emergency procedures.

20

Music 'live'

There are three questions that a producer should ask before becoming involved in the production of any music.

First, is the material offered relevant to programme needs? The technically minded producer, audiophile or engineer can easily create reasons for recording or broadcasting a particular music occasion other than for its value as good programming. It may represent an attractive technical challenge, or be the sort of concert which at the time seems a good idea to have 'in the can'. Alternatively, the desire to record a particular group of performers might outweigh considerations of the suitability of what they are playing. Sometimes musicians are visually persuasive but aurally colourless, as with many club performers dependent on an atmosphere difficult to reproduce on radio. To embark on music with little prospect of actually using it is a speculative business and the radio producer should not normally commit resources unless it's known, perhaps only in outline, how the material is to be aired.

The second question asks whether the standard of performance is good enough for broadcasting. There cannot be a single set of objective criteria for standards since much depends on the programme's purpose and context. A national broadcaster will undoubtedly demand the highest possible standards. Regionally and locally there is an obligation to broadcast the music-making of the area, the standard of which will almost certainly be of less than international excellence. The city orchestra, college swing band, and amateur pop group all have a place in the schedules, but for broadcasting to a general audience, as opposed to a school concert which is directed only to the parents of performers, there is a lower limit below which the standard must not fall. In identifying this minimum level, the producer has to decide whether what is being broadcast is primarily the music, the musician or the occasion. Certainly one should not go ahead without a clear indication of the likely outcome – new groups should be auditioned first, preferably 'live' rather than from a submitted recording. If the technical limits of performance are apparent, musicians should be persuaded to play items that lie within their abilities. Simple music well played is infinitely preferable to the firework display that does not come off.

And finally, is the recording or broadcast within the technical capability of the station? Even at national level there are limits to what can be expended on a single programme. Numbers and types of microphones, the best specification for stereo connections, special circuits, engineering facilities at a remote OB site for a 'live' broadcast, the availability of more than one echo source and so on – these are considerations that affect what the listener will hear and will therefore contribute to an appreciation of the performer. The broadcaster has an obvious obligation not only to encourage local artists but also to present them in the most appropriate manner without the intervention of technical limits. This presupposes that the producer knows exactly what is involved in any given situation and, for example, understands the implications of a 'live' concert where the members of a pop group sing as they play, where public address relay is present, or hypercardioid, variable pattern or tie-clip mics are required to do justice to a particular sound balance. Small programme-making units should avoid taking on more than they can adequately handle, and instead should stay within the limits of their equipment and expertise. Much better for a local station to refuse the occasional music OB as beyond its technical scope than it should broadcast a programme that it knows could have been improved, given better equipment. Alternatively if a station is not equipped or staffed to undertake outside music recording, it could consider contracting the work out to a specialist facilities company, with the station retaining editorial control.

The remainder of this chapter is directed to help producers in their understanding of some of the technical factors of music recording.

The philosophy of music balance is divided into two main groups – first, the reproduction of a sound that is already in existence, and second, the creation of a synthetic overall balance which exists only in the composer's or arranger's head and subsequently in the listener's loudspeaker.

Reproduction of internal balance

Where the music created results from a carefully controlled and self-regulating relationship between the performers, it would be wrong for the broadcaster to alter what the musicians are trying to achieve. For example, the members of a string quartet are sensitive to each other and adjust their individual volume as the music proceeds. They produce a varying blend of sound that is as much part of the performance as the notes they play or the tempo they adopt. The finished product of the sound already exists

Choir – 34 performers = 1 mic

Vocals

Pop Group – 4 performers = 12 mics

Figure 20.1 The technical complexity of broadcasting live music is not related to how many performers there are. The producer must decide whether adequate programme standards can be achieved with the available resources

and the art of the broadcaster is to find the place where the microphone(s) can most faithfully reproduce it. Other examples of broadly internally balanced music are symphony orchestras, concert recitalists, choirs and brass bands. The dynamic relationship between the instruments and sections of a good orchestra is under the control of its conductor; the broadcaster's task is to reproduce this interpretation of the music and not create a new sound by boosting the woodwind or unduly accentuating the trumpet solo. Since the conductor controls the internal balance by what he or she hears, it is in this area that one searches initially for the 'right' sound. Using a 'one-mic balance' or stereo pair, the rule of thumb is to place it with respect to the conductor's head – 'three metres (10 feet) up and three metres (10 feet) back'.

Similarly the conductor of a good choir will listen, assess, and regulate the balance between the soprano, contralto, tenor and bass parts. If the choir is short of tenors and this section needs reinforcing, this adjustment can of course be made in the microphone balance. But this immediately creates an enhanced – even artifical – sound, not made simply by the musicians, and raises interesting questions about the lengths to which broadcasters may go in repairing the deficiencies of performers.

It is possible to 'improve' a musician's tone quality, to clarify the diction of a choir or to correct an unevenly balanced group. With multi-track recording it is even possible to correct a soloist's late entry by moving that track forward in time. The ultimate in such cosmetic treatment is to use the techniques of the recording studio to so enhance a performer's work that it becomes impossible for them ever to appear 'live' in front of an audience, a not-unknown situation. But while it is entirely reasonable to make every possible adjustment in order to obtain the best sound from a school orchestra, it would be unthinkable to tamper with the performance of an

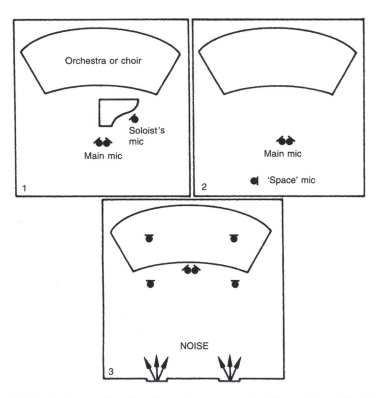

Figure 20.2 Basic consideration for mic placement with an internally balanced group. 1. Mic balance suitable under low noise conditions in a good acoustic. A soloist's mic or occasional 'filler' mic may be added. 2. In too 'dry' an acoustic, a space mic may added to increase the pick-up of reflected sound. 3. Under conditions of unacceptable ambient noise, the microphones are moved closer to the source and increased in number to preserve the overall coverage. It may be necessary to add artificial reverberation

artist who had a personal reputation. It is in the middle ground, including the best amateur musicians, where the producer is required to exercise this judgement – to do justice to the artist but without making it difficult for the performer in other circumstances.

The foregoing assumes that the music is performed in a hall that has a favourable acoustic. Where this is not the case, steps have to be taken to correct the particular fault. For example, in too 'dry' an acoustic, some artificial reverberation will need to be added, preferably on-site at the time of recording, although it is possible for it to be added later. Alternatively, a 'space' microphone can be used to increase the pick-up of reflected sound. If the hall is too reverberant or the ambient noise level, from an air-conditioning plant or outside traffic, is unacceptably high, the microphone should be moved closer to the sound source. If this tends to favour one part at the expense of another, further microphones are added to restore full coverage. The best sound quality from most instruments and ensembles is to be found not simply in front of them, but above them. A good music studio will have a high ceiling and the microphone for an internally balanced group will be kept well above head height. As mentioned previously, the standard starting procedure is to place a main microphone 'three metres up and three metres back' from the conductor's head, with a second microphone perhaps further back and a little higher. The outputs of the two microphones are then compared. The one giving the better overall blend without losing the detail then remains and the other microphone is moved to another place for a second comparison. An alteration in distance of even a few centimetres can make a significant difference to the sound produced. Hence the development of the Decca 'tree' which consists of three omni mics, placed in a horizontal equilateral triangle two metres apart above the conductor's head. They are panned left, centre and right. This helps to avoid the effects of either the reinforcement or the phase cancellation of the reflected sound, whereby mono microphones are best placed asymmetrically within a hall, off the centre line. Similarly one avoids the axis of a concave surface such as a dome.

The process of mic combining and comparison continues until no further improvement can be obtained. It is essential that such listening is done with reference to the actual sounds produced by the musicians. It is important that producers and sound balance engineers do not confine themselves to the noises produced by their monitoring equipment, good though they may be, but listen in the hall to what the performers are actually doing.

Having achieved the best placing for the main microphone, soloists', 'filler' or 'space' mics may be added. These additional microphones must not be allowed to take over the balance, nor should their use alter the 'perspective' during a concert. For a group that is balanced internally the only additional control is likely to be a compression of the dynamic range as the music proceeds. An intelligent anticipation of variations in the dynamics of the sound is required and although a limiter/compressor will avoid the worst excesses of overload, it can also iron out all artistic subtlety in the performance. There is no substitute for a manual control that takes account of the information from the musical score and is combined with a sensitive appreciation of what comes out of the loudspeaker.

The aim is to broadcast the music together with whatever atmosphere may be appropriate. To secure the confidence of the conductor, band-master or leader, it is good practice to invite them into the sound mixing and monitoring area to listen to the balance achieved during rehearsal. It should be remembered that since conductors normally stand close to the musicians, they are used to a fairly high sound level and the playback of rehearsal recordings should take this into account. Recordings should not as a rule be played back into the room in which they were made since this can lead to an undue emphasis of any acoustic peculiarity. A final balance should also be monitored at low level, at what more closely resembles domestic listening conditions. This is particularly important with vocal work when the clarity of diction may suffer if the end result is only judged at high level.

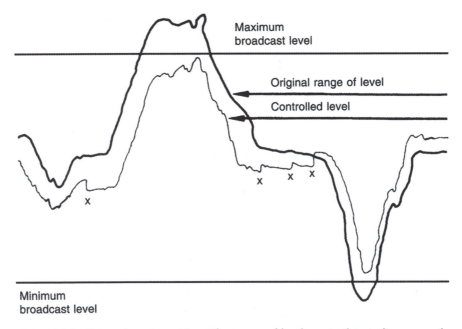

Figure 20.3 Dynamic compression. The range of loudness in the studio can easily rise above the maximum, and fall below the minimum, acceptable to the broadcasting system. By anticipating a loud passage, the main fader is used to reduce the level before it occurs. Similarly, the level prior to a quiet part is increased. Faders generally work in small steps, each one introducing a barely perceptible change – at X

Creation of a synthetic balance

Whereas the reproduction of an existing sound calls for the integration of performance and acoustic using predominantly one main microphone, the creation of a synthetic balance, which in reality exists nowhere, requires the use of many microphones to separate the musical elements in order to 'treat' and reassemble them in a new way. For example, a concert band arrangement calling for a flute solo against a backing from a full brass section would be impossible unless, for the duration of the solo, the flute can be specially favoured at the expense of the brass. Achieving this relies on the ability to separate the flute from everything else in such a way that it can be individually emphasised without affecting the sound from other instruments. The factors involved in achieving this separation are: studio layout, microphone types, source treatment, mixing technique and recording technique.

Studio layout

The physical arrangement of a music group has to satisfy several criteria, some of which may conflict:

1   To achieve 'separation', quiet instruments and singers should not be too close to loud instruments.
2   The spatial arrangement must not inhibit any stereo effect required in the final balance.
3   The conductor or leader must be able to see everyone.
4   The musicians must be able to hear themselves and each other.
5   Certain musicians will want to see other musicians.
6   If an audience is present, the group will need to play to 'the front'.

The producer should not force players to adopt an unusual layout against their will since the standard of performance is likely to suffer. Any difficulty of this kind has to be resolved by suggesting alternative means of meeting the musical requirements. For example, a rhythm section of piano, bass, drums and guitar needs to be tightly grouped together so that they can all see and hear each other. If there is a tendency for them to be picked up on the microphones of adjacent instruments, then working closely to highly directional mics may improve the separation, or they might need to be screened off. If this in turn affects their or other musicians' ability to see or hear, the sight-line can be restored by the use of screens with inset glass panels, and aural communication maintained by means of headphones – or a foldback speaker on the floor near them – fed

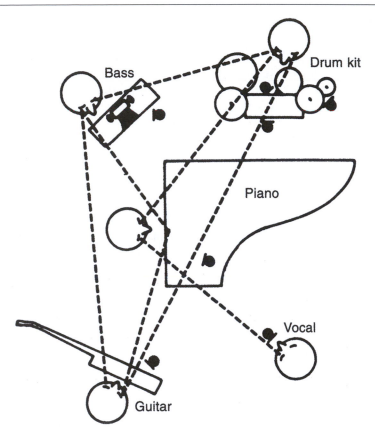

Figure 20.4 Studio layout. The musicians need to see each other, and the microphones placed so as to avoid undue pick-up of other instruments

with whatever mix of other sources they need. It's worth noting that working simply from a floor-plan may inhibit thinking in three dimensions. The solution to a sight-line problem might be to use rostra or staging to raise, say, a brass section – this can improve the separation by keeping it off other mics, which are pointing downwards.

Left to themselves, musicians often adopt the layout usual to them for cabaret or stage work. This may be unsuitable for broadcasting and it is essential that the producer knows the instrumentation detail beforehand in order to make positive suggestions for the studio arrangements. It's obviously important to know whether bass and guitar players have an electric or acoustic instrument. Also how much 'doubling' is anticipated – that is, one musician playing more than one instrument, sometimes within the same musical piece, and whether the players are also to sing. This information is also useful in knowing how many music stands to provide.

It is desirable to avoid undue movement in the studio during recording or transmission but it may be necessary to ask a particular musician to move to another microphone, for example for a solo. It is also usual to ask brass players to move into the microphone slightly when using a mute. This avoids opening the fader to such an extent that the separation might be affected.

Microphones for music

As an aid in achieving separation, the sound recordist's best friend is the directional microphone. The physical layout in the studio is partly arrived at in the light of what microphones are available. A ribbon microphone, with its figure-of-eight directivity pattern, is most useful for its lack of pick-up on the two sides. Used horizontally just above a flute or piano, it is effectively dead to other instruments in the same plane. A cardioid microphone gives an adequate response over its front 180° and will reject sounds arriving at the back – good for covering a string section. A hyper-cardioid microphone will narrow the angle of acceptance, giving an even more useful area of rejection. Condenser or electret types of microphone with variable directivity patterns are valuable for the flexibility they afford in being adjustable – often by remote control – to meet particular needs.

Mics for singers working very close to them will need additional windshield or 'pop-stoppers' to prevent blasts of breath causing audible noise. Even mics with a built-in anti-blast basket, such as the ubiquitous Shure SM58, are not completely pop-proof. Don't be afraid to use a further foam pop shield over the metal gauze. Improving any slight loss of clarity is easily achieved at the mixer with a gentle boost of the treble frequencies.

The presenter of a live music show with an audience will probably need his or her mic to be fed to a public address system as well as to the broadcaster's mixer. Careless PA suffers from the all too familiar howlround problem, even with a mixer desk that provides an anti-howl facility by tuning out the most susceptible frequency. However, a personal radiomic is a great help – provided that the sound mixer can see the stage. PA mixing is best done in the hall itself so that the operator hears what the live audience hears and can take immediate action in the event of the onset of a 'howl'. If this is not possible, a video link is extremely useful in allowing both the PA and the broadcast mixers to see what the performers are doing. The producer should also decide if the presenter needs talkback via an earphone.

The wide operatic platform can be covered by a suspended stereo pair, or a row of three panned, front-of-stage mics. The low 'boundary effect' mic is particularly effective in this context, although prone to footstep sounds.

The closer a microphone is placed to an instrument, the greater the relative balance between it and the sounds from other sources. However, while separation is improved, other effects have to be considered:

1   The pick-up of sound very close to an instrument may be of inferior musical quality. It can be uneven across the frequency range or sound 'rough' due to the reproduction of harmonics not normally heard.
2   There may be an undue emphasis of finger noise or the instrument's mechanical action.
3   The volume of sound may produce overload distortion in the microphone, or in the subsequent electronics.
4   Movement of the player relative to the microphone becomes very critical, causing significant variations in both the quantity and quality of the sound.

Figure 20.5   A high-quality Neumann condenser speech/vocal mic with a fine gauze 'pop-stopper' to prevent breath blasts causing audible pops

5   Close microphone techniques require more microphones and more mixing channels, thus increasing the complexity of the operation and the possibility of error.

The choice and placing of microphones around individual instruments is a matter of skill and judgement by the recording engineer. But no matter how complex the technical operation, the producer must also be aware of other considerations. Technicalities such as the changing of microphones, alterations to the layout, and the running of cables or audio feeds, should not be allowed to interfere unduly with the music-making or human relations aspects. It is possible to be technically so pedantic as to inhibit the performance. If major changes in the studio arrangements are required, it is much better to do them while the musicians are given a break. Under these circumstances the broadcaster has an obvious additional responsibility to safeguard instruments left in the studio.

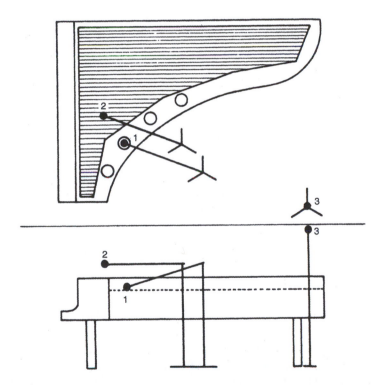

Figure 20.6 Piano balance. Typical microphone positions for different styles of music. 1. Emphasises percussive quality for pop music and jazz. 2. Broader intermediate position for light music. 3. Full piano sound for recitals and other serious or classical music. For an upright piano, the mic is generally placed behind, or even just inside the instrument, at the top end

The ultimate in separation is to do away with the microphone altogether. This is possible with certain electronic instruments where their
output is available via an appropriately terminated lead, as well as acoustically for the benefit of the player and other musicians. The lead should be
connected through a 'direct inject' (DI) box to a normal microphone input
cable. Used particularly in conjunction with electric guitars, the box obvi-
ates the hum, rattles and resonances often associated with the alternative
method, namely placing a microphone in front of the instrument's loudspeaker. Note that certain instruments require specific types of DI box.
Acoustic guitars, for example, often require very high impedance connections if a dramatic loss of clarity and detail is not to be experienced.
Commonly these days the guitar amplifier or 'combo' may also provide a
direct output for recording purposes.

A further point in the use of any electrically powered instrument is
that it should be connected to a studio power socket through a mains
isolating transformer. This will protect the power supply in the event
of a failure within the instrument and will exclude the possibility of an
electrocution accident – so long as the studio equipment and the individual instrument are correctly wired and properly earthed. Faulty wir-
ing or earthing arrangements can cause an accident, for example in the
event of a performer touching simultaneously two 'earths' ('grounds') –
such as an instrument and a microphone stand – which in fact are at a
different potential. While comparatively rare it is a matter that needs
attention in the broadcasting of amateur pop groups using their own
equipment. The broadcasting organisation is legally responsible for the
safety of performers and contributors while on their premises or under
their supervision. Many stations insist that all outside equipment is
safety tested by a qualified Portable Appliance Tester before they can
be plugged in.

Having turned the acoustic sound into electrical output, a number of
treatments may now be applied to individual sources, some of which can
affect the separation. These include frequency control, dynamic control
and echo.

Frequency control

The tone quality of any music source can be altered by the emphasis or
suppression of a given portion of its frequency spectrum. Using a graphic
equaliser, the EQ on the mixer channel, or other frequency discriminating
device, a singer's voice is given added 'presence', and clarity of diction is
improved, by a lift in the frequency response in the octave between 2.8 kHz
and 5.6 kHz. A string section can be 'thickened' and made 'warmer' by an
increase in the lower and middle frequencies, while brass is given greater

Software EQ

Rack-mounted Hardware EQ

Figure 20.7 In addition to EQ on each channel, it may be desirable to apply EQ to the overall sound through a graphic equaliser, selecting the gain for each band of frequencies. In this case there is some emphasis of the top middle with a rolling off of the bass. EQ can be applied via software or hardware, each device giving a graphical representation of how the frequencies are being adjusted

'attack' and a sharper edge by some 'top lift'. It should be noted, however, that in effectively making the microphone for the brass section more sensitive to the higher frequencies, spill-over from the cymbals is likely to increase and separation in this direction is therefore reduced. Fairly savage control is often applied to jazz or rhythm piano to increase its percussive quality. It is also useful on a one-microphone balance, particularly on an outside broadcast, to reduce any resonance or other acoustic effect inherent in the hall.

Dynamic control

This can be applied automatically by inserting a compressor/limiter device in the individual microphone chain. On a digital desk this facility, together with a noise gate, is likely to be found on each channel. Once set, the level obtained from any given source will remain constant – quiet passages remain audible, loud parts do not overload. It becomes impossible for the flute to be swamped by the brass. Because of the economics of popular music, commercial recording companies have attained a high degree of sophistication in the use of dynamic control. This is unlikely to be reached by most broadcasters who are not able to go to such lengths in recording their music. However, devices of this type can save studio time and their progressive application is likely. Variations include a 'voice-over' facility that enables one source to take precedence over another. Originally intended for DJs, its obvious use is in relation to singers and other vocalists but it can be applied to any source relative to any other. Digital desks can be programmed to store, in a memory, the changing mix required throughout a performance. In the final take the basics can be left to look after themselves.

Echo

The echo effect, more correctly described as reverberation, used to be generated in an empty room but today's digital effects equipment is so versatile that any kind of echo can be created electronically through a number of controls and presets which affect the duration, type, mix and decay characteristic of the reverberation. Modern desks not only offer echo and delay, but facilities may also include dynamic compression, equalisation and noise reduction – even of a specific frequency – making it especially useful in controlling the howl-round of PA. A digital mixer might have all these facilities on every channel. While it's not normal to add reverberation to a band's rhythm section, in order to keep it 'tight', a useful tip is to increase the echo return channel when a source – for example, the string section of a band – is itself being increased. This gives the sound of 'soaring strings' for the duration of this particular passage. Especially in light music, the echo channel is not set and fixed but is varied as the arrangement demands. Echo variation may also be appropriate for a solo piano and for a singer. Incidentally, when echo is added to singers' voices it is wise to let them listen on headphones, since they will almost certainly want to adjust their phrasing to suit the new reverberant sound.

An alternative to a digital echo device is the mechanical plate or spring, made to vibrate by the signal from the desk. These vibrations travel through the device to a transducer which converts them back into electrical energy, returning as echo to the mixing desk. Equivalent to a two-dimensional room, the reverberation time is adjustable depending on the mechanical damping applied. Some such devices are in portable form and are often preferred for 'retro' OB applications.

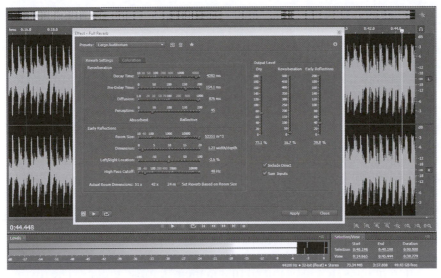

Software Reverberation

Rack-mounted Hardware Reverberation

Figure 20.8 Artificial reverberation. In addition to applying reverberation to individual channels, it is often necessary to add it to the overall sound. Here it is added to simulate a large concert hall with a general reverberation time of 2.3 seconds. It is also set for the early reflections – from the nearest wall – to take 1.2 seconds

Channel delay

The clever way of overcoming the effect of altering the perspective when opening up an individual 'spot' mic is to delay the output of the mic by the same amount as the sound of that source coming through the main mic. For example, if an orchestral clarinet is 40 feet from the main mic, the time taken for its sound to reach that microphone is some 40 milliseconds (sound travels at 1,100 feet/sec or about 1 foot/ms). Therefore, with a delay of 40 m/s on the close clarinet mic, its sound is heard as if it were still 40 feet away. On opening up its own mic, the perceived effect is that the instrument is 'louder but not closer'. This is yet another facility that a digital mixer may offer on every channel.

Mixing technique

Before mixing a multi-microphone balance, each channel should be checked for delivering its correct source, with adequate separation from its neighbours, having the desired amount of 'treatment', and producing a clear distortion-free sound. It is a sensible procedure to physically label the channel faders with the appropriate source information – solo vocal, piano, lead guitar and so on. For a stereo balance it is necessary to be clear about the placing of each instrument in the stereo picture. Stereo microphones are physically adjusted to spread their output across the required width of the picture, and the mono microphone 'pan-pots' set so that their placing coincides with that provided by any stereo microphone covering the same source. It is possible to create a stereo balance using only monophonic microphones by 'spotting' them across the stereo picture, but it is an advantage to have at least one genuine stereo channel, even if it is only the echo source return.

Microphones should be mixed first in their 'family' groups. The rhythm section, backing vocals, or brass is individually mixed to obtain an internal balance. The sections are then added to each other to obtain an overall mix. Music desks allow their channels to be grouped together so that sections can be balanced relative to each other by the operation of their 'group' fader without disturbing the individual channel faders. Successful mixing requires a logical progression, for if all the faders are opened to begin with, the resulting confusion can be such that it is very difficult to identify problems. It is important to arrive at a trial balance fairly quickly since the ear will rapidly become accustomed to almost anything. The overall control is then adjusted so that the maximum level does not exceed the permitted limit.

The prime requirement for a satisfying music balance is that there should be a proper relationship between melody, harmony and rhythm. Since the group is not internally balanced, listening on the studio floor will not indicate the required result. Only the 'hands-on' person manipulating the faders and listening to the loudspeaker is in a position to arrive at a final result. The melody will almost certainly be passed around the group, and this will need very precise operation of the faders. If the lead guitar takes the tune from an up-beat, that fader must be opened further *from that point* – no earlier otherwise the perspective will alter. At the end of that section the fader must be returned to the normal position to avoid loss of separation. As is noted above, it is probable that alterations to a source making use of echo will need a corresponding and simultaneous variation in the echo return channel. It is a considerable help, therefore, to have an elementary music score or 'lead sheet' which indicates which instrument has the melody at any one time. If this is not provided by the musical

Figure 20.9 A lead sheet. This gives the mixer operator precise information about when particular instruments or sections take the lead, and their mics need to be opened up – and taken back again

director, the producer should ensure that notes are taken at the rehearsal –
for example, how many choruses there are of a song, at what point the
trumpets put their mutes in, and perhaps more importantly, when they take
them out again, when the singer's microphone has to be 'live' and so on.
A professional music balance operator quickly develops a flair for reacting
to the unexpected but, as with most aspects of broadcasting, some basic
preparation is only sensible.

It is important that faders that are opened to accentuate a particular
instrument are afterwards returned to their normal setting. Unless this is
done it will be found that all the faders are gradually becoming further and
further open with a compensating reduction of the overall control. This is
clearly counter-productive and leads to a restriction in flexibility as the
channels run out of 'headroom'.

As with all music balance work, the mixing must not take place only
under conditions of high-level monitoring. The purpose of having the loudspeaker turned well up is so that even small blemishes can be detected
and corrected. But from time to time the listening level must be reduced
to domestic proportions to check particularly the balance of vocals against
the backing, and the acceptable level of applause. The ear has a logarithmic
response to volume and is far from being equally sensitive to all the frequencies of the musical spectrum. This means that the perceived relationship
between the loudness of different frequencies heard at a high level is not the
same as when heard at a lower level. Unless the loudspeaker is used with
domestic listening in mind, as well as for professional monitoring, there will
be a tendency as far as the listener at home is concerned to underbalance the
echo, the bass and the extreme top frequencies. For this reason many professional studios check the final mix on a simple domestic loudspeaker.

Recording technique

The successful handling of a music session requires the cooperation of
everyone involved. There are several ways of actually getting the material
recorded and the procedure to be adopted should be agreed at the outset.

The first method is obviously to treat a recording in the same way as a
live transmission – start when the red light goes on and continue until it
goes out. This will be the procedure at a public concert when, for example,
no retakes are possible.

Second, a studio session with an audience present. Minor faults in the
performance or mixing will be allowed to pass but the producer must
decide when something has occurred that necessitates a retake. The audience may be totally unaware of the problem but someone will have lightly
and briefly to explain the existence of 'a gremlin' and quickly brief the
musical director on the need to retake from point A to point B.

Third, without an audience the musicians can agree to use the time as thought best. This may be to rehearse all the material and then to record it, or to rehearse and record each individual number. In either case, breaks can be taken as and when decided by the producer in consultation with the musical director. A music producer must of course be totally familiar with any agreements with the performers' unions that are likely to affect how a session is run.

Fourth, to record the most material in the shortest time or to accommodate special requirements, the music may be 'multi-tracked'. This means that instead of immediately arriving at a final mix, the individual instruments or groups of channels are recorded as separate tracks. Digital recording in this way offers a wide range of signal manipulation, monitoring, auto-location and editing. Some tracks can be set to play while others record, so enabling musicians to record their contribution without the necessity of having all the performers present simultaneously. It also permits 'doubling', that is, the performing of more than one role in the same piece of music. This double-tracking is most often used for an instrumental group who also sing. The musicians first play the music – either conventionally recorded or multi-tracked – and when this is played back to them on headphones, they add the vocal parts. These are recorded on separate tracks or mixed with the music tracks to form a final recording. This process can be repeated to enable singers to sing with themselves as many times as is required to create the desired sound. An added advantage

Figure 20.10 A 4-track digital recorder which records to an SD card in a choice of WAV or MP3 formats

of multi-tracking is that any mistake does not spoil the whole performance – a retake may only be required to correct the individual error. The
subsequent 'mixing down' or 'reduction' of many tracks is a relatively
straightforward process, giving the producer the opportunity to tidy up the
performance. Many is the recording session when it was said – 'we'll sort
it out in post-production'.

Since recording at different times represents total separation, multi-
tracking techniques can solve other problems. For instance the physical
separation or screening of loud instruments such as a drum kit can be
avoided by recording the rhythm section separately – usually first, in order
to keep subsequent performers in sync. It is an especially useful technique
to apply to a music recording which has to be done in a small non-purpose-
built studio, and even working in four tracks can help to produce a more
professional sound.

Figure 20.11 A multi-tracked computer-based recording studio. Recording,
arranging and post-production effects can all be controlled by the software. In
addition to recording real acoustic instruments, the on-screen mixer can also control
and mix MIDI tracks which are recorded via the MIDI output of the digital keyboard.
The iPad, which can just be seen in the top right corner, is used to display the music
and lyrics of the song during the session

A digital audio workstation (DAW) can act as the heart of a music studio and is similar to a multi-track recorder and mixer, but working in the digital domain on a personal computer. The DAW has features that can only be performed in the digital domain, and in addition to recording audio via microphones can also use music in data form from MIDI sources such as keyboards and sequencers.

Production points

Essentially the producer's job is to obtain a satisfactory end product within the time available, and above all to avoid running into overtime which with professional musicians is expensive.

Neither do musicians appreciate an oppressively hard taskmaster so that the session ends with time to spare, and everyone is relieved to get out. The producer must remember that all artists need encouragement and that 'production' has a positive duty in helping them give the best performance possible.

Finally, some general points.

- Avoid using the talkback into the studio to make disparaging comments about individual performers. Instead, direct all remarks to the conductor or group leader via a special headphone feed, or better still by going into the studio for a personal talk.
- Often acting as the mediator between the performers and the technical staff, explain delays and agree new timescales.
- Anticipate the need for breaks and rest periods. During these times it is often sensible to invite the MD or group leader into the control cubicle to hear the recorded balance and discuss any problems.
- Avoid displaying a musical knowledge which, while attempting to impress, is based on uncertain foundations. 'Can we have a shade more "arco" on the trombones?' is a guaranteed credibility loser.
- Make careful rehearsal notes for the recording or transmission, and take details of the final timings and any retakes for subsequent editing.
- When an audience is present, decide how it is best handled and what kind of introduction and warm-up is appropriate – and get them (and the musicians) to switch off their cell phones. This is often the producer's job.
- Resolve the various conflicts that arise – whether the air conditioning or heating is working properly, whether the lights are too bright or whether the banging in another part of the building can be stopped. Remember other needs – drinking water, lavatories, car parking times.

- When a retake is required, make a positive decision and communicate it quickly to the musical director. The producer exercises the broad-caster's normal editorial judgement and is responsible for the quality of the final programme.
- Ensure that the musicians are paid – those performers who signed a contract and made the music and, through the appropriate agencies, those who wrote, arranged, and published the work still in copyright.
- After the session say thank you – to everyone.

21

Drama – principles

We all love stories and the radio medium has a long and distinguished history of turning thoughts, words and actions into satisfying pictures within the listener's mind by using the techniques of drama. But there is no need for the producer to think only in terms of the Shakespeare play – the principles of radio drama apply to the well-made 30-second commercial, a programme trail, dramatised reading, five-minute serial or two-minute teaching point in a programme for schools. The size and scope of the pictures created are limited only by the minds that devise and interpret them – for example, the piece for young children on the website. Since the medium in its relationship to drama is unique, any radio service is the poorer for not attempting to work in this area.

As an illustration of the effective simplicity of the use of sound alone, listen now to the celebrated example by Stan Freberg who gave it as part of his argument for selling advertising time on radio.

MAN:	Radio? Why should I advertise on radio? There's nothing to look at . . . no pictures.
GUY:	Listen, you can do things on radio you couldn't possibly do on TV.
MAN:	That'll be the day.
GUY:	Ah huh. All right, watch this. (Clears throat) OK, people, now when I give you the cue, I want the 700-foot mountain of whipped cream to roll into Lake Michigan which has been drained and filled with hot chocolate. Then the Royal Canadian Air Force will fly overhead towing the 10-ton maraschino cherry which will be dropped into the whipped cream, to the cheering of 25,000 extras. All right . . . cue the mountain . . .
SOUND:	Groaning and creaking of mountain into big splash.
GUY:	Cue the Air Force!
SOUND:	Drone of many planes.
GUY:	Cue the maraschino cherry . . .

```
SOUND:    Whistle of bomb into bloop! of cherry hitting whipped
          cream.
GUY:      Okay, twenty-five thousand cheering extras . . .
SOUND:    Roar of mighty crowd. Sound builds up and cuts off sharp!
GUY:      Now . . . you wanta try that on television?
MAN:      Well . . .
GUY:      You see . . . radio is a very special medium, because it
          stretches the imagination.
MAN:      Doesn't television stretch the imagination?
GUY:      Up to 21 inches, yes.
```

(Courtesy of Freberg Ltd)

We like stories, partly because a story can offer a framework for the understanding – or at least an interpretation – of life's events. Often, a mirror in which we can see ourselves – our actions, motives and faults – and the outcomes and results can contribute to our own learning. Drama is essentially about conflict and resolution, relationships and feelings and people being motivated by them, both driving and driven by events. What happens should be credible, the people believable, and the ending have a sense of logic however unusual and curious, so that the listener does not feel cheated or let down.

The aim with all dramatic writing is for the original ideas to be recreated in the listener's mind and, since the end result occurs purely within the imagination, there are few limitations of size, reality, place, mood, time or speed of transition. Unlike the visual arts, where the scenery is provided directly, the listener to radio supplies personal mental images in response to the information given. If the 'signposts' are too few or of the wrong kind, the listener becomes disorientated and cannot follow what is happening. If there are too many, the result is likely to be obvious – 'cheesy' and 'corny'. Neither will satisfy. The writer must therefore be especially sensitive to how the audience is likely to react – and since the individual images may stem largely from personal experience, of which the writer of course knows nothing, this is not easy. But it is the ageless art of the storyteller – saying enough to allow listeners to follow the thread but not so much that they do not want to know what is to happen next or cannot make their own contribution.

The writer must have a thorough understanding of the medium and the production process, while the producer needs a firm grasp of the writing requirements. If they are not one and the same person, there must be a strong collaboration. There can be no isolation, but if there is to be a dividing line, let the writer put everything down knowing how it is to sound, while the producer turns this into the reality of a broadcast knowing how it is to be 'seen' in the mind's eye.

The component parts with which both writer and producer are working are speech, music, sound effects and silence.

Figure 21.1   Drama rehearsal in a TWR studio in Johannesburg

Adapting for radio

Rewriting an existing work for radio sets a special kind of challenge. Staying faithful to the original so as not to upset those who already know and love the book or play, yet conveying it in this different medium, and probably compressing it in time, requires a distinctive writing skill. Translation from another language is generally easier than working in the same language since it's then necessary to use entirely new forms of speech, whereas a great question in adaptation concerns the need to use the same words and phrases as the original. But how about keeping to the same style and pace? What is to be sacrificed if a lengthy original is to be a half-hour play? Radio works well with speech in dialogue form, but if there is little conversation, will a narrator do? Radio also effectively conveys 'internal' thoughts, thinking and talking to oneself. What does the original have of 'mood'? The essential character should still be present – torment, depression, pressure, conflict, excitement, expectation, achievement or resolution, and above all the relationships – their strengths, weaknesses and doubts. The adapting writer should care for the original while analysing it, and preserve its essential features in the new medium. If the work is still in copyright the original publisher – who actually owns the work – will need to be consulted for permission to adapt, and may well have views regarding a radio treatment.

The idea

Before committing anything to paper, it is essential to think through the basic ideas of plot and form – once these are decided, a great deal follows naturally. The first question is to do with the material's suitability for the target audience, the second with its technical feasibility.

Assuming that the writer is starting from scratch and not adapting an existing work, what is the broad intention? Is it to make people laugh, to comment on or explain a contemporary situation, to convey a message, to tell a story, to entertain? How can the writer best enable the listener to 'connect' with this intention? Is it by identifying with one of the characters? Should the basic situation be one with which the listener can easily relate?

The second point at this initial stage is to know whether the play has to be written within certain technical or cost limitations. To do something simple and well is preferable to failing with something complicated. There seems little point in writing a play that calls for six simultaneous sound effects, echo, a variety of acoustics, distorted voice-over, and a crowd chanting specific lines of script, if the studio facilities or staff are not able to meet these demands. Of course, with ingenuity even a simple studio can provide most if not all these devices. But the most crucial factor is often simply a shortage of time. There may be limitations too in the capabilities of the talent available. The writer, for example, should be wary of creating a part that is emotionally exceptionally demanding only to hear it inadequately performed. Writing for the amateur or child actor can be very rewarding in the surprises which their creative flowering may bring, but it can also be frustrating if you automatically transfer into the script the demands and standards of the professional stage.

The best brief advice is that writers should write about things they know about, and about characters they care about.

Thus the writer must know at the outset how to tailor the writing for the medium, in what form to put over a particular message or effect, how the audience might relate to the material, and whether this is technically and financially possible. From here there are three possible starting points – the story, the setting or the characters.

Story construction

The simplest way of telling a story is to:

1   explain the situation
2   introduce 'conflict'
3   develop the action
4   resolve the conflict.

Of course there may be intriguing complications, mystery and sub-plots, twists and surprises at several levels. This is an essential aspect of the multi-strand soap opera where the listener is invited to relate to several different characters. In an absorbing story there will be small personal struggles to be resolved as well as the big issues. However, the essence of the thing is to find out 'what happens in the end'. Who committed the crime? Were the lovers reunited? Did the cavalry arrive in time? The element that tends to interest us most is the resolution of conflict and since this comes towards the end, there should be no problem of maintaining interest once into the 'rising action' of the play. And in the final scene it is not necessary to tie up all the loose ends – to dot every 'i' and cross every 't' in a neat and tidy conclusion. Life seldom works that way. It is often better to have something unsaid, leaving the listener still with a question, an issue or a motivation to think about. Parables are stories that deliberately do not go straight from A to B, but take a parabolic route leaving the hearer to work out the implications of it all. This is one of the fascinations of the story form.

In radio, scenes can be much shorter than in the theatre, and intercutting between different situations is a simple matter of keeping the listener informed about where we are at any one time. This ability to move quickly in terms of location should be used positively to achieve a variety and contrast which itself adds interest. The impact of a scene involving a group of fear-stricken people faced with impending disaster is heightened by its direct juxtaposition with another, but related, group unaware of the danger. If the rate of intercutting becomes progressively faster and the scenes shorter, the pace of the play increases. This sense of acceleration, or at least of movement, may be in the plot itself, but a writer can inject greater excitement or tension simply in the handling of scene length and in the relation of one scene to another. Thus the overall shape of the play may be a steady development of its progress, heightening and increasing. Or it may revolve around the stop-go tempo of successive components. Interest through contrast can be obtained by a variety of means, for example:

1   change of pace: fast/slow action, noisy/quiet locations, long/short scenes;
2   change of mood: tense/relaxed atmosphere, angry/happy, tragic/dispassionate, good/bad, storm/calm, right/wrong;
3   change of place: indoors/open air, crowded/deserted, opulent/poverty-stricken, town/country, different geography or accents, earthly/heavenly;
4   change of time: present/past, flashback/imagined future.

The radio writer is concerned with images created by sound alone. If the effect wanted is of colour and mood, then this must be painted with the words and music used, and through locations that are aurally evocative.

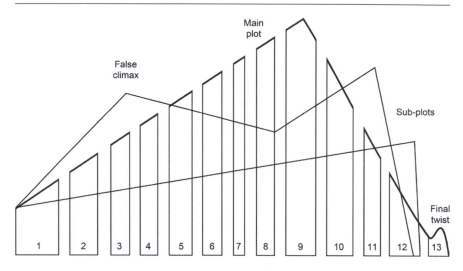

Figure 21.2   A plot diagram to illustrate the shortening scene length as the action rises to its climax

1–2	Introduction, setting and context, characterisation established;
3–5	conflict, events arising from characters reacting to the situation;
6–8	rising action, sub-plot complication, suspense;
9	maximum tension, crisis, climax;
10–11	falling action, resolution;
12–13	denouement, surprise, twist.

On a personal note, to illustrate the impact of contrast within a play, the authors recall a moment from a drama on the life of Christ. The violence and anger of the crowd demanding his execution is progressively increased, the shouting grows more vehement. Then we hear the Roman soldiers, the hammering of nails and the agony of the crucifixion. Human clamour gives way to a deeper, darker sense of tragedy and doom. Christ's last words are uttered, a crash of thunder through to a climax of discordant music gradually subsiding, quietening to silence. A pause. Then slowly – birdsong.

How wrong it would have been to spoil that contrast by using the narrator, as elsewhere in the play.

The setting

Situation comedy drama often begins with a setting – an office, prison, hospital, shop, or even a radio station – which then becomes animated with characters. The storyline comes later, driven by the circumstances, generally a series of predicaments in which the characters find themselves. Fortunately radio can provide almost any setting at will – a royal household in ancient Egypt, a space capsule journeying to a distant planet, or a ranch house in outback Australia. The setting, plus one or two of the principal characters, may be the key theme holding a series of plays – or advertisements – together. It is important that the time and place are well researched, especially if it is historical, so that credibility is maintained for those listeners who are expert in the particular situation.

Within a play the setting can obviously vary considerably and one of the devices used to create interest is to have a strong contrast of locale in adjacent scenes – for example, to move from an opulent modern office peopled by senior managers to a struggling rural hospital affected by their decisions. Changes of location are very effective when run in parallel with changes in disposition or mood.

Characterisation

One of Britain's best-known advisers in this area, Bart Gavigan – often referred to as a 'script doctor' – says that there are three questions to be answered for a compelling story:

1   Who is the hero – or heroine?
2   What does he, or she, want?
3   Why should I care?

This last question emphasises the need for at least main characters to develop a rapport with the listener. They must be more than thinly described cardboard cutouts – they should appear as credible people with whom the listener can identify and whose cause the listener can about. The writer has to create a play – or 30-second advertisement – so that the human motivation and behaviour is familiar. This does not mean that the setting has to be the same as the domestic circumstances of the audience – far from it. For, whatever the setting, the listener will meet people he or she can recognise – fallible, courageous, argumentative, greedy, fearful, compassionate, lazy and so on. Characterisation is a key ingredient and many writers find it important to sketch out a pen portrait of each character. This helps to stabilise them as people and it's easier to give them convincing dialogue. Here are some headings:

- their age, sex, where they live and how they talk;
- height, weight, colouring and general appearance;
- their social values, sense of status, beliefs of right and wrong;
- the car they drive, what they eat and drink, the clothes they like, their wealth – or not;
- their family connection, friends – and enemies;
- the jokes they make – or not, how trusting they are, how perceptive;
- their moods, orientations, preferences and dislikes.

Characters have faults, they fall apart in crises; they will have internal contradictions of values and behaviour, and appear illogical because their words and deeds won't always coincide. They will reveal themselves in what they say, but even more in what they do. Someone may profess honesty but fiddle their income tax. One of the necessary tensions in compelling drama is the inner conflict that exists in human beings – the inconsistencies between what I want to do, what I ought to do and what I actually do. Saints will have their failings and even the worst sinners may have their redeeming features under certain circumstances. Characters that are genuinely engaging have credible substance. They convey real human complexity and the writer cannot accurately portray them until he or she knows them. When characterisation is fully in place the writer may let the characters almost write themselves, since they know how they are likely to react to a set of events driven by the story. Writers and producers should tell actors as much as they can about the characters – their personality, typical behaviour and disposition.

Dialogue

'Look out, he's got a gun.' Lines like this, unnecessary in film, television or theatre where the audience can see that he has a gun, are essential in radio as a means of conveying information. The difficulty is that such 'point' lines can so easily sound contrived and false. All speech must be the natural colloquial talk of the character by whom it is uttered. In reproducing a contemporary situation, a writer can do no better than to take a notebook to the marketplace, restaurant or party and observe what people actually say, and their manner of saying it. Listen carefully to the talk of shoppers, eavesdrop on the conversation on the bus. It is the stuff of your reality. And amid the talk, silence can be used to heighten a sense of tension or expectation. Overlapping voices convey anger, passion, excitement or crisis. Not only do radio characters say what they are doing, but people reveal their inner thoughts by helpfully thinking aloud, or saying their words as they write a letter. These are devices for the medium and should be used with subtlety if the result is to feel true to life.

Producers should beware writers who preface a scene with stage directions: 'The scene is set in a lonely castle in the Scottish Highlands. A fire is roaring in the grate. Outside a storm is brewing. The Laird and his visitor enter.'

Such 'picture setting' designed for the reader rather than the listener should be crossed out and the dialogue considered in isolation. If the words themselves create the same scene, the directions are superfluous; if not, the dialogue is faulty:

```
LAIRD:      (Approaching) Come in, come in. It looks like a storm is
            on the way.

VISITOR:    (Approaching) Thank you. I'm afraid you're right. It was
            starting to rain on the last few miles of my journey.

LAIRD:      Well come and warm yourself by the fire; we don't get
            many visitors.

VISITOR:    You are a bit off the beaten track, but since I was in
            the Highlands and've always been fascinated by castles
            I thought I would call in – I hope you don't mind.

LAIRD:      Not at all, I get a bit lonely by myself.

VISITOR:    (Rubbing hands) Ah, that's better. This is a lovely room –
            is this oak panelling as old as it looks?

            etc., etc.
```

The producer is able to add considerably to such a scene in its casting, in the voices used – for example, in the age and accents of the two characters, and whether the mood is jovial or sinister.

In addition to visual information, character and plot, the dialogue must remind us from time to time of who is speaking to whom. Anyone 'present' must either be given a line or be referred to, so that they can be included in the listener's mental picture:

```
ANDREW:     Look, John, I know I said I wouldn't mention it but . . .
            well something's happened I think you should know about.

JOHN:       What is it, Andrew? What's happened?
```

The use of names within the dialogue is particularly important at the beginning of the scene.

Characters should also refer to the situation not within the immediate picture so that the listener's imagination is equipped with all the relevant information:

```
ROBINS:     And how precisely do you propose to escape? There are
            guards right outside the door, and more at the entrance
            of the block. Even if we got outside, there's a barbed
            wire fence two storeys high – and it's all patrolled by
            dogs. I've heard them. I tell you there isn't a chance.

JONES:      But aren't you forgetting something – something rather
            obvious?
```

An obvious point which the writer does not forget is that radio is not only blind, but unless the drama is in stereo, it is half-deaf as well. Movement and distance have to be indicated, either in the acoustic or other production technique, or in the dialogue. Here are three examples:

```
(Off-mic) I think I've found it, come over here.

Look, there they are – down there on the beach. They must be half a
mile away by now.

(Softly) I've often thought of it like this – really close to you.
```

We shall return later to the question of creating the effects of perspective in the studio.

To achieve a flow to the play, consecutive scenes can be made to link one into the other. The dialogue at the end of one scene points forward to the next:

```
VOICE 1:    Well, I'll see you on Friday - and remember to bring the
            stuff with you.
VOICE 2:    Don't worry, I'll be there.
VOICE 1:    Down by the river then, at 8 o'clock - and mind you're
            not late.
```

If this is then followed by the sound of water, we can assume that the action has moved forward to the Friday and that we are down by the river. The actual scene change is most often through a fade-out of the last line – a line incidentally like the second half of the one in this last example, words we can afford to miss. There is a moment's pause, and a fade-up on the first line of the new scene. Other methods are by direct cutting without fades, or possibly through a music link. The use of a narrator will almost always overcome difficulties of transition so long as the script avoids clichés of the 'meanwhile, back at the ranch' type.

A narrator is particularly useful in explaining a large amount of background information which might be unduly tedious in conversational form or where considerable compressions have to be made, for example in adapting a book as a radio play. In these circumstances the narrator can be used to help preserve the style and flavour of the original, especially in those parts which have a good deal of exposition and description but little action:

```
NARRATOR:   Shortly afterwards, Betty died and John, now destitute
            and friendless, was forced to beg on the streets
            in order to keep himself from starvation. Then one
            afternoon, ragged and nearing desperation, he was
            recognised by an old friend.
```

When in doubt, the experienced writer will almost certainly follow the simplest course, remembering that the listener will appreciate most what can be readily understood.

Drama from personal story

Dr Bill Dorris, a Lecturer in Communications at Dublin City University, encourages his students to base drama writing on their real experience. As he explains:

Pick a topic which will trigger strong memories, memories that will evoke feelings and images. Make it a topic that would have grabbed your attention when you experienced it – the night before surgery, your first step dance competition, or maybe just that summer trip out to your uncle's farm. The topic can be commonplace, your first competition, a family vacation, etc. The key thing is that the event triggered emotion and grabbed your attention at the time. Then jot down whatever bits you remember. Don't try to edit or organise or make a story out of them. Just jot them down. Remember where you were, what you were doing, who was doing it with you, what scared you, surprised you, made you feel sad, what was rattling round your head. Cheat if it helps – pull out the map, journal, photos, hum the tune, phone a friend, Google the web . . .

The example used here is called 'Rattlesnakes' – produced by two of my students, Kim Cahill and Fiona McArdle. Here we consider only the first, third and final drafts, although typically in such work there will be four or five rewrites from original to final draft.

First draft

I was seven and went with my family to visit my Aunt and Uncle at their dream farmhouse in southern Canada.

Wraparound porches and many windows with beautiful rural countryside views in any direction were some of the old house's main features. So were the snakes.

We'd sit out in the long grass toasting marshmallows on a crackling fire, whilst Uncle Mike and Dad skinned the trout we had caught earlier that day, the nervous jump when one of the many green little grass snakes would venture a little too near for comfort .What would those little snakes think of my Uncle Mike skinning one of the larger members of the snake family . . . a rattlesnake?!

I was so nervous holding the skin, really believing it would come back to life! I was fascinated by the crisp little rattle of its tail and had plenty of amusement rattling the skin around chasing after my little brother!

I had my first cup of coffee. My Aunt Helen and I were up early and watched the sun rise on the old rocking chairs on the front porch. In imitation of her I curled up with this giant cup of strong smelling coffee.

After you've got several of these bits, eight or ten of them, read them over and see what they suggest for possible 'stories'. What potential conflicts are there that you could rough a 'story' around? In the above draft, for example there's 'the first cup of coffee', and 'the rattlesnake's – and the trout's? – revenge'. But the one that leaps out as having the most potential and most already written and suggested, including a punchline – 'rattling my little brother' – is 'snakes and rattlers'.

So the next step is to rough out such a story and organise, elaborate, even make up bits to give some surge and flow in the tension around the core conflict, i.e. kids' fears and fun with snakes – all set in that memory of 'skinning trout' and 'toasting marshmallows' over the 'crackling campfire' just off those 'wraparound porches'.

Most people have a tough time with this at first, that is with making a story where there was none. It helps to bring in someone who's done this before and who's outside the original experience. Even then in the early drafts there is often an emphasis on descriptive narrative, the where, when, who, etc., rather than what's needed, which is to build the story around the conflict and where it leads.

Third draft

One of my earlier memories of travelling is staying in my aunt and uncle's old farm house out in southern Canada. It had wraparound porches with grand windows looking out over the beautiful rural countryside.

After a day's fishing, we'd sit out there off the porch, skin the trout and cook it over the campfire . . . baked potatoes, corn on the cob, roasting marshmallows and . . .

'AAAh! my LEG!! It's on me! Get it off! Get it off!!'

Suddenly I'm panicking, jumping, crying . . . it's one of those little green grass snakes. They're all laughing, and my little brother's the loudest.

At least it wasn't one of those prairie rattlers, the ones my Uncle Mike skins and hangs on the barn door. They make a crisp little rattle when you shake them in your hand. Like when you're walking slowly across the porch . . . up behind my brother!

Now we've got the makings of a short evocative piece of drama. If you compare this draft with the final one below, the main changes will reflect the guidelines that need to be considered in getting from here to the final draft. Briefly they are:

1 No general narrative – 'one of my earliest memories . . . ' in the third draft. Scene setting leads straight into the conflict – 'soon there'll be the marshmallows . . . ' in final draft.

2 Told from the point of view of the narrator as she is experiencing the story, with a flux of information and emotion in her words after she screams in final draft.

3 No asides. Everything is written so as to pull us into the drama through the use of dialogue and active verbs. There are changes between drafts in how

the information about how Uncle Mike's rattlesnake skins are used as 'little warning bells', etc.

4   All prose and Fx are evocative, informative and distinctive, and pared down to the bare minimum. All 'summary adjectives' like 'beautiful', 'old', and 'rural', and redundant information like 'panicking . . .' are cut. Only one extremely evocative Fx is used.

<u>Final draft</u>

Wraparound porches, tall rattling windows, clear summer wheat-fields . . . we're sitting round the campfire . . . just off the porch of my aunt and uncle's old farmhouse out in southern Canada . . . grandpa's skinned the trout, potatoes, corn on the cob. Soon there'll be the marshmallows . . .
   'AAAAh!! My leg! My leg! Get it off me! Get it off me! off me!'

They're all laughing. Brian, my little brother . . . it's just a little green, wriggly, harmless garden snake. He's laughing. It's frantic. It's . . .

   <u>*FX – rattlesnake sound low, fade out at 'warning bells', then up again at 'fitting in'*</u>

. . . not one of my Uncle Mike's prairie rattlesnake skins, hanging from the barn door. Those rattlesnakes with their little warning bells, used to let you know . . . those little warning bells fitting in your hand when you're sneaking across the porch . . . just behind . . . my little brother . . .

'AAAAh!!'

Truth vs. drama

In 'Rattlesnakes', memory was used like a book might be for developing a film script – as source material for a dramatic adaptation. The focus here is on creating radio that grabs the listener, not on communicating an accurate account of what actually happened. What is important here concerning 'truth' is that the core emotional experience is communicated – that sense of kids on vacation, fishing and marshmallowing round the campfire, rattlesnaking each other right off the porch.

In other applications – interviewing or commentary – the same issues about the balance of 'truth' vs. drama will inevitably arise. The outcome in each case will obviously depend on what 'truth' you are trying to communicate.

That, from Bill Dorris and his students, was a different approach to drama, more like poetry perhaps than the regular play. It is a process that can work well writing dialogue for a factual documentary where no actual record exists of what was said.

Script layout

Following the normal standard of scripts intended for broadcast use, the page should be typed on one side only to minimise handling noise, the paper being of a firm 'non-rustle' type. The lines should be triple-spaced to allow room for alterations and actors' notes, and each speech numbered for easy reference. Directions, or details of sound effects and music, should be bracketed, underlined or in capitals, so that they stand out clearly from the dialogue. The reproduction of scripts should be absolutely clear and there should be plenty of copies so that spares are available.

An example of page layout is shown below.

```
1   (INDOOR ACOUSTIC)

2   BRADY:        Why isn't Harris here yet? – You people at the
                  Foreign Office seem to think everyone has got time
                  to waste.

3   SALMON:       I don't know, Colonel, it's not like him to be
                  late.

4   BRADY:        Well it's damned inconsiderate, I've a good mind to
                  . . .

5   Fx:                       KNOCK AT DOOR

6   SALMON:       (RELIEVED) That's probably him. (GOING OFF)
                  I'll let him in.

7   Fx:                       DOOR OPENS OFF

8   HARRIS:       (OFF) Hello, John.

9   SALMON:       (OFF) Thank goodness you've arrived. We've been
                  waiting some time. (APPROACH) Colonel Brady I
                  don't think you've met Nigel Harris. He's our
                  representative . . .

10  BRADY:        (INTERRUPTING) I know perfectly well who he is,
                  what I want to know is where he's been.

11  HARRIS:       Well, I've been trying to get us out of trouble.
                  It's bad news I'm afraid. The money we had ready
                  for the deal has gone, and Holden has disappeared.

12  BRADY:        This is preposterous! Are you suggesting he's taken
                  it?

13  HARRIS:       I'm not suggesting anything, Colonel, but we know
                  that last night he was at Victoria Station – and
                  bought a ticket for Marseilles.

14  SALMON:       Marseilles? By train?

15  HARRIS:       By train. At this moment I should think he, and the
                  money, are halfway across Europe.

16  Fx:           TRAIN WHISTLE APPROACH AND ROAR OF TRAIN PASSING.

17  Fx:           CROSSFADE TO INTERIOR, TRAIN RHYTHM CONTINUES
                  UNDERNEATH.

18  STEWARD:      (APPROACHING) Take your last sitting for lunch
                  please.
                  (CLOSER) Last sitting for lunch. Merci, Madame.
```

1	Fx:	COMPARTMENT DOOR SLIDES OPEN
2	STEWARD:	(CLOSE) Last sitting for lunch, Sir. Excusez-moi, Monsieur – will you be wanting lunch? Monsieur?
		(TO SELF) C'est formidable. Quite a sleeper.
		(LOUDER) Excuse me, Sir – allow me to remove the newspaper.
3	Fx:	PAPER RUSTLE
4	STEWARD:	Will you be . . . (GASP)
		Oh . . . Terrible . . . Terrible.
5	Fx:	TRAIN NOISE PASSING AND FADES INTO DISTANCE.
6	Fx:	PHONE RINGS. RECEIVER PICKED UP. RINGING STOPS.
7	BRADY:	Hello.
8	VOICE:	(DISTORT) Is that Colonel Brady?
9	BRADY:	Yes. Who's that?
10	VOICE:	(DISTORT) It doesn't matter but I thought you ought to know he's dead.
11	BRADY:	Who's dead? Who is this?
12	VOICE:	(DISTORT) Oh you know who's dead all right – and I've got the money.
13	BRADY:	You've got what money? Who are you?
14	VOICE:	You'll know soon enough. I'll be in touch . . .
15	Fx:	TELEPHONE CLICK. DIALLING TONE.
16	BRADY:	Hello, hello . . . oh blast it.
17	Fx:	RECEIVER SLAMMED DOWN.
18	Fx:	OFFICE INTERCOM BUZZER.
19	SECRETARY:	(DISTORT) Yes, Sir?
20	BRADY:	Joan, I want you to get hold of Salmon and Harris – can you do that?
21	SECRETARY:	(DISTORT) Yes, Sir – they went back to the Foreign Office.
22	BRADY:	Well I want them – at once. And get me on tonight's plane – to Marseilles.
23	CD:	MUSIC TO END.

The actors

Casting a radio drama, whether it is a one-hour play or a short illustration, will nearly always end by being a compromise between who is suitable and who is available. Naturally, the producer will want the best performers but this is not always possible within the constraints of money. It is also difficult to assemble an ideal cast at one place, and possibly several times, for rehearsal and recording, and this to coincide with the availability of studio space. Again, it may be that two excellent players are available, but their voices are too similar to be used in the same piece. So there are several factors that will determine the final cast.

Actors new to radio have to recognise the limitations of the printed page, which is designed to place words in clearly readable lines. It cannot overlay words in the same way that voices can, and do:

```
VOICE 1:    The cost of this project is going to be 3 or 4 million -
            and that's big money by anyone's reckoning.
VOICE 2:    But that's rubbish, why I could do the job for . . .
VOICE 1:    (Interrupting) Don't tell me it's rubbish, anyway
            that's the figure and it's going ahead.
```

Although this is what might appear on the page, voice two is clearly going to react to the cost of the project immediately on hearing the figures – halfway through Voice 1's first line. The scriptwriter may insert at that point (react) or (intake of breath) but it is generally best left to the imagination of the actor. Actors sometimes need to be persuaded to act, and not to become too script-bound. Voice 1 (interrupting) does not mean waiting until Voice 2 has finished the previous line before starting with the 'Don't tell me . . . '. Voice 1 starts well before Voice 2 breaks off, say on the word 'could'. The two voices will overlay each other for a few words, thereby sounding more natural. Real conversation does this all the time.

On the matter of voice projection, the normal speaking range over a conversational distance will suffice – intimate and confidential, to angry and hysterical. As the apparent distance is increased, so the projection also increases. In the following example, the actress goes over to the door and ends the line with more projection than was being used at the start. The voice gradually rises in pitch throughout the speech. This is on the website.

```
VOICE:      Well I must go. (DEPARTS) I shan't be long but there
            are several things I must do.
            (DOOR OPENS, OFF) I'll be back as soon as I can.
            Goodbye. (DOOR SHUTS)
```

In moving to the 'dead' side of a bi-directional microphone, the actor's actual movement may have been no more than a metre. The aural impression given may be a retreat of at least five metres. It is important that such 'moves' are made only during spoken dialogue otherwise the actor will appear to have 'jumped' from a near to a distant position. Of course, moving off-mic, which increases the ratio of reflected to direct sound, can only serve to give an impression of distance in an interior scene.

When the setting is in the open air, there is no reflected sound and the sense of distance has to be achieved by a combination of the actor's higher voice projection and a smaller volume derived from a low setting of the microphone channel fader. By this means it is possible to have a character shouting to us from 'over there', having a conversation

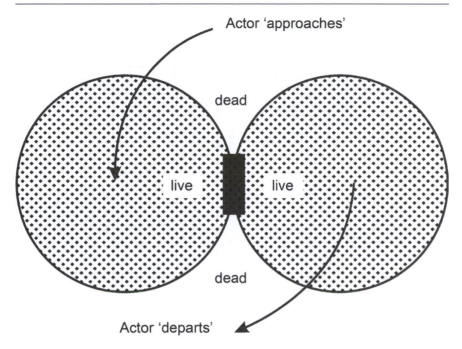

Actor 'approaches'

dead

live live

dead

Actor 'departs'

Figure 21.3 Movement on-mic. The shaded circles are the normal areas of pick-up on both sides of a bi-directional microphone. An actor 'approaching' speaks while moving from the dead side to the live side, also reducing in voice projection. Effective for indoor scenes

with another person 'in the foreground' who is shouting back. Such a scene requires considerable manipulation of the microphone fader with no overlay of the voices. A preferred alternative is to have the actors in separate rooms with their own microphone, each with a headphone feed of the mixed output.

The acoustic

In any discussion of monophonic perspective, distance is a function that separates characters in the sense of being near or far. The producer must always know where the listener is placed relative to the overall picture. Generally, but not necessarily, 'with' the microphone, the listener placed in a busily dynamic scene will need some information which helps in following the action by moving through the scene rather than simply watching it from a static position.

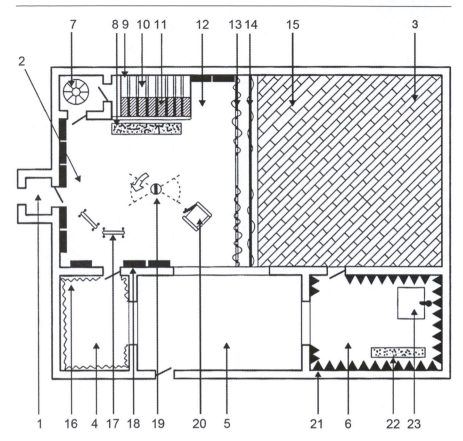

Figure 21.4 Features of a drama studio. 1. Entrance through sound lobby. 2. Studio 'dead' end. 3. Studio 'live' end. 4. Narrator's studio. 5. Control cubicle with windows to all other areas. 6. Dead room – outdoor acoustic. 7. Spiral staircase – iron. 8. Gravel trough. 9. Sound effects staircase. 10. Cement tread stairs. 11. Wooden tread stairs. 12. Carpet floor. 13. Soft curtain drapes. 14. 'Hard' curtain – canvas or plastic. 15. Wood block floor. 16. Curtains. 17. Movable acoustic screen. 18. Acoustic absorber – wall box. 19. Bi-directional microphone. 20. Sound effects door. 21. Acoustic wedges – highly absorbent surface. 22. Sand or gravel trough. 23. Water tank

Part of this distinction may be in the use of an acoustic which itself changes. Accompanied by the appropriate sound effects, a move out of a reverberant courtroom into a small ante-room, or from the street into a telephone box, can be highly effective.

There are five basic acoustics comprising the combinations of the quantity, and duration of the reflected sound:

1	No reverberation	Outdoor	Created by fully absorbent 'dead' studio
2	Little reverberation but long reverberation time	Library or large well-furnished room	Bright acoustic or a little reverberation added to normal studio
3	Mid reverberation amount and mid reverberation time	Courtroom or stage rehearsal	Adjust amount/time to achieve realism
4	Much reverberation but short reverberation time	Telephone box, bathroom	Small enclosed space of reflective surfaces – 'boxy' acoustic
5	Much reverberation and long reverberation time	Cave, 'Royal Palace', concert hall, cathedral	Artificial echo added to normal studio output

A digital reverberation device, such as that illustrated in Figure 21.5, can reproduce virtually any acoustic, from a train station to the Grand Canyon. Additional frequency discrimination applied to its output will create any distinctive coloration required. The typical characteristic of a drama studio with a 'neutral' acoustic would have a reverberation time of about 0.2 second. Associated with this would be a 'bright' area with a reverberation time of, say, 0.6 second, a separate cubicle for a narrator, and a 'dead' area for 'outdoor' voices. The key factor is flexibility so that by using screens, curtains and carpet, a variety of acoustic environments may be produced. The specialist drama studio illustrated in Figure 21.4 is, of course, very expensive. However, many of the facilities it offers can be reproduced quite cheaply as demonstrated by the 'Acoustic Walk' on the website.

The use of personal tie-clip mics for actors avoids any pick-up of studio acoustic and may be used to give an 'outdoor' sound. It also allows considerable freedom of movement. Crossfading between a studio mic and a tie-clip mic with a suitable drop in vocal projection also provides a very credible 'think piece' in the middle of dialogue.

Sound effects

When the curtain rises on a theatre stage the scenery is immediately obvious and the audience is given all the contextual information it requires for

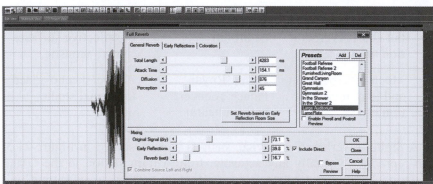

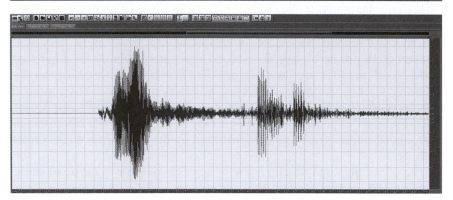

Figure 21.5 A digital reverberation device. The top image shows speech recorded in a dry acoustic. The centre image illustrates the reverberation options being set in the selection screen, and the bottom image has the speech with added reverberation. Note how, as a result of the added reverberation, the words now merge together

Figure 21.6   A sound effects door with a lock and bolt, and stairs – metal and wood

the play to start. So it is with radio, except that to achieve an unambiguous impact the sounds must be refined and simplified to those few which really carry the message. The equivalent of the theatre's 'backdrop' are those sounds that run throughout a scene – for example, rain, conversation at a party, traffic noise or the sounds of battle. These are most likely to be pre-recorded and reproduced from a stereo CD or digital source. The 'incidental furniture' and 'props' are those effects which are specially placed to suit the action – for instance, a ringing telephone, pouring a drink, closing a door, or firing a gun. Such sounds – called 'spot' effects – are best made in the studio at the time of the appropriate dialogue, if possible by the actors themselves – for example, lighting a cigarette or taking a drink – but by someone else if hands are not free due to their holding a script. Going upstairs is a quite different sound from walking on the flat, hence the flight of stairs in a drama studio with different treads – iron, concrete, wood, etc.

There is a temptation for a producer new to drama to use too many effects. While it is true that in the real world the sounds we hear are many and complex, radio drama in this respect purveys not what is real but what is understandable and convincing. It is possible to record genuine sounds which, divorced from their visual reality, convey nothing at all. The sound of a modern car drawing up has very little impact, yet it might be required to carry the dramatic turning point of the play. In the search for clear associations between situation and sound, radio over the years has developed conventions with generally understood meanings. The urgently stopping car virtually demands a screech of tyres, slamming doors and running footsteps. It becomes a little larger than life. Overdone, it becomes comical.

Some other sounds that have become immediately understood are:

1   passage of time – clock ticking;
2   night time – owl hooting;
3   on the coast – seagulls and seawash;
4   on board sailing ship – creaking of ropes;
5   early morning – cock crowing;
6   urban night time – distant clock chime, dog barking;
7   out of doors, rural – birdsong.

A particularly convincing sound to run behind an outdoor scene is a 'rumble' at low level. Creating the feeling of 'atmosphere' it is especially effective on coming indoors when it is suddenly cut. This can be heard on the website.

The convention for normal movement is to do without footsteps. These are only used to underline a specific dramatic point.

Background sounds may or may not be audible to the actors in the studio, depending on the technical facilities available. It is important, however, for actors to know what they are up against, and any background sounds should be played over to them to help them visualise the scene and judge their level of projection. This is particularly important if the sounds are noisy – in the cockpit of a light aircraft, a fairground or battle. If an actor has to react to a pre-recorded sound a headphone feed of the output will be needed, or through a loudspeaker. The normal studio speaker is of course cut when any mic channel is opened, but an additional loudspeaker fed from the effects source can remain on, unaffected by this muting. This facility is called *foldback*, and has the advantage not only that the cast can hear the effects, even though their own mics are live, but that any sounds reproduced in this way and picked up on the studio microphones will share the same acoustic as the actors' voices.

Producers working in drama will soon establish their own methods of manufacturing studio sounds. The following are some that have saved endless time and trouble:

1   walking through undergrowth or jungle – a bundle of old recording tape rustled in the hands;
2   walking through snow – a roll of cotton wool squeezed and twisted in the hands, or two blocks of salt rubbed together. Alternatively, a wooden dowel crunching on a bowl of corn starch makes wonderful walking in snow!;
3   horses' hooves – halved coconut shells are still the best from pawing the ground to a full gallop. They take practice though. A bunch of keys will produce the jingle of harness;

4 pouring a drink – put a little water into the glass first so that the sound starts immediately the pouring begins;

5 opening champagne – any good sound assistant ought to be able to make a convincing 'pop' with their mouth, otherwise blow a cork from a sawn-off bicycle pump. A little water poured on to Alka-Seltzer tablets or fruit salts close on-mic should do the rest;

6 a building on fire – cellophane from a biscuit pack rustled on-mic plus the breaking of small sticks;

7 a thin stick or bamboo can make a splendid 'whoosh';

8 marching troops – a marching box is simply a cardboard box approximately $20 \times 10 \times 5$ cm, containing some small gravel. Held between the hands and shaken with precision it can execute drill movements to order;

9 in the case of a costume drama, some silk or taffeta material rustled near the mic occasionally is a great help in suggesting movement;

10 creaks – rusty bolts, chains or other hardware are worth saving for the appropriate aural occasion. A little resin put on a cloth and pulled tightly along a piece of string fastened to a resonator is worth trying.

It's useful to keep a store of things that reliably squeak, pop, clatter, jangle and scrape, for Fx use. The essential characteristic for all electronic or acoustic sound-making devices is that they should be simple and consistent. And if the precise effect cannot be achieved, it's worth remembering that by playing a sound through a digital converter, it can be altered in 'size' or made unrelatable to the known world. Hence the fantasy sounds varying from dinosaurs to outer space. Voices can be made 'unhuman', lighter or deeper without affecting their speed of delivery. Time spent exploring the possibilities of digital effects will pay dividends for the drama producer.

In making a plea for authenticity and accuracy, it is worth noting that such attention to detail saves considerable letter-writing to those among the audience who are only too ready to display their knowledge. Someone will know that the firing of the type of shot used in the American War of Independence had an altogether characteristic sound, of course certain planes used in the Australian Flying Doctor services had three engines not two, and whoever heard of an English cuckoo in February? The producer must either avoid being too explicit, or be right.

Music

An ally to the resourceful producer, music can add greatly to the radio play. However, if it is overused or badly chosen, it becomes only an irritating distraction. Avoiding the obvious or over-familiar the producer must decide in which of its various roles music is to be used:

1   As a 'leitmotif' to create an *overall style*. Opening and closing music, plus its use within the play as links between some of the scenes, will provide thematic continuity. The extracts are likely to be the same piece of music, or different passages from the same work, throughout.

2   Music chosen simply to *create mood* and establish the atmosphere of a scene. Whether it is 'haunted house' music or 'a day at the races', music should be chosen that is not so well-known that it arouses in the listener personally preconceived ideas and associations. In this respect it pays the producer to cultivate an awareness of the more unfamiliar works in the various production 'mood music' libraries, some of which are non-copyright.

3   Reiterative or relentless music can be used to mark the *passage of time*, thus heightening the sense of passing hours, or seconds. Weariness or monotony is economically reinforced.

In using music to be deliberately evocative of a particular time and place, the producer must be sure of the historical period and context. Songs of the First World War, or ballads of Elizabethan England – there is sure to be at least one expert listening ready to point out errors of instrumentation, words or date. To use a piano to set the mood of a time when there was only the harpsichord and virginals is to invite criticism.

The drama producer must not only search the shelves of a music library but should sometimes consider the use of specially written material. This need not be unduly ambitious or costly – a simple recurring folk song or theme played on a guitar, harmonica or flute can be highly effective. There are considerable advantages in designing the musical style to suit the play, and having the music durations exactly to fit the various introductions and voice-overs.

Production technique

Producers will devise their own methods and different plays might demand an individual approach, as will working with children or amateurs as opposed to professional actors. However, the following is a practical outline of general procedure:

1   The producer works with the writer, or on the script alone rewriting for the medium, and making alterations to suit the transmission time available.

2   Cast the play, issue contracts, clear any copyright issues, distribute copies of the script, arrange rehearsal or recording times.

3   The producer or the sound team assemble the sound effects, book the studio, arrange for any special technical facilities or acoustic requirements and choose the music.

4   The cast meets, not necessarily in the studio, for a read-through. The actors may need to be briefed on the character they are playing – personality and behaviour. Awkward wordings may be altered to suit individual actors. The producer gives points of direction on the overall structure and shape of the play and the range of emotion required. This is to give everyone a general impression of the piece. Scripts are marked with additional information such as the use of cue lights.

5   In the studio, scenes are rehearsed on-mic, with detailed production points concerning inflection, pauses, pace, movement, etc. The producer should be careful not to cause resentment by 'over-direction', particularly of professional actors. The producer may by all means say what effect a particular line should achieve, but is likely to undermine an actor by going as far as to say, or even further to demonstrate, precisely how it should be delivered. Sound effects, pre-recorded or live in the studio, are added. The producer's main task here is to encourage the actors and to listen carefully for any additional help they might need or for any blemish that should be eradicated.

6   As each scene is polished to its required perfection, a recording 'take' is made. Are the pictures conjured up at the original reading of the script being brought to life? Is the atmosphere, content and technical quality exactly right? Necessary retakes are made and the script marked accordingly.

7   The recording is edited using the best 'takes', removing fluffs and confirming the final duration.

8   The programme is placed in the transmission system and the remaining paperwork completed.

There are many variations on this pattern of working. Here are some alternative approaches:

● Do away with the studio. If the play is suitable then, for example, the recording can be made with portable recorders out of doors among the back alleys and railway tracks of the city – a *radio verité*? Away from any studio, a 10-hour production of Tolstoy's *War and Peace* with a cast of 60 was recorded for the BBC entirely on location, often using handheld boom mics as if it were film. The point of doing this was to get a variety of authentic acoustics. The battle scene effects were recorded at an actual reconstruction of the Battle of Austerlitz, and the play was broadcast across a single day.

- There is no need to think only in terms of the conventional play. A highly effective yet simple format is to use a narrator as the main storyteller with only the important action dialogue spoken by other voices. Few but vivid effects complement this radio equivalent of the strip cartoon – excellent as a children's serial.

- Dialogue on its own can be a simple and powerful means of explaining a point – two farmers discussing a new technique for soil improvement – or illustrating a relationship, such as a father and teenage daughter arguing about what she should wear. Given the right people it doesn't even have to be scripted – give them the basic idea and let them ad-lib it.

- The use of monologue. It sounds dull perhaps, but writers – such as Alan Bennett with his *Talking Heads* – enable a single character to unravel a richly riveting stream of internal thinking, motivation, musings and real or imagined behaviour which, spoken in the idiom of the listener, creates immediate identification. Perhaps with a little music and effects added, radio monologue creates an intense world of two – the character and the listener. One such example was the award-winning *Spoonface Steinberg* – Lee Hall's hour-long internal monologue by a seven-year-old autistic girl, dying of cancer. Just her voice, thinking of the music she loves and why. Imaginative, inspiring, insightful and memorable – a lovely use of the medium.

Any producer finding their way into the drama field is well advised to listen to as many radio plays and serials as possible. He or she will gather ideas and recognise the value of good words simply spoken.

In realising the printed page into an aural impression, it is not to be expected that the visual images originally conceived at the start will be exactly translated into the end result. Actors are not puppets to be manipulated at will, they too are creative people and will want to make their own individual contribution. The finished play is an amalgam of many skills and talents; it is a 'hand-made', 'one-off' product which hopefully represents a richer experience than was envisaged by any one person at the outset.

22

Documentary and feature programmes

These terms are often used as if they were interchangeable and there is some confusion as to their precise meaning. But here are exciting and creative areas of radio and because of the huge range they cover, it is important that the listener knows exactly what is being offered. The basic distinction is to do with the initial selection and treatment of the source material. A documentary programme is wholly fact, based on documentary evidence – written records, attributable sources, contemporary interviews and the like. Its purpose is essentially to inform, to present a story or situation with a total regard for honest, balanced reporting. The feature programme, on the other hand, need not be wholly true in the factual sense, it might include folk song, poetry or fictional drama to help illustrate its theme. The feature is a very free form, where the emphasis is often on portraying rather more indefinable human qualities, atmosphere or mood.

The distinctions are not always clear cut and a contribution to the confusion of terms is the existence of hybrids – the feature documentary, the semi-documentary, the drama documentary, faction and so on.

It is often both necessary and desirable to produce programmes that are not simply factual, but are 'based on fact'. There will certainly be times when, through lack of sufficient documentary evidence, a scene in a true story will have to be invented – no actual transcript exists of the conversations that took place during Columbus's voyage to the New World. Yet through his diaries and other contemporary records, enough is known to piece together an acceptable account which is valid in terms of reportage, often in dramatic form. While some compromise between what is established fact and what is reasonable surmise is understandable in dealing with the long perspective of history, it is important that there is no blurring of the edges in portraying contemporary issues. Fact and fiction are dangerous in combination and their boundaries must be clear to the listener. A programme dealing with a murder trial, for example, must keep to the record; to add fictional scenes is to confuse, perhaps to mislead. Nevertheless, it is a perfectly admissible programme idea to interweave serious fact, even a court case, with contrasting fictional material, let us say songs and nursery rhymes; but it must then be called a feature not a documentary. Ultimately what is important is not the subject or its treatment, but that we all understand what is meant by the terms used. It is essential that the listener knows the purpose of the broadcaster's programme – essentially the difference between what is true and what is not. If the producer sets out to provide a balanced, rounded, truthful account of something or someone – that is a documentary. If the intention is not to feel so bound to the whole truth but to give greater rein to the imagination, even though much of the source material is real – that is a feature.

The documentary

Very often subjects for programmes present themselves as ideas that suddenly become obvious. They are frequently to do with broad contemporary issues such as race relations, urban development, pollution and the environment, or medical research – attempting to examine how society copes with change. Alternatively, a programme might explore in detail a single aspect of one of these subjects. Other types of documentary deal with a single person, activity or event – the life of Abraham Lincoln, the discovery of radium, the building of the Concorde aeroplane, the causes of child obesity, or the work of a particular factory, theatre group or school.

Essentially these are all to do with people, and while statistical and historical fact is important, the crucial element is the human one – to underline motivation and help the listener understand the prevailing social climate, why certain decisions were made, and what makes people 'tick'.

The main advantage of the documentary approach over that of the straightforward talk is that the subject is made more interesting and colourful. It's brought alive by involving more people, more voices and a greater range of treatment. It should entertain and tell a story while it informs and provokes further thought and concern.

Planning

Following on the initial idea is the question of how long the programme should be. It may be that the brief is to produce something for a 30-minute or one-hour slot, in which case the problem is one of selection, of finding the right amount of material. Given a subject that is too large for the time available, a producer has the choice either of dealing with the whole area fairly superficially or reducing the topic range and taking a particular aspect in greater depth. It is, for example, the difference between a 20-minute programme for schools on the life of Chopin, and the same duration or more devoted to the events leading up to Chopin's writing of the *Revolutionary Study*, directed to a serious music audience. For a large subject area, it's always possible, of course, to make a series of related programmes.

Where no overall duration is specified, simply an intent to cover a given subject, the discipline is to contain the material within a stated aim without letting it become diffuse, spreading into other areas. For this reason it is an excellent practice for the producer to write a programme brief in answer to the questions 'What am I trying to achieve?', 'What do I want to leave with the listener?' Later on, when deciding whether or not a particular item should be included, a decision is easier in the light of the producer's own statement of intent. This is not to say that programmes cannot change their shape as the production proceeds, but a positive aim helps to prevent this happening without the producer's conscious knowledge and consent.

At this stage the producer is probably working alone, gradually coming to terms with the subject, exploring it at first-hand. Many producers find it useful to make a 'mind map' – a visual diagram of how different aspects of the issue are related. Even more possibilities may be opened up by brainstorming the subject with colleagues. This initial research, and in particular listing those topics within the main subject which must be included, leads to decisions on technique – how each topic is to be dealt with. From this emerges the running order in embryo. Very often the title comes much later – perhaps from a significant remark made within the programme. There is no formally recognised way of organising this programme planning; each producer has a preferred method. By committing thoughts to paper and seeing their relationship one to another – where the emphasis should be and what is redundant – the producer is more likely to finish up with a tightly constructed, balanced programme.

Let's take an example. Here are the first planning notes for a local radio programme. This radio station serves a coastal region where the trawler fleet has been seriously affected by the loss of fishing rights in international waters:

Working title: 'The Return of the Trawlermen'

Aim: To provide the listener with an understanding of the impact which changes in the deep-sea fishing industry over the last 20 years have had on the people who work in it.

Duration: 30 minutes.

Information: Annual figures for ships in use, shipping tonnage, men employed, fish landed, turnover and profit, investment, changes in the legal requirements of the fishing industry, etc.

Content: Historical account of development in the 20 years:

- technological change, searching, catching, freezing methods;
- economic change, larger but fewer ships – implication for owners in the increase in capital cost;
- social change resulting from fewer jobs, higher paid work, longer voyages, and better conditions on board;
- political change, new European regulations relating to international waters, size of nets, offshore limits. Impact of other fishing fleets.

Key questions: What has happened to the men and the ships that used to work here in large numbers?
                     What has happened to those areas of the city where previously whole streets were dependent on fishing as a livelihood? Family life, local shop-keepers, trade, etc., loss of comradeship?
                     How has the price of fish changed relative to the cost of running the new hi-tech ships? What has happened to fish stocks? What is the scientific evidence for this? Sources of conflict.
                     How significant are the political factors affecting fishing rights in distant waters or was the industry in any case undergoing more fundamental change?
                     What do the men, and their families, feel about the industry now? Where is any animosity directed? Will the trends continue into the future?

Interview sources: National Federation of Fishermen's Organisations.
                     Docks and Harbour Board? Shipbuilder?
                     Fleet owner.
                     Skippers and fishermen – past and present.
                     Sea Fish Industry Authority.
                     Representative of fish processing industry – oils – frozen food.

	Government – ministry official; members of parliament. Wives of seamen, etc.
Reference sources:	Newspaper cuttings. Library – shipping section. Relevant websites Government White Paper. Trawler company reports. Magazine – *Fishing News International*. Fred Jones (another producer who did a programme on the docks some time ago).
Actuality:	Bringing in the catch, nets gear running, ship's bridge at sea, engine noise, radio communications, shoal radar, etc. Unloading – dockside noises. Auction house sale, fish market.

By setting out the various factors that have to be included in the programme, it is possible to assess more easily the weight and duration that should be given to each, and whether there are enough ideas to sustain the listener's interest. It probably becomes apparent that there is a lot to get in. It would be possible to do a programme that concentrated solely on the matter of international fishing rights, but in this case the brief was broader and the temptation to dwell on the latest or most contentious issue, such as safety at sea, must be resisted – that's another programme.

A final point on planning. A producer's statement of intent should remain fixed, but how that aim is met may change. Initial plans to reach the goals in a certain way may be altered if, in the course of production, an unforeseen but crucial line of enquiry opens up. The programme material itself will influence decisions on content.

Research

Having written the basic planning notes, the producer must make the programme within the allocated resources of time, money, people, etc. The key decision is whether to call on a specialist writer or to write one's own script. Depending on this will rest the matter of further research – perhaps it is possible to obtain the services of a research assistant or reference library. The producer who is working to a well-defined brief knows what is wanted and in asking the right questions will save both time and money.

The principle with documentary work is always, as far as possible, to go back to sources, the people involved, eye witnesses, the original documents, and so on.

Structure

The main structural decision is whether or not to use a narrator. A linking, explanatory narrative is obviously useful in driving the programme forward in a logical, informative way. This can provide most of the statistical fact and the context of the views expressed, and also the names of various speakers. A narrator can help a programme cover a lot of ground in a short space of time, but this is part of the danger; it may give the overall impression of being too efficient, too 'clipped' or 'cold'. The narrator should *link* and *not interrupt*, and there will almost certainly not be any need to use a narrative voice between every contribution. There are styles of documentary programme that make no use at all of links but each item flows naturally from one to the next, pointing forward in an intelligible juxtaposition. This is not easy to do but can often be more atmospheric.

Collecting the material

Much of the material will be gathered in the form of location interviews, if possible while at sea during a fishing trip. If it has been decided that there will be no narrator, it is important to ensure that the interviewees introduce themselves – 'speaking as a trawler owner . . . ' or 'I've been in this business now for 30 years . . . '. They may also have to be asked to bring out certain statistical information. This may be deleted in the editing but it is wise to have it in the source material if there is no obvious way of adding it in a linking script.

It must be decided whether the interviewer's voice is to remain as part of the interviews. It may be feasible for all the interviewing to be done by one person, who is also possibly the producer, and for the programme to be presented in the form of a personal investigative report. Pursuing this line further, it is possible for the producer to hire a well-known celebrity to make a programme as a personal statement – still a documentary, but seen from a particular viewpoint that is known and understood. Where the same interviewer is used throughout, he or she becomes the narrator and no other linking voice is needed. Where a straightforward narrator is used, the interviewer's questions are removed and the replies made to serve as statements, the linking script being careful to preserve them in their original context. What can sound untidy and confusing is where in addition to a narrator, the occasional interviewer's voice appears to put a particular question. A programme should be consistent to its own structure. But form and style are infinitely variable and it is important to explore new ways of making programmes – clarity is the key.

Impression and truth

The purpose of using actuality sounds is to help create the appropriate atmosphere. More than this, for those listeners who are familiar with the subject, recognition of authentic backgrounds and specific noises increases the programme's authority. It may be possible to add atmosphere by using recorded sound effects. These should be used with great care since a sound only has to be identified as 'not the genuine article' for the programme's whole credibility to suffer. The professional broadcaster knows that many simulated sounds or specially recorded effects create a more accurate impression than the real thing. The producer concerned not simply with truth but with credibility may use non-authentic sounds only if they give an authentic impression.

The same principle applies to the rather more difficult question of fabrication. To what extent may the producer create a 'happening' for the purpose of the programme? Of course it may be necessary to 'stage manage' some of the action. If you want the sound of ship's sirens, the buzzing of a swarm of angry bees, or children in a classroom reciting poetry, these things may have to be made to happen while the recorder is running. Insofar as these sounds are typical of the actual sounds, they are real. But to fabricate the noise of an actual event, for example a violent demonstration with stones thrown, glass breaking, perhaps even shots being fired; this could too easily mislead the listener unless it is clearly referred to as a simulation. Following the work of broadcasters in war zones it is probably true that unless there are clear indications to the contrary, the listener has a right to expect that everything heard in a documentary programme is genuine material to be taken at face value. It is not the documentary producer's job to deceive, or to confuse, for the sake of effect.

Even the reconstruction of a conversation that actually happened, using the same individuals, can give a false impression of the original event. Like the 'rehearsed interview', it simply does not feel right. Similarly it is possible to alter a completely real conversation by switching on a recorder – a house builder giving a quotation for a prospective purchaser is unlikely to be totally natural with a 'live' microphone present!

Faced with the possibility that 'reality' simply won't happen, either in an original recording or by a later reconstruction, the documentary producer may be tempted to obtain material by secretive methods.

An example would be to use a concealed recorder to get a conversation with an 'underground' dealer for a programme on drug smuggling.

This is a difficult area that brings the broadcaster into conflict with the quite reasonable right of every individual to know when they are making a statement for broadcasting. Certainly the BBC is opposed in general to the use of surreptitious production techniques as being an undue invasion of personal liberty (see the ACMA Code on p. 22). It is a question that producers, staff or freelances, should not take upon themselves. If such a method is used, it is as a result of a decision taken at a very senior level.

Of course, if the subject is historical, it is an understood convention that scenes are reconstructed and actors used. Practice in other countries differs, but in Britain a documentary on even a recent criminal trial must of necessity employ actors to reconstruct the court proceedings from the transcripts, since the event itself cannot be recorded. No explanation is necessary other than a qualification of the authenticity of the dialogue and action. What is crucial is that the listener's understanding of what is broadcast is not influenced by an undisclosed motive on the part of the broadcaster.

Music

Current practice makes little use of music in documentary programmes, perhaps through a concern that it can too easily generate an atmosphere, which should more properly be created by real-life voices and situations. However, producers will quickly recognise those subjects that lend themselves to special treatment. Not simply programmes that deal with musicians, orchestras or pop groups, but where specific music can enhance the accuracy of the impression – as background to youth club material, or to accompany reminiscence of the depressed 1930s. A line from a popular song will sometimes provide a suitably perceptive comment, and appropriate music can certainly assist the creation of the correct historical perspective. Again, as with drama, one of the many specialist MCPS 'mood music' libraries can help. However, be warned that many complaints come from listeners about music in these circumstances being 'intrusive', 'too loud', or 'distracting'.

Compilation

Having planned, researched, and structured the programme, written the basic script and collected material, the producer must assemble it so as to meet the original brief within the time allotted. First, a good opening. Two suggestions that could apply to the earlier example of the programme on the fishing industry are illustrated by page one of the following script:

Example one

```
1   Sound effects:     Rattle of anchor chain.
                       Splash as anchor enters water.

2   Narrator:          The motor vessel Polar Star drops anchor for
                       the last time. A deep-sea trawler for the
                       last 24 years, she now faces an uncertain
                       future.

                       Outclassed by a new generation of freezer
                       ships and unable to adapt to the vastly
                       different conditions, she and scores of
                       vessels like her are now tied up - awaiting
                       either conversion or the scrap yard.

                       In this programme we look at the causes of
                       change in the industry and talk to some of
                       the men who make their living from the sea.
                       Or who, like their ships, feel that they
                       too have come to the end of their working
                       life . . . etc.
```

Example two

```
1   Skipper Matthews:  I've been a trawler skipper for 18 years -
                       been at sea in one way or another since I was
                       a lad. Never thought I'd see this. Rows of
                       vessels tied up, just rusting away - nothing
                       to do. We used to be so busy here. I never
                       thought I'd see it.

2   Narrator:          The skipper of the Grimsby Polar Star. Why
                       is it that in the last few years the fishing
                       fleet has been so drastically reduced? How
                       have men like skipper Matthews adapted to
                       the new lives forced on them? And what does
                       the future look like for those who are left?
                       In this programme we try to find some of the
                       answers . . . etc.
```

The start of the programme can gain attention by a strong piece of sound actuality, or by a controversial or personal statement carefully selected from material that is to be heard within the programme. It opens 'cold' without music or formal introduction, preceded only by a time check and station identification. An opening narration can outline a situation in broad factual terms, or it can ask questions to which the listener will want the answers. The object is to create interest, even suspense, and involve the listener in the programme as soon as possible.

The remainder of the material might consist of interviews, narrator's links, actuality, vox pop, discussion and music. Additional voices may be used to read official documents, newspaper cuttings or personal letters. It is better if possible to arrive at a fairly homogeneous use of a particular

technique, not to have all the interviews together, and to break up a long
voice piece or statement for use in separate parts. The most easily understood progression is often the chronological one, but it might be desirable
to stop at a particular point in order to counter-balance one view with its
opposite. And during all this time the final script is being written around
the material as it comes in – cutting a wordy interview to make the point
more economically in the narration, leaving just enough unsaid to give the
actuality material the maximum impact, dropping an idea altogether in
favour of a better one, always keeping one eye on the original brief.

Programme sequence

There are few rules when it comes to deciding the programme sequence.
What matters is that the end result makes sense – not simply to the producer, who is thoroughly immersed in the subject and knows every nuance
of what was left out as well as what was included, but to the listener who is
hearing it for the first time. The most consistent fault with documentaries
is not with their content but in their structure. Examples of such problems
are insufficient 'signposting', the re-use of a voice heard sometime earlier
without repeating the identification, or a change in the convention regarding the narrator or interviewer. For the producer who is close to the material
it is easy to overlook a simple matter which may present a severe obstacle
to the listener. The programme maker must always be able to stand back
and take an objectively detached view of the work as its shape emerges.

The ending

To end, there are various alternatives:

1   to allow the narrator to sum up – useful in some types of educational
    programme or where the material is so complex or the argument so
    interwoven that some form of clarifying resumé is desirable;
2   to repeat some of the key statements using the voices of the people
    who made them;
3   to repeat a single phrase that appears to encapsulate the situation;
4   to speculate on the future with further questions;
5   to end with the same voice and actuality sounds as those used at the
    opening;
6   to do nothing, leaving it to the listener to form an assessment of the
    subject. This is often a wise course to adopt if moral judgements are
    involved.

Contributors

The producer has a responsibility to those asked to take part. First, tell them as much as possible of what the programme is about. Provide them with the overall context in which their contribution is to be used. Second, tell them, prior to transmission, if their contribution has had to be severely edited, or omitted altogether. Third, whenever possible let contributors know in advance the day and time of transmission. These are simple courtesies and the reason for them is obvious enough. Whether they receive a fee or not, contributors to documentary programmes generally take the process extremely seriously, often researching additional material to make sure their facts are right. They frequently put their professional or personal reputation at risk in expressing a view or making a prediction. The producer must keep faith with them in keeping them up to date as to how they will appear in the final result.

What the producer cannot do is to make the programme conditional upon their satisfaction with the end product. Contributors cannot be allowed access to the edited material in order to approve it before transmission. Not only would there seldom be a programme because contributors would not agree, but it would be a denial of proper editorial responsibility. The programme goes out under the producer's name and that of the broadcasting organisation. That, the listener understands, is where praise and blame attaches and editorial responsibility is not to be passed off or avoided through undisclosed pressures or agreements with anybody else.

Programmes in real time

A great use of the medium is to unravel the telling of a true story across the same time-span as the original. It might be a hospital emergency, a court case, a tropical storm or a rescue operation. An excellent example is Len Deighton's *Bomber*, a Second World War documentary broadcast by BBC Radio 4. It was on the anniversary of a bomber raid over Germany, but it was done in real time, in chunks over an afternoon and evening. It took us from the initial lunchtime briefing, the take-off and some of the flight to the raid itself – by which time everything had gone horribly wrong. It was interwoven with the personal details and voices of the people in the air and on the ground, on both sides, who had done this for real 50 years before. It was immensely complex, adding drama, effects and actuality. You had to stay up until midnight to hear the planes – some of them – returning home. Anguish, death, hope, despair, it was innovative as well as authentic – and tremendously compelling radio.

A BBC Radio 4 documentary in the *Farming Today* series on the theme of 'North Sea Trawlermen' appears on the website: www.focalpress.com/cw/mcleish.

The feature

Whereas the documentary must distinguish carefully between fact and fiction and have a structure that separates fact from opinion, the feature programme does not have the same formal constraints. Here all possible radio forms meet, poetry, music, voices, sounds – the weird and the wonderful. They combine in an attempt to inform, to move, to entertain or to inspire the listener. The ingredients may be interview or vox pop, drama or discussion, and the sum total can be fact or fantasy. A former Head of BBC Features Department, Laurence Gilliam, described the feature programme as:

> a combination of the authenticity of the talk with the dramatic force of a play, but unlike the play, whose business is to create dramatic illusion for its own sake, the business of the feature is to convince the listener of the truth of what it is saying, even though it is saying it in dramatic form.

It is in this very free and highly creative style that some of the most memorable and innovative radio has been made. On the subject of North Sea fishing, Charles Parker for the BBC produced a 'Radio Ballad' – 'Singing the Fishing' – with folk singers, instrumental music, sound effects and actuality voices, recounting the dangers and satisfactions of being a North Sea fisherman. The overall effect is not reasoned argument in documentary form, but a collage of colourful songs, human experience and endeavour, which after many years is still memorable.

The possible subject material for the feature ranges more widely than the documentary since it embraces even the abstract: a programme on the development of language, a celebration of St Valentine's day, the characters of Dickens, a voyage among the stars. Even when all the source material is authentic and factually correct, the strength of the feature lies more in its impact on the imagination than in its intellectual truth. Intercut interviews with people who served in the Colonial Service in India mixed with the appropriate sounds can paint a vivid picture of life as it was under the British Raj – not the whole truth, not a carefully rounded and balanced documentary report, it is too wide and complicated a matter to do that in so short a time, but a version of the truth, an impression. The same is true of a programme dealing with the music of the slave trade, the countryside in summer, the life of Byron, or the romance of the early days of aviation. The feature deals not so much with issues but with events and emotions, and at its centre is the ancient art of telling a story.

The production techniques and sequence are the same as for a documentary – statement of intent, planning, research, script, collection of material, assembly, final editing. In a documentary the emphasis is on the collection of the factual material. Here, the work centres on the creative writing of the script – a strong storyline, clear visual images, the unfolding of a sequence of events with the skill of the dramatist, the handling of known facts but still with a feeling of suspense. Some of the best programmes have come from the producer/writer who can hear the end result begin to come together even while doing the research. Only through immersion in the subject comes the qualification to present it to the rest of us.

Once again, because of the multiplicity of treatment possible and the indistinct definitions we use to describe them, an explanatory subtitle is often desirable.

> 'A personal account of . . . '
> 'An examination of . . . '
> 'The story of . . . '
> 'Some aspects of . . . '
> 'A composition for radio on . . . '

Thus, the purpose of the finished work is less likely to be misconstrued. For the final word on the documentary and feature area of programming, Laurence Gilliam again:

> It can take the enquiring mind, the alert ear, the selective eye, and the broadcasting microphone into every corner of the contemporary world, or into the deepest recess of experience. Its task, and its destiny is to mirror the true inwardness of its subject, to explore the boundaries of radio and television, and to perfect techniques for the use of the creative artist in broadcasting.

23

The work of the producer

So what does the producer actually do?

Ideas

First and foremost he or she is essentially creative – has ideas – ideas for programmes, or items, people to interview, pieces of music or subjects for discussion – new ways of treating old ideas, or creating a fresh approach to the use of radio. New ideas are not simply for the sake of being different, they stimulate interest and fresh thought, so long as they are relevant. But ideas are not the product of routine, they need fresh inputs to the mind. The producer therefore must not stay simply within the confines of the world of broadcasting, but must get involved physically and mentally in the community being served. It is all too easy for media people to stay in their ivory towers and to form an elite, not quite in touch with the world of the listener. Such an attitude is one of a broadcasting service in decline. Ideas for programmes must be rooted firmly in the needs and language of the audience they serve; the producer's job is to assess, reflect and antici-pate those needs through a close contact with his potential listeners. If the audience is geographically far away, then it's important to read their newspapers, magazines, emails, tweets and posts – to talk to the returning traveller, carefully study the incoming mail, and visit their country as and when possible. The producer will carry a small notebook to jot down the fleeting thought or snatch of conversation overheard. And if it is difficult to think of genuinely new ideas, then act as a catalyst for others, stimulat-ing and being receptive to their thoughts, and at least recognise an idea when one arises. Only then may the experiential producer retreat to the quiet of an office to think – to shape and develop ideas into draft outlines.

There is, however, a great deal of difference between a *new idea* and a *good idea* and any programme suggestion has to be thought through on a number of criteria. An idea needs distilling in order to arrive at a workable form. It has to have a clarity of aim so that all those involved know what they are trying to achieve. It has to be seen as relevant to its target audience, and it must be practicable in terms of resources. Is there the talent available to support the idea? Is it going to be too expensive in people's time? Does it need additional equipment? What will it cost? Is there sufficient time to plan it properly? Any new programme idea has to be thought through in relation to the four basic resources – people and their skills, money, technical equipment and time. It might be depressing to have to modify a really good idea in order to make it work with the resources available, but here's another of the producer's most important tasks – to reconcile the desirable with the possible.

The audience

Given the initial programme idea in a practical format, the producer might have to persuade the head of department, programme controller or schedule 'gatekeeper' – the boss – that the proposal is the best thing that could happen to the broadcast output. Further, not only will the programme not fail, but it will enhance the manager's reputation, as well as provide a memorable programme. While the producer will see a project as intrinsically worthwhile or personally creative, the manager of the service might be much more concerned with competitive ratings. His or her first question is likely to be: 'What will it do for the audience?' There are two possible answers: 'satisfy it' or 'increase it'. A good programme may do both. In allocating a transmission slot, the time of day selected and the material preceding it can be crucial to the success of the programme. It is no good putting out a programme for children at a time when children are not available, nor is it helpful to broadcast an in-depth programme at a time when the home environment is busy and the necessary level of concentration is unlikely to be sustained. This is where a knowledge of the target audience is essential. Farmers, industrial workers, housewives, teenagers, doctors will all have preferred listening times which will vary according to local circumstances.

The fairly superficial news/information and 'current affairs plus music' type of continuous output, where all the items are kept short, may be suitable for the general audience at times when other things are happening – such as meal times or at work. But the timing of the more demanding documentary, drama or discussion programme can be critical and will depend on individual circumstances. Factors to be considered when assessing audience availability may include weekday/weekend work and leisure patterns, the potential car listenership which can represent a significant 'captive' audience, television viewing habits, digital/FM/MW usage, and so on. The producer has to be involved in marketing the product and normal consumer principles apply whether or not the radio service is commercially financed. We look at this further in the next chapter.

Resource planning

Having agreed on a time slot for the broadcast, the producer must ensure there is reasonable time available for its preparation. Is it to be next week or in six months' time? No producer will say that there is sufficient time for production work but there is much to commend working within definite deadlines, for such pressure can lend creative impetus to the work.

The programme idea is now accepted and the transmission date and time allocated. At this stage the producer draws up a detailed budget and obtains authorisation for any additional resources needed – money for research effort, scriptwriting, contributors or a music group. Fees may have to be negotiated. Now check the availability of appropriate studio facilities and make arrangements for any necessary engineering or other staff support. Obtain the clearance of any copyright work to be used. Conditions for the broadcasting of material in which usage rights are owned by someone outside the broadcasting service vary widely. In the case of literary copyright, books, poems, articles, etc., the publisher will normally be the point of reference, but if the work is not published, the original author (or if dead, his or her estate) should be consulted. Under British copyright law such rights of ownership exist for a period of 70 years from the date of publication or from the death of the author, whichever is the later. The rules vary according to the law of the country in which the broadcast is to be made and in cases of doubt it is well worth taking specialist legal advice – discussing copyright fees after the broadcast is, to say the least, a weak negotiating position.

One other set of rules that has to be considered is the risk assessment referred to on p. 249. Is there any kind of hazard attached to the programme? As the person in charge of the project, safety is the producer's responsibility.

Preparation of material

Programme requirements may be very simple and the producer able to fulfil them without any help – interview material, music selected from the library, and some crisply written and well-presented links might be all that is required. Good ideas are often simple in their translation into radio and can be easily ruined by 'over-production'. On the other hand, it might be necessary to involve a lot more people, such as a writer, 'voices', actors, musicians, specialist interviewer or commentator. The interpretation of the original idea might call for specially written music or the compilation of sound effects, electronic or actuality. Again, it is easy to get carried away by an enthusiasm for technique, which is why the original brief is such an important part of the process. It should serve as a reference point throughout the production stages.

After selecting the contributors, agreeing fees and persuading them to share the programme's objectives, the producer's task is basically to stay in close touch with them. Remember also that the producer must always be on the look-out for new voices and fresh talent. The next task is to revise the draft scripts and clarify individual aims and concepts so that when everybody comes together in the studio they all know what they are doing and can work together to a common goal. What is needed here is a great sense of timing so that everything integrates at the same moment – the broadcast or recording. Above all the producer gives encouragement. The making of programmes has to be both creative and businesslike. There is a product to be made, restrictions on resources and constraints of time to be observed. But it also calls on people to behave uniquely, to write something they have never written before, to give a new public performance, to play music in a personal way. The producer is asking them to give something of themselves. Contributors, artists and performers of all kinds generally give their best in an atmosphere of encouragement, not uncritical, not complacent, but with a recognition that they are involved in the process of creative giving. To an extent it is self-revealing, and this leaves the artist with a feeling of vulnerability that needs to be reassured by a sense of succeeding in this attempted communication. The producer's role is to provide feedback in whatever form it is required. Therefore, be watchful, perceptive and supportive of contributors, whether they are professional or amateur.

During this time while material is being gathered and ordered there may be a number of permissions to seek. Broadcasters have no rights over and above those of any other citizen, and to interview someone or to make recordings in a home, hospital, school, factory or other non-public place requires the approval of appropriate individuals. In the great majority of cases it is not withheld, and indeed it is most often only an informal verbal clearance that is required. It does not do, however, to record on-site without the knowledge or consent of the legitimate owner or custodian of the property. But neither is it acceptable to be given permission subject to certain conditions, for example to undertake to play back the material recorded and not to broadcast any of it without the further permission of the person concerned. In response to a request to record, the producer must accept only a 'yes' or a 'no' and not be tempted to accept conditional answers. The listening public has the right to believe that the programme they hear is what the producer whose name attaches to it wants them to hear, and is not the result of some secret deal imposed by an outside party. Accountability for the programme rests with the producer: it can seldom be shifted elsewhere.

So the programme takes shape – promising ideas are developed, poor material discarded, items and thoughts explained, put into the listener's context, juxtaposed to give variety, impact, chronology or other meaningful structure. In its design the producer must remember to engage the listener's attention *at the start of the programme* and continue to do so throughout. Now comes the time for it to be broadcast or recorded.

The studio session

Consider the pre-recording of a special speech/music programme. Here again the producer must combine a talent for shrewd business with a yearning for artistic creativity. There are limited resources, particularly of time, and yet people want to give of their best. Some might be in unfamiliar surroundings and are possibly tense, almost certainly nervous. So, arriving in good time, the first task is to set them at their ease and create the appropriate atmosphere. There is no single 'right' way of doing this since people and programmes are all different – the atmosphere in a news studio needs to be different from that of a drama production. A music recording session will be different again from a talk or group discussion. It might be a case of providing coffee all round or even a *small* quantity of something stronger. Any lavish hospitality of this kind is generally much better left until after the programme. But there are two points that a producer must observe. First, make any necessary introductions so that people know who everyone else is and what they are doing, including any technical staff. Second, run over the proposed sequence of events so that individual contributors know their own place in the timescale. These two practices help to reassure and provide some security for the anxious. There is nothing worse for a contributor than standing around wondering what is going on or even whether he or she has come to the right place.

The producer has brought everything needed for the programme: prerecorded inserts correctly labelled and timed with optional out cues; music inserts timed; plenty of copies of the script; coloured pens for making changes; stopwatch; and so on. Everyone in the studio should have a script or running order, and know what is required of them. They should know of any breaks in the session and be sufficiently acquainted with the building as to be able to find their own way to the source of coffee or to the lavatory. There should be enough chairs. Rehearsal, recording or transmission can begin.

Whatever the attitude and approach of the producer, it will find its way into the end product. To get the job done, there has to be a certain studio discipline. 'That's not quite right yet, let's take it again from the beginning' – this is the signal for a new and better concerted effort. 'Everyone check the running order' – means *everyone*. 'Let's start again in 20 minutes', cannot mean people wandering back in half an hour. The producer needs to control, to drive the process forward, to maintain the highest possible quality with the time and talent available. It is generally a compromise. Too strict a control can be stifling to individual creativity, anxiety increases, the studio atmosphere becomes formal and inflexible. On the other hand, lack of control can mean a drifting timescale, an uncertainty as to what is going on and a lowering of morale. The appropriate balance is developed with experience, but the following points apply generally when managing a studio full of people:

1   Use general talkback for announcements to studio participants sparingly. Such use should be brief and should be overall praise or straightforward administration. Never use talkback from the control cubicle into the studio for individual feedback, let alone criticism.
2   Listen to suggestions from contributors for alterations but be positive in making up your mind as to what will be done.
3   In the studio provide plenty of individual feedback to contributors.
4   Keep in mind the needs of the technical, operational or other broadcasting staff – they also want to feel that they are contributing their skills to the programme.
5   Watch the clock, plan ahead for breaks, recording or transmission deadlines. Avoid a last-minute rush.
6   Mark the script as a recording proceeds for any retakes needed or editing required.
7   If rehearsing for a live programme, work out and write down any critical timings for particular items. The 'must be finished by' times are most critical.
8   If the programme is live and is under-running or over-running – and you can do little about it, for example a concert – tell other people who need to know as far in advance as possible and agree what is to happen.
9   Be encouraging. Be communicative. Keep calm. Keep control.

When rehearsing a straight talk, it might be necessary for the producer to sit in the studio opposite the speaker in order to persuade him or her that they are actually talking to someone. The effect of knowing that there is an intent listener is likely to make the vocal delivery much more natural. Moreover, any verbal pedantry or obscure construction in the script is the signal for the producer to ask for clarification. Since it is given in conversational form, this can then become the basis of the suggested rewrite. Almost always, constructive suggestions for simplification, professionally given, are gratefully accepted, often with relief. Producers should remember, however, that their role is not to create in their contributors imitations of themselves. In making suggestions for script changes, or how an actor might tackle a certain line, the producer must be visualising not how they themselves would do it but how that particular performer can be most effective.

In the presence of a live mic, or through a glass window, the producer's non-verbal language is characterised by the following most universal hand signals:

1   the cue for someone to start. The hand is brought from the raised position to point directly at the person to speak. This is also used for handovers from one broadcaster to another;

2   to keep an item going, e.g. to lengthen an interview. The hands are slowly moved apart as if stretching something between them;

3   to start winding up. The index finger describes slow vertical circles in the air, getting faster as the need to stop becomes more urgent;

4   to come to an immediate end, to cut. The hand is passed swiftly across the throat – often with an anguished facial expression.

So much for the producer's responsibility to the other people involved in making the programme, but of course his prime responsibility is towards the listener. Is the programme providing a clear picture of what it is intending to portray? Are the facts correct and in the right order? Is it legally all right? Is it of good technical quality? Is it interesting? Most of these questions are self-evident and so long as they are borne in mind will answer themselves as the programme proceeds. Some questions, however, may require a good deal of searching – for example, is it in good taste?

Taste

Taste: The ability to discern what is of good quality or of high aesthetic standard. Conformity to a specified degree with generally held views on what is appropriate or offensive.

(Concise Oxford Dictionary)

Quality and standards are discussed elsewhere, but what about 'generally held views'? How can we proceed when there is such a huge range of what is acceptable? Nevertheless, it is crucial to know the generally accepted social tenor of the time and the cultural flavour of the place in order to succeed with the general audience and avoid giving unwitting offence. It's possible to decide that a particular programme is only directed to the bawdy revellers in the marketplace with little thought for those who would be shocked at such goings on. Or one could design a programme simply for a cultural or intellectual elite whose acceptable standards are 'more advanced' than those of 'ordinary' folk. So be it, but either way the radio casting of such programmes is, by definition, broad rather than narrow; others will hear and their reaction too must be calculated as part of the overall response. This is especially true when the station content is streamed on the Internet and consequently heard by people with very different cultural backgrounds and values. How then do you interpret 'good taste'? It is virtually impossible, but nevertheless should be borne in mind. On questions of content, material will be designed for a *specific target audience* but the matter of acceptable taste is a much broader issue that the radio producer must try to sense accurately. In stepping outside it, there is a considerable social risk. In deciding on a style of language, or the inclusion of a particular joke which raises the question of good or bad taste, there is one simple rule: 'would I say this to someone I did not know very well in a face-to-face situation?' If so, it is fine for broadcasting. If not, the producer must ask whether the microphone is being used as a mask to hide behind. Simply because the studio appears to be isolated from contact with the audience it is sometimes tempting to be daring in one's assumed relationship with the individual listener. The seeming separation is not a cause for bravado but a reason for sensitivity. The matter of taste in broadcasting so often resolves itself in a recognition of the true nature of the medium.

If in real doubt over a knife-edge decision, it is wise to talk to a senior, more experienced colleague. For this reason the matter of taste is discussed further in Chapter 24.

Ending the session

After a recording and while the contributors are still present, it is often possible to put together some additional material for on-air trailing and promotional use. A specially constructed 30-second piece will later pay dividends in the attention which it can attract.

The producer has a responsibility to professional colleagues who use the same technical facilities. This finds expression in a number of ways:

1   *Studio cleanliness:* the smaller the radio station, the more it operates on a 'leave the place as you would wish to find it' basis. You are probably not required to do the clearing up in detail but it should be left in its 'technically normal' and usable state.

2   *Fault reporting system:* every studio user must contribute to the engineering maintenance by reporting any equipment faults that occur. It is extremely annoying for a producer to be seriously hampered by a studio 'bug' only to find that someone else had the same trouble a few days previously but did nothing about it.

3   *Return of borrowed equipment:* a radio centre is a communal activity; its facilities are shared. An additional microphone, or even a chair, taken from one place to another for a specific programme should be returned afterwards. It might not be the producer who actually does this but it is the responsibility of the borrower to ensure that some other user is not inconvenienced.

If the programme was 'live', the contributors have been thanked and the occasion suitably rounded off. This might mean the dispensing of some 'hospitality' or simply a discussion of 'how it went'. It is generally unhelpful to be too analytical at this stage: most people know anyway whether as a programme it was any good.

Post-production

If the programme was recorded, the producer might have to do some editing. At this stage the running order or script should be marked in detail with the necessary edits, and also with any additional cuts to be made in the light of the overall timing. At a final editing session all the material is heard and a final judgement made on what is to be included. There might still be time here for second thoughts. Should the music tracks be re-mixed? Are the speech/music levels or the stereo balance correct? Does the sound need to be enhanced in any way – by adding echo or special effects treatment? This is probably the last opportunity to hear the programme in its final form to check that what the listener hears is what the producer intends.

Programme administration

The finished programme, together with the necessary paperwork, is then deposited within 'the approved system' so that it finds its way satisfactorily on-air. Often, a producer, while excellent as a creative impresario, artistic director or catalyst in the community, can have a total blind-spot when it comes to simple programme administration. One may be fortunate enough to have an assistant to look after much of this; nevertheless, it is a production responsibility to see that such things are done. The following is a summary of the likely tasks:

1 the completion of a recording or editing report and other details such as library numbers which will enable the programme to get on the air in accordance with the system laid down;
2 the writing of introductory on-air announcements, cues and other presentation material detailing the transmission context of the programme;
3 the initiation of payment to contributors, or at least thanks, giving the transmission details if these were not known at the time of the recording;
4 the supply of programme details covering the use of music or other copyright material. Depending on local circumstances these items will need to be reported to the various copyright societies so that the original performers and copyright holders can receive their proper payment;
5 the issue of a publicity handout, press release or programme billing, for use by newspapers, or programme journal published by the broadcasting organisation. The placing of on-air trails or promos drawing attention to the programme;
6 the reply to any form of feedback generated by the broadcast. While not necessarily representative of listener reaction as a whole, an understanding response forms an important part of a producer's public accountability. Apart from the PR value to the particular radio service, such enquiries and expressions of praise or criticism constitute a consumer view which should not be treated lightly.

Technician, editor, administrator, and manager

It will be obvious that a producer has to be a wizard of multitasking. In summary, the job is in four parts: the technical and operational, the editorial, the administrative and the managerial. The technical part is to do with the proper operational use of the tools of the trade, knowing when and how to use programme-making equipment. The editorial function is about ideas and decisions. It is to do with making judgements about what is and is not appropriate and legal for a particular programme. It is about backing hunches and taking risks, about choosing and commissioning material. The administrative part is procedural – following agreed systems of paperwork to do with contracts, running orders and scripts, expenses and payments, overtime, leave applications, studio bookings, copyright returns, logging transmissions, verifying traffic, reporting faults, requisitioning music and Fx, replying to letters, texts and emails, etc. But the producer is also a manager, managing projects called programmes. This means setting objectives for other people, monitoring their progress, controlling, organising and motivating them in their work. This will be the person who disciplines the habitual latecomer, who resolves conflict between contributors, and encourages the new and uncertain.

The journalist and the DJ, the presenter and the performer, frequently regard themselves as the pre-eminent component in a mixed sequence. The producer as manager must create the team where each is sufficiently confident to support the other. As manager, the producer recognises the financial responsibility of the job – agreeing the budget, monitoring expenditure and taking action to remain within the allocation. On occasion, it might be necessary to argue the case for more, but the editorial and managerial aspects cannot be separated. Editorial decisions *are* resource decisions. Like any other manager, he or she is primarily in charge of the quality of what happens; the bottom line is the standard of the programme – this is the person who ultimately says what is good enough and what isn't. At the end of the day the producer decides and communicates what is to be done, to what standard, by when, by whom, at what cost. That's editorial management.

Having completed the programme, the producer is already working on the next. For some it is a constant daily round to report new facts and discover fresh interests. For others it might be a painstaking progress from one epic to another. Unlike the purely creative artist, the producer cannot remain isolated, generating material simply from within. The role is that of the communicator, the interpreter who attempts to bring about a form of contact that explains the world a little more. For the most part it is an ephemeral contact leaving an unsubstantial trace. Radio works very much in the present tense; reputations are difficult to build and even harder to sustain. The producer is rarely regarded as any better than his or her last programme.

24

The executive producer

Senior producer, programme manager, controller, organiser or director, the producer's boss comes in a wide range of guises and titles. His or her essential job is not so much to make programmes as to make programme makers. Whether responsible for a programme strand, a station, or a network, this senior editorial figure is there to listen to the output and provide feedback on it to the producers. A handwritten note left on the producer's desk – a habit of former BBC Managing Director of Radio, David Hatch – is particularly effective. This head of programmes will also lead the programme meetings where there is a systematic discussion of the output – there is more detail on this procedure in the next chapter. The purpose of this is continual improvement through professional encouragement and evaluation, critique, discussion of ideas, programme development, and so on. Without this, the output stagnates.

This senior role also has the job of 'gatekeeper' – the person who says yes or no to programme suggestions, who takes risks with new ideas, who spots talent and encourages new presenters – or gets rid of them. In his or her mind all the time are questions such as:

- What is the purpose of this station?
- What are my success criteria?
- Are the current programmes achieving these?
- If not, how can I improve – the programmes?
  – the presenters?
- What stops the output being better in my terms?
- How can I overcome this?
- What is my competition doing?

Finding answers to these questions is what should be driving the station forward. This doesn't have to be a lonely quest, for everyone can be involved. Perhaps the first lesson for any executive is that while they might make the major decisions, they are not the only person with good ideas.

Station management

The executive producer has to combine the two aspects of the job – the people and the task. To concentrate solely on the people might make for good relationships but what is the quality of the work? On the other hand, to put the whole emphasis on the task may create an unduly work-driven atmosphere in which no one really thrives. The balance is seen in a management style that bears fruit in the creative work done by motivated people who enjoy being part of a team.

The primary tasks are fourfold, to do with (a) maintaining all the necessary relationships with people outside the station, (b) finding, recruiting and training new staff and talent, (c) keeping to a budget of income and expenditure, and (d) running the station – actually keeping the show on the road.

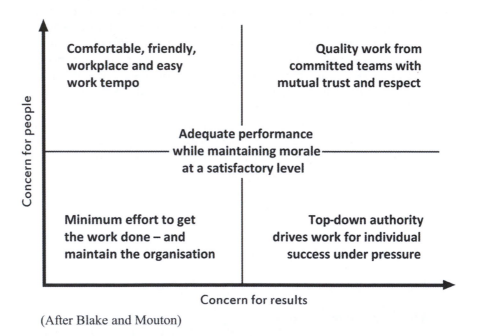

(After Blake and Mouton)

Figure 24.1 A form of the Blake–Mouton Grid illustrating the likely outcome of management concentrating on the tasks as opposed to an emphasis on the people. To do both fully can result in highly motivated teams

Something of the complexity of the first of these is illustrated on p. 19, where management is shown as a coordinating focus. Finding good programme people is often difficult with risks to be taken, but minimised through proper training. An additional problem is that when you have good creative people, they so often want to spend more than you can afford!

A useful shorthand of what management entails can be set out in four steps.

1   *Decide what you want to achieve.* Have a clear objective in line with your overall purpose for the station. This requires a sense of direction and some clarity about priorities.
2   *Decide the means of reaching your objective.* The process of getting to your goal will depend on the resources available – money, time, people and their skills, equipment. It requires planning and organising ability and perhaps some persuasion or negotiation with others involved.
3   *Integrate the effort of the team.* Lots of communication so that everyone knows what's expected of them, and by when. Team building and motivation are important.
4   *Monitor progress and provide feedback.* Keep across things and evaluate how the task, the money and the people are going. Resolve any conflict, change things if necessary, and encourage success.

This is not by any means always easy. Time pressures alone may mean that insufficient attention is given to any part of the job. Clear priorities and good delegation certainly help.

Staff development

Inherent in the business of management is the whole area of developing the people you are responsible for – a matter very much in the remit of the executive producer. This will include the process of annual appraisal – taking time formally to review a person's work over a period of time to assess the problems, difficulties and successes, asking how it could be made better, more rewarding or challenging. This can lead to a personal development plan for each individual. But improving people performance should also happen all the time in many more informal ways. For example, giving feedback on programmes, as discussed in the next chapter, and holding meetings to analyse specific problems or to extend the knowledge and understanding of current issues. Some stations form groups that meet to confront problems together and share 'best practice'. A scheme called SONAR – Sharing Opportunities in Nations and Regions – operates among groups of linked BBC stations as a forum to discuss solutions to particular questions. Visiting another station and, if necessary, buying in specific expertise for a day can contribute much to personal learning and confidence. Not as formal as the skills courses discussed in Chapter 26, such experience adds to the process of making sure that your staff are increasingly better equipped to do the job.

Scheduling

Some would describe the drawing up of a good programme schedule as perhaps the highest art form in radio. Certainly there are schedulers who seem to have just the right knack of placing programmes and people where the target audience most appreciates them.

The basis of a schedule that works is (1) deciding the role and purpose of the station – *what do I want to say, to whom, and with what effect*? and (2) knowing the needs, likes and dislikes, habits, work patterns and availability of the intended audience.

The first of these comes from perceiving a gap in the market and supplying what is needed – rock music, pop, classical, news, speech-based programming, phone-in chat, community services, etc. No station can do everything and deciding the limits of programme provision is crucial. The second is discovered by audience research in whatever form possible. This may be available from the national regulator – FCC, Ofcom, etc. – or there may be a specialist organisation set up to do it, from whom you can commission or buy the information needed. In the absence of such a facility, it is possible to infer the information from other sources, e.g. government statistics, newspaper circulation figures, population breakdown by socio-economic groups, etc. The scheduler ideally knows what the target audience is doing at different times of the day – and night – and at the weekend; what time the intended audience wakes up and goes to bed; when the home of the intended audience is busy – typically at breakfast time – so as to keep items short, with plenty of time checks; when the home is quiet and the listener available for more extended listening; when the typical listener is likely to be travelling – by car – 'drivetime'; when the time is right for news or relaxation; how to capitalise on a popular programme by following it with something really appropriate for the inherited audience.

Audience habits and therefore listening habits are very different at the weekends. Broadcasters to Muslim communities know that this mostly applies to Friday and Saturday, while Sunday is generally a normal working day. The schedule also needs to recognise special days – anniversaries, holidays, the New Year, Christmas and Easter, Hanukkah, Ramadan and other religious festivals.

RADIO 3 SCHEDULE

Saturday	Sunday	Monday	Tuesday	Wednesday	Thursday	Friday
Morning						
07:00 Breakfast	07:00 Breakfast	06:30 Breakfast	06:30 Breakfast	06:30 Breakfast	06:30 Breakfast	06:30 Breakfast
09:00 CD Review	09:00 Sunday Morning	09:00 Essential Classics	09:00 Essential Classics	09:00 Essential Classics	09:00 Essential Classics	09:00 Essential Classics
12:15 Music Matters	12:00 Private Passions	12:00 Composer of the Week	12:00 Composer of the Week	12:00 Composer of the Week	12:00 Composer of the Week	12:00 Composer of the Week
Afternoon						
13:00 Lunchtime concert	13:00 Lunchtime concert	13:00 Lunchtime concert	13:00 Lunchtime concert	13:00 Lunchtime concert	13:00 Lunchtime concert	13:00 Lunchtime concert
14:00 Saturday Classics	14:00 The Early Music Show	14:00 Afternoon on 3	14:00 Afternoon on 3	14:00 Afternoon on 3	14:00 Afternoon on 3	14:00 Afternoon on 3
	15:00 Choral Evensong			15:30 Choral Evensong		
16:00 Sound of Cinema	16:00 The Choir	16:30	16:30	16:30	16:30	16:30
17:00 Jazz Record Requests		In Tune	In Tune	In Tune	In Tune	In Tune
	17:30 Words and Music					
Evening						
18:00 Jazz Line-Up	18:45 Sunday Feature	18:30 Composer of the Week	18:30 Composer of the Week	18:30 Composer of the Week	18:30 Composer of the Week	18:30 Radio 3 Live in Concert
19:30 Radio 3 Live in Concert	19:30 Radio 3 Live in Concert	19:30 Opera on 3	19:30 Radio 3 Live in Concert	19:30 Radio 3 Live in Concert	19:30 Radio 3 Live in Concert	19:30 Radio 3 Live in Concert
21:45 Between the Ears	22:00 Drama on 3		22:00 Free Thinking	22:00 Free Thinking	22:00 Free Thinking	22:00 The Verb
22:30 Hear and Now		22:45 The Essay	22:45 The Essay	22:45 The Essay	22:45 The Essay	22:45 The Essay
		23:00 Jazz on 3	23:00 Late Junction	23:00 Late Junction	23:00 Late Junction	23:00 World on 3
00:00 Geoffrey Smith's Jazz						
Night						
01:00 Through the Night	00:30 Through the Night	00:30 Through the Night	00:30 Through the Night	00:30 Through the Night	00:30 Through the Night	01:00 Through the Night

Figure 24.2 A simple method of illustrating the schedule – for publicity purposes – when the daily programmes are stripped across the week, with variations at the weekend (Courtesy of BBC Radio 3)

Television schedules make much use of library material – feature films or repeats. Radio may also be able to do this, depending on the format. For example, a live programme recorded off-air in the evening may be usefully repeated for a different audience in the morning, or during the night, in the same week. Schedulers have to remember that habitual listeners to any programme do not like change – a change of presenter for a holiday break is understandable, but a new presenter may take months to be accepted. A change of timing for a programme might mean that the regular listener cannot then hear it at all. This should, therefore, only be done after extensive and conclusive research. Continual tinkering with the schedule simply irritates, giving rise to the cry from the audience – 'who do they think this station belongs to?' – a good question.

Rescheduling

Sooner or later an event occurs that will cause the cancellation of the planned schedule. This may happen in one of three ways.

First, in the short term, following the death of a national leader or other eminent person it might be appropriate to play 'solemn music'. This is especially so when the death is unexpected or is violent. But what does such music mean within the context of the station's format? How should the station react to the assassination of a head of state as opposed to the anticipated natural death of an ageing president? Standby music of an appropriate kind should always be available for such emergencies. However, with the best will in the world it may turn out to be quite wrong – like the music station which, on the death of the Pope, found that its first standby track was a non-vocal version of 'Arrivederci Roma'!

The second, longer-term rescheduling applies either when a death requires a longer period of mourning or in the event of a major national disaster such as a serious terrorist incident or outbreak of war. This becomes not simply a matter of solemn music, but of revising the whole output, perhaps to allow for additional news slots, and of changing programmes. This was certainly the case in the days following the catastrophe of 9/11, on the death of Diana, Princess of Wales, in India after the assassination of Prime Minister Gandhi, and on many occasions during the Iraq and Syria wars. It is also necessary to advise presenters on how to be sensitive to the public mood, for it is here that off-the-cuff remarks or anarchic humour can cause great offence among the audience.

The third circumstance for rescheduling requires special vigilance. It most frequently occurs after a one-off event of significant national or local impact – a tragedy at sea or an air crash, a particularly brutal killing or terrorist bomb. In itself the event might not justify the cancellation or suspension of the planned schedule – but these things have a habit of coinciding with something in the output that is quite inappropriate in the new context. Programme titles – even song titles – have to be scrutinised to avoid the unfortunate juxtaposition.

Strategic planning

So much management time is given to short-term crises – 'putting out fires'. But it is essential also to take a broader view, for one of the roles of senior editorial staff is to give a sense of movement to the output – that this year is not the same as last, but things have moved on. The station is not only up to date in the sense of being topical, but it is also advancing in its technology, its skills and its relationship with the listener – this in particular is a current preoccupation – to become more *personal* with the individual listener.

Strategic planning requires a response to such questions as:

- Where do I want this station to be in five years' time?
- What do I want to be known for?
- What should we be doing that's new?

It is armed with a sense of vision, that the strategic planner begins to glimpse new programme ideas and, in discussion with others, slowly turns them into reality. A nice definition of 'vision' is: a compelling picture of the future that is better than today. So, we have a three- to five-year plan not simply in the business sense but one that envisions programme development.

Even with a single programme strand it is possible to think strategically. For example, an educational series is an obvious case of leading the listener to a point at the end that was not the same as the beginning. The idea is to cover a body of knowledge or skill that the listener acquires – understanding a period of history, or learning a language. But this concept of 'taking the listener somewhere' need not be confined to formal education, it applies to all leadership, i.e. the idea of having a vision of where you want to get to, and enabling the listener to get there. Competitive quiz programmes begin with initial rounds and semi-finals, culminating in a final. DJ programmes can be designed so that they move towards a live pop concert or festival.

In such ways the output, although consistent, is not the same week after week and year after year. It is changing and developing, and extending the audience as it does so – not in a haphazard, accidental way but as part of a planned strategy. Even the public servant may exercise such leadership.

Commissioning programmes

A radio organisation might not produce all its programmes in-house. It may approach an outside production facility or programme supplier to fill part of its schedule. How then does it ensure that it gets the programming it wants – that the programme will be appropriate for the audience listening to the station at that time?

The commissioning of programmes varies widely in different parts of the world – not least in the matter of money, of who pays whom. Some organisations will specify what is wanted and pay a freelance producer or independent production facility to supply it. Others will offer airtime to a programme maker who then buys it in order to reach the audience which the broadcaster is targeting – the producer recouping the cost from a sponsor or other donor agency. The danger here is that the schedule could become a patchwork of unrelated, independent programmes with little or no continuity. However, even when airtime is sold, it is possible to create a coherent schedule through a positive commissioning procedure.

The Programme Commissioning Form illustrated on pp. 352–53 is a simple basis for the procedure used in an African context.

Retaining editorial control means that the executive producer has to specify with some precision what is wanted. Programmes are commissioned within the framework of a known and agreed editorial policy. The purpose of the programme or series has to be stated, not in great detail, but sufficiently to evaluate the result in terms of its intended audience and desired response.

For example, a farming programme in East Africa might be specified as follows:

> A 30-minute programme in the Hausa language designed for the working community to enable them to make more efficient use of the land, and to promote self-sufficiency through improved crops by the better use of simple tools, seeds and other agricultural products, in line with government policy. The speech-based programme format may include scripted talk, interview, discussion or drama. No commercial sponsor shall dictate programme content.

The specification for a comedy programme in the UK could be:

> The purpose of this late Saturday evening programme is to amuse and entertain especially through the satirising of the events of the previous week in the world of politics and publicly reported news. The format may include talk, sketches, songs, dialogue, etc. using a variety of voices. The programme must be self-regulating in matters of libel, good taste, etc.

Essential details regarding cost, technical specification and programme delivery must also be negotiated. To ensure that the agreement is as watertight as possible, a Programme Commissioning Form is drawn up. This will be given to the programme supplier, together with any other statements of editorial policy and a legal contract.

The BBC, when commissioning programmes for Radio 4, is very keen that programme suppliers describe what the listener will hear – not simply the proposed programme technique and content, but how it will sound on radio. To help with correctly identifying the target audience, the Commissioning Guidelines include audience profiles for different times of the day and each slot has a detailed commissioning brief, with a profile of the audience at that time – the age, proportion of men and women, social groups, and what they are likely to be doing – together with a broad description of the genre and some detail of the programme wanted. Audience need is the key factor, linking what radio can offer to listener lifestyle. Full details of the BBC commissioning process are online.

A possible danger of commissioning exists when the process is applied to too much of a station's output. The station begins to lose its own in-house expertise of production, and simply becomes the broadcaster of other people's programmes.

Codes of Practice

The executive producer is the guardian of the programme output, ensuring that it does not fall foul of the legal necessities placed on broadcasters. But it is not simply libel and race relations and the other laws of the land that have to be considered. We have mentioned before the mass of regulatory constraints or Codes of Practice – some voluntary, some not – which come with a station's licence to broadcast.

As we have seen, the national regulator will lay down its own Codes of Guidance – on Fairness and Privacy, on Taste and Decency and general programme standards. These are useful documents of which all broadcasters should be aware (see pp. 81 and 385–86).

The UK Broadcasting Act (1990) says

> nothing shall be included in programmes which offends against good taste or decency or is likely to encourage or incite to crime or to lead to disorder or to be offensive to public feeling.

The problem, of course, is in the interpretation of this requirement – it is difficult to challenge or provoke without offending someone. No one wants blandly soporific radio. The Codes usefully distinguish between

XYZ RADIO   PROGRAMME COMMISSIONING FORM

Date: _____

Programme to be supplied by:_____

Address: _____   Phone:_____

            _____   Email:_____

            _____

Programme Title:_____ No. of Programmes:_____

Service Block:_____ Language:_____

Transmission: Day_____ Time_____ Duration_____

Purpose of Programme:

Primary target audience:

Programme format:

Specification of content:

Limits of commercial sponsorship:

Cost / Budget and Payment details:

Goals for listener response:

Procedure for follow-up, correspondence etc:

Technical specification – peak levels, file details:

Programme delivery:

Copyright details:

Other requirements:

This programme commission to be reviewed on_____

Signed for XYZ Radio_____

Signed for programme supplier_____

rules and advice in attempting to define and pinpoint likely matters of contention:

- sexual explicitness;
- blasphemy;
- the portrayal of violence;
- bad language, especially in programmes for younger listeners;
- the danger of misleading people through simulated news bulletins and reconstructions;
- the interviewing of children or criminals;
- appeals for donations;
- the whole area of the occult, including horoscopes and hypnotism;
- rules for lotteries, competitions and so on.

The Codes also raise issues such as the need to avoid suffering and distress when interviewing people involved in personal tragedy, the use of hidden microphones, about not causing offence against religious sensibilities – or the bereaved. They outline the care needed in areas such as drugs, alcohol, race, gender stereotypes, people with disabilities or mental health problems, etc. They draw attention to the lyrics of songs that glamorise crime or invite aggression.

There is always a balance to be struck between the broadcasters' right to freedoms of speech and expression and the listeners' right not to have their own liberties infringed. While programmes should be creative and perhaps experimental, challenging convention, they also need to recognise their responsibility to different audiences. Radio is not for on-air graffiti. It is not an anonymous communication, but always signed and attributed. Suffice it to say that since ignorance of the law is no defence, senior programmers should essentially acquaint themselves with this part of the professional infrastructure – designed as much for the protection of broadcasters as for the public. Knowing where the boundaries are means that it is possible, when necessary, to test and challenge them.

Complaints

When the broadcaster is thought to be 'wrong' the listener tends to complain in one of three categories:

1   on hearing something that affects the listener personally, such as what they believe to be an incorrect news report in which they were involved, or a lack of balance in reporting a particular argument. They are aggrieved at being treated unfairly;

2   a historical inaccuracy, a wrong date, or a misquote – an error of fact;
3   when they are offended by something said – bad language, overtly
     sexual comment, inappropriate humour – a matter of taste.

While all media attract the inveterate letter writer or tweeter, especially seeming eccentrics, a station that takes its audience seriously should
regard what a listener says with some importance. People like a feeling
of having some control – at least that their complaints are taken note of,
that their views matter. They therefore need a reply, preferably on the air
in the same time slot as the item that caused their offence. An error of
fact requires an apology and a correction; a difference of view requires an
explanation while, at the same time, respecting the other opinion.

Research indicates that listeners and viewers have clear images in their
minds about the channels and stations they use. Many complaints arise
because a programme was at odds with their expectation of it – a family
programme was found to be offensive or shocking, perhaps because it was
broadcast at an inappropriate time of day, or a provocative programme
was too tame and did not sufficiently challenge or provoke. The complaint
is often not in the deed but in the unexpected or uncharacteristic. A station, a programme, must live up to its promise. It must satisfy – if possible
exceed – the expectations the listener has of it. In the event of serious
complaint, the executive role is to calm the anger or allay the fears of the
listener; it is to defend the producer, and deal internally with the cause.

Website

Senior station staff have to decide what kind of presence they want on the
Internet and what resources to devote to it. Undoubtedly, the dissemination of the audio output in this way is a part of broadcasting, as is distribution by cable – many stations, of course, use the Internet as their only
means of transmission.

The complexity of a site is governed not so much by the amount of
information provided or the number of linked pages but by the rate at
which page content is changed. Categories of increasing complexity and
therefore input effort and cost are as follows:

1   a single static page with details of station name, logo, frequency,
     address, coverage area and purpose statement. Seldom needs updating;
2   a page that is updated periodically, e.g. to show variations in the
     weekly schedule. It may also carry advertising;
3   page(s) that are changed continually to give the latest news and weather
     together with other information, using text, pictures and symbols;

Figure 24.3 A station website illustrating the channel with programme details and pictures, together with a 'Listen Live' button and links to other parts of the organisation (Courtesy of BBC Radio 3)

4   a website that provides an audio output of the most recent news bulletin and news updates, plus station promos and announcements;
5   provision of the continuous live audio output;
6   a fully interactive site that offers access to live camera output of the studios, the ability to re-hear recent programmes, and keyboard text discussion with users in real time.

The possibilities for website design are virtually limitless. They include links to presenter pages, sound clips of recent news output – with pictures, podcasts, a tour of the station, webcams looking out of the window at the traffic or weather, special announcements, coming events, classified ads – jobs and cars, chart music listings, details of music played, interactive opinion-seeking questions, advice pages, etc. In order to attract new viewers to your pages, they should be 'search friendly'. That is, their main title should avoid obscure or descriptive words, and instead use the key words that people will use when searching. This is called Search Engine Optimisation – SEO – attracting search engines. Headlines that begin with the proper names of people in the news are an obvious starting point.

The more complex the design of the home page, especially with regard to pictures and animated graphics, the longer it will take to download, and therefore the greater the potential for frustration and the user not completing the connection. When this occurs, the casual surfer is unlikely to return to the site. There is therefore much to be said for making the first entry page relatively straightforward, giving the basic information and a menu of links, designed in bold colours with succinct text reversed out on a dark background.

Inputting data – text, graphics and audio – is a relatively cheap process, but the greater the rate at which it is changed, the faster and more expensive must be the data connection with the service provider. Before starting a website, the webmaster should explore other radio websites, be prepared to negotiate over costs and if necessary seek specialist advice.

A list of some useful and interesting radio websites is given on p. 387.

Archival policy

- What of the output should be kept?
- In what form?
- How should it be catalogued?

Decisions regarding the archives inevitably fall to the senior programme staff – producers themselves have a tendency to want to keep everything.

It may be that the whole of the station/network output is automatically kept on a digital logger. But why keep archives anyway? This will depend considerably on the nature of the output. A music station may want to keep very little; a news station, potentially at least, will want to keep most things in the short term since any story has the capacity to grow. Keeping news items for a week or two is sensible, either to follow a developing story or to produce a pull-together of the week's news. When archives are stored on computer, storage space limitations may mean it is necessary to transfer less frequently accessed material to some other medium, such as local high-capacity storage devices, or 'the Cloud'. This transfer of material will also act as a reliable 'back-up' to the valuable archive resource. But of course, the indexing must be detailed and thorough – by date, subject, title and producer/presenter name – to allow for efficient searches and speedy retrieval of specific items when required in the future. It should not be forgotten that the purpose of an archive is not storage but retrieval.

In the past, many archives built up a substantial number of programmes recorded on quarter-inch tape which could progressively be transferred to digital format. If this is impractical, special care is needed in the long-term storage of tape. The enemies of any magnetically sensitive medium, including audio, DAT and cassettes, are excessive heat, damp, vibration or banging, and external electrical or magnetic fields, for example from some microphones. For the long-term storage of tape the recommended temperature is 10°C (50°F).

All archived material should ideally carry the following information:

- subject matter;
- name and address of speaker, musicians, performers, etc.;
- location of recording;
- date of recording;
- details of copyright material;
- duration;
- catalogue number or category.

Items designed for the archives should be labelled in this way before archiving as a matter of routine, whatever the recording medium.

Without encouraging the hoarding of unnecessary clutter, stations should also consider the keeping of certain categories of paper – schedules, concert programmes, posters, publicity leaflets, public appeals and so on. It is simply another way of keeping track of where the station has been.

Good decisions about archives and a sensible system will mean that repeats, year-end programmes, anniversaries and other retrospectives are easy to produce and satisfyingly representative. The station then deals not only with the ephemera of today but becomes the repository for the community or nation – the guardian of its oral history.

25

Programme evaluation

A crucial activity for any producer is the regular evaluation of what he or she is doing. Programmes have to be justified. Other people may want the resources or the airtime. Station owners, advertisers, sponsors and accountants will want to know what programmes cost and whether they are worth it. Above all, conscientious producers striving for better and more effective communication will want to know how to improve their work and achieve greater results. Broadcasters talk a great deal about quality and excellence, and rightly so, but these are more than abstract concepts, they are founded on the down-to-earth practicalities of continuous evaluation.

Programmes can be assessed from several viewpoints. We shall concentrate on three:

- Production and quality evaluation
- Audience evaluation
- Cost evaluation.

Production evaluation

Programme evaluation carried out among professionals is the first of the evaluative methods and should be applied automatically to all parts of the output. However, it is more than simply a discussion of individual opinion, for a programme should always be evaluated against previously agreed criteria.

First, the basic essential of the proper technical and operational standards. This means there is no audible distortion, that intelligibility is total, that the sound quality, balance and levels are correct, the fades properly done, the pauses just right and the edits unnoticeable.

Second, what is the programme for – what does it set out to do? A statement of purpose should be formulated for every programme so that it has a specific direction and aim. Without such an aim, any programme can be held to be successful. So, what is the 'target audience'? What is the programme intending to do for that audience? How well does it set about doing it? (Whether it actually succeeds in this is an issue for audience evaluation.)

Third, is a professional evaluation of content and format. Were the interviews up to standard? The items in the best order? The script lucid, and the presenter communicative? These questions are best discussed at a regular meeting of producers run by the senior editorial figure. It is important at these meetings that everyone has heard the programme under review – either off-air or played at the start of the meeting. It is good to have some positive feedback first – what did we like about it? In what ways did we feel it appealed to the target audience? What could be improved? Is there a follow-up? Everyone should be encouraged to take part in this evaluation, including of course the programme's producer. The focus of the discussion should be on constructive improvement rather than finding fault. When producers are first involved in playback and discussion sessions, they are bound to show some initial defensiveness and sensitivity over their work. This has to be understood. It is best minimised by focusing the discussion on the programme, not on the programme maker. The process is essentially about problem-solving and creatively seeking new ideas in pursuit of the programme aim.

Programme quality

Quality is a much overused word in programme making. Is it only something about which people say 'I know it when I see it or hear it, but I wouldn't like to say what it is'? If so, it must be difficult to justify the judges' decisions at an awards ceremony. Of course there will be a subjective element – a programme will appeal to an individual when it causes a personal resonance because of experience, preference or expectation. But there must also be some agreed professional criteria for the evaluation of programme excellence – quality will mean that at least some of the following eight components will be in evidence.

First, *appropriateness*. Irrespective of the size of the audience gained, did the programme actually meet the needs of those for whom it was intended? Was it a well-crafted piece of communication that was totally appropriate to its target listeners, having regard to their educational, social or cultural background? Programme quality here is not about being lavish or expensive; it is about being in touch with a particular audience, in order exactly to serve it, providing with precision the requirements of the listener.

Second, *creativity*. Did the programme contain those sparks of newness, difference and originality that are genuinely creative, so that it combined the science and logic of communication with the art of delight and surprise? This leaves a more lasting impression, differentiating the memorable from the dull, bland or predictable.

Third, *accuracy*. Was it truthful and honest, not only in the facts it presented and in their portrayal and balance within the programme, but also in the sense of being fair to people with different views? It is in this way that programmes are seen as being authoritative and reliable – essential of course for news, but necessary also for documentary programmes, magazines and, in its own way, for drama.

Fourth, *eminence*. Quality acknowledges known standards of ability in other walks of life. A quality programme is likely to include first-rate performers – actors or musicians. It will make use of the best writers and involve people eminent in their own sphere. This of course extends to senior politicians, industrial leaders, scientists, sportsmen and women – known achievers of all kinds. Their presence gives authority and stature to the programme. It is true that the unknown can also produce marvels of performance, but quality output cannot rely on this and will recognise established talent and professional ability.

Fifth, *holistic*. A programme of quality will certainly communicate intellectually in that it is understandable to the sense of reason, but it should appeal to other senses as well – the pictorial, imaginative or nostalgic. It will arouse emotions at a deeper and richer level, touching us as human beings responsive to feelings of awe, love, compassion, sadness, excitement – or even the anger of injustice. A quality programme makes contact with more of the whole person – it will surely *move* me.

Sixth, *technical advance*. An aspect of quality lies in its technical innovation, its daring – either in the production methods or the way in which the audience is involved. Technically ambitious programmes, especially when 'live', still have a special impact for the audience.

Seventh, *personal enhancement*. Was the overall effect of the programme to enrich the experience of the listener, to add to it in some way rather than to leave it untouched – or worse to degrade or diminish it? The end result may have been to give pleasure, to increase knowledge, to provoke or to challenge. An idea of 'desirable quality' should have some effect which gives, or at least lends, a desirable quality to its recipient. It will have a 'Wow' factor.

Eighth, *personal rapport*. As a result of a quality experience, or perhaps during it, the listener will feel a sense of rapport – of closeness – with the programme makers. One intuitively appreciates a programme that is perceived as well researched, pays attention to detail, achieves diversity or depth, or has personal impact. In short, is *distinctive*. The listener identifies not only with the programme and its people, but also with the station. Programmes which take the trouble to reach out to the audience earn a reciprocal benefit of loyalty and sense of ownership.

Combining accuracy with appropriateness, for example, means providing truthful and relevant news in a manner that is totally understandable to the intended audience at the desired time and for the right duration. Quality news will also introduce creative ways of fairly describing difficult issues, so leaving listeners feeling enriched in their understanding of the world.

Programme quality requires several talents. It takes time to think through and is less likely to blossom if the station's primary requirement is quantity rather than excellence. It cannot be demanded in every programme, for creativity requires experiment and development. It needs the freedom to take risks and therefore occasionally to make mistakes. Qualitative aspects of production are not easy to measure, and it may be that this is why an experienced programme maker determines them intuitively rather than by logic alone. Nevertheless they have to be present in any station that has quality on the agenda, or aspires to be a leading broadcaster.

Quality allied to programming as a whole – especially that thought of as public service – takes us back to criteria described on p. 15. Quality in this sense will mean a diversity of output, meeting a whole range of needs within the population served. It will reflect widely differing views and activities, with the intention of creating a greater understanding between different sections of the community. Its aim is to promote tolerance in society by bringing people together – surely always the hallmark of quality communication.

The cynic will say that this is too idealistic and that broadcasting is for self-serving commercial or even propagandist ends – to earn a living, and provide music to ease the strain of life for listeners. If this is the case, then simply evaluate the activity by these criteria. The many motivations for making programmes and the values implicit in the work are outlined on p. 17. What is not in doubt is the need to evaluate the results of what we do against our reason for doing it.

Audience evaluation

Formal audience research is designed to tell the broadcaster specific facts about the audience size and reaction to a particular station or to individual programmes. The measurement of audiences and the discovery of who is listening at what times and to which stations is of great interest not only to programme makers and station managers but also to advertisers or sponsors who buy time on different stations. Audience measurement is the most common form of audience research, largely because of the importance attached to obtaining this information by the commercial world.

Several methods of measurement are used and, in each, people are selected from a specific category to represent the target population:

1   people are interviewed face to face, generally at home;
2   interviews are conducted by phone.

In both 1 and 2, respondents are asked about listening 'yesterday'.

3   respondents complete a listening diary;
4   a selected sample wear a small personal meter.

The more detailed the information required – the number of people who heard part or all of a given programme on a particular day, and what they thought of it – the more research will cost. This is because the sample will need to be larger, requiring more interviews (or more diaries) and because the research has to be done exclusively for radio. If the information required is fairly basic – how many people tuned to Radio 1 for *any* programme last month and their overall opinion of the station – the cost will be much less since suitable short questions can be included in a general market survey of a range of other products and services.

It should be said that constructing an adequately representative sample of the target population is a process requiring some precision and care. For example, we know that the unemployed are likely to be heavy users of the media generally, yet as a category they are especially difficult to represent. Nevertheless, an adequate sample should cover all demographic groups and categories in terms of age, gender, social or occupational status, ethnic culture, language and lifestyle, e.g. urban and rural. It should reflect in the proper proportions any marked regional or other differences in the area surveyed. This pre-survey work ensures, for example, that the views of Hindi-speaking students, male and female, are sought in the same ratio as the over-65s living in rural areas as these two categories exist in the population as a whole. Only when the questioning is put to a correctly constructed sample of the potential audience will the answers make real sense.

A further important definition relates to the meaning of the word 'listener'. Many sequence programmes are two or three hours long – do we mean a listener is someone who listened to it all? If not, to how much? In Britain, RAJAR – Radio Joint Audience Research – is a Joint Industry Committee or JIC, comprising a not-for-profit agency formed by all UK radio stations, together with all the major advertisers and agencies. This is the body that conducts surveys of all radio listening using a self-completion diary. Here a listener is defined as someone who listens for a minimum of five minutes within a 15-minute time segment. The figure for *weekly reach* is the total number of such listeners over a typical week, expressed as a percentage of the population surveyed. The resulting figures have even more value over a period of time as they indicate trends in programme listening, seasonal patterns and changes in a specific audience, such as car drivers. This allows comparisons to be made between different kinds of format and schedule. Research, therefore, helps us to make programme decisions, as well as providing the figures for managers to justify the cost of airtime. For comparison between stations, researchers talk about audience *share* (see pp. 3, 398, 399).

Personal meters

The first two methods of radio audience measurement listed above depend on listeners' memories and on whether they actually know what radio stations they have been listening to, while the third method, using diaries, depends on the respondents' discipline and commitment. With all three methods, a lot depends on the respondents' own self-evaluation of their listening activity. However, there are electronic methods that get round these limitations. For example, a selected number of people agree to wear a small personal programme meter that records what stations the wearer has been 'in the audible presence of' during his or her waking hours. There are two types.

One is a small pager-sized meter that relies on all radio and TV stations having an embedded but inaudible identity code that it can recognise and record. The other system involves wearing a wristwatch that records and compresses four seconds of sound during every minute that the watch is worn. These compressed digital clips are then compared with the output of all radio and TV stations during the period being measured. At the end of a period – daily or weekly – the data contained in the device is sent to a

Figure 25.1   A wristwatch worn as a listening meter (Courtesy of GfK)

computer via a telephone line for analysis. Such systems have been in regular use in several countries. One major advantage is that they can provide comparable data for both radio and TV from the same source. But there are also several disadvantages and these are very likely to bring the use of these personal people meters to an end. One is the high cost involved, and another is that these meters produce figures that are rather different from those produced by diaries and questionnaires, partly at least because they record unintended listening. But perhaps the most important disadvantage is that neither of the two types of meter can effectively capture listening on headphones, an activity which is growing in importance.

Research panels

Another method of research is through listening panels scattered throughout the coverage area. Such groups can, by means of a questionnaire, be asked to provide qualitative feedback on programmes. Panel members will be in touch with their own community and therefore may be chosen to be broadly representative of local opinion. Once a panel has been established, its members can be asked to respond to a range of programme enquiries which, over time, may usefully indicate changes in listening patterns.

Such panels are also appropriate where the programme is designed for a specific minority such as farmers, the under-fives, the unemployed, hospital patients, adult learners or a particular ethnic or language group. Here, the panel might meet together to discuss a programme and provide a group response. Visited by programme makers from time to time, a panel can sustain its interest by undertaking responsible research for the station. But beware any such sounding-board which is too much on the producer's side. Too close an affiliation creates a desire to please, whereas programme makers must hear bad news as well as good. Indeed, one of the key survey questions is always to find out why someone *did not* listen to my programme.

Panels have become very much more important in the age of the Internet and email. These ways of making contact with listeners have dramatically reduced the costs of research. It is now possible and inexpensive to gather detailed and comprehensive feedback on programmes, services, particular issues and changes in output through the use of self-completion questionnaires sent out to selected respondents. A radio station can build up a panel representing the general listenership and can also build special panels to respond on questions to do with particular genres or programmes designed for particular target groups. Computer technology, the Internet and email make possible research activity that was frequently too expensive to do very often. The challenge, however, is twofold. One is to keep in check the amount one uses new technologies to do research. And the other is to remember that not all listeners are online. What proportion of your audience are 'nonliners', as they are sometimes called? You need to know – and when you do research, make sure that you have some way of adding them into your research coverage.

The simple email/post audience research questionnaire on p. 367 is designed for individual members of a listening panel. The station completes sections 1 and 2 before distribution.

Questionnaires

When designing a research questionnaire ensure that the concepts, words and format are appropriate to the person who will be asked to complete it. A trained researcher filling in the form while undertaking an interview can cover greater complications and variables than a form to be completed by an individual listener on their own. Before large-scale use, any draft questionnaire should be tested with a pilot group to reveal ambiguities or misunderstandings. Here are some design criteria:

- Decide exactly what information you need, and how you will use it.
- Do not ask for information you don't need – redundant questions only complicate things.
- Write the introduction to indicate who wants the information, and why, and what will be done with it. Establish the level of confidentiality.
- Number each question for reference.
- The layout should be in lines rather than boxes or columns. This enables it to be completed either by typing or longhand.
- The information you may want is in three categories:
  - facts: name, age, family, address, job, newspapers read;
  - experience: listening habits, reception difficulties, use of TV/ videos;
  - opinion: views of programmes, presenters, of competitor stations.
- Questions here come in four categories:
  - Where the answer is either yes or no: Are you able to receive Radio XYZ?   Yes/No
  - Where you offer a multiple-choice question: How difficult is it to tune your radio to Radio XYZ? Very difficult/Moderately difficult/Fairly easy/Very easy
  - Where you provide a numerical scale for possible answers: On a scale of 0–5 how difficult is it to tune your radio to Radio XYZ? Very difficult 0 . . . 1 . . . 2 . . . 3 . . . 4 . . . 5 Very easy
  - Where you invite a reply in prose: What difficulties have you had in receiving Radio XYZ?

RADIO XYZ   Research Questionnaire

We would ask you please to reply to the following questions -

1. Programme Title..

2. Date & Time of broadcast...

3. Do you listen to this Station (circle)   every day   most days   once a month   never

4. Do you listen mostly on (circle)   MW/FM radio   Digital radio   Online   Podcast

5. Did you hear this particular programme? (circle)   Yes   No

6. If 'no' go to question 12.
   If 'yes' did you hear (circle)   all of it   most of it   parts of it   a little of it

7. What did you think of the programme ? (circle)   excellent   good   fair   poor

8. What did you like most about it ?

9. What did you dislike about it ?

10. Is the programme broadcast at a suitable time for you ? (circle)   Yes   No

11. If 'no' what would be a better time   The same day at.................................
    A different day.. at.................................

12. If you did not hear the programme why was this?..

13. If you did hear the programme did you do anything as a direct result ?..................................
....

Some information about yourself is very useful since it enables us to contact you for follow-up if necessary and it helps us to know more about our audience. However, this section is optional and you may leave it blank if you prefer.
Any information you give is for programme evaluation only and is strictly confidential.

Name:...Sex: M / F

Address:...Married / Single

....

Age: (circle)   under 15   16–24   25–39   40–59   60+

Occupation...

Hobbies and Interests...

Other radio stations listened to...

- Yes/No, tick box, multiple-choice and numerical scale questions take up little space and can be given number values and so are useful in producing statistics.
- Add 'other' to appropriate multiple-choice questions to allow for responses you have not thought of.
- Prose answers can be difficult to decipher but give good insights and usable quotes.
- Avoid imprecise terms – 'often', 'generally', 'useful' – other than in a multiple-choice sequence.
- Avoid questions that appear to have a right or preferred answer.
- Keep the questionnaire simple, and as short as possible.

Written programme response

Informal audience research – anecdotal evidence, press comment and immediate feedback – often has an impact on the producer that is out of proportion to its true value. Probably the most misleading of these – in relation to the listenership as a whole – is the written response. Several studies have shown that there is no direct correlation between the number of letters, emails or texts received and the size or nature of the audience. News is often the most listened-to part of the output, yet generates comparatively little feedback.

Written correspondence will indicate something about the individuals who are motivated to write – where they live, their interests perhaps, what triggered them to pick up a pen or type their thoughts, or what they want in return. But it is wrong to think that each writer represents a thousand others – they might do, but you don't know that and cannot assume it. Broadcasting to 'closed' countries, or where mail of any kind is subject to interference, frequently results in a lack of response which by no means necessarily indicates a small audience. Low literacy, interruptions in the electricity supply, or a genuine inability to pay the postage are other factors that complicate any real accuracy in attempting audience evaluation through the correspondence received. It might give some useful indicators, and raise questions worthy of feeding back into programmes – for each individual respondent has to be taken most seriously – but they are self-selecting along patterns that are likely to have more to do with education, income, available time and personal motivation than with any sense of the audience as a whole.

A method that partly overcomes the unknown and random nature of written response is to send with every reply to a correspondent a questionnaire (together with a stamped addressed return envelope, or your own email address) designed to ask about the writer's listening habits, to your own and to other services. It also asks for information about the person. Over a period of several months it is possible to build up some useful demographic data. It still only relates to those who write, but it can be compared with official statistical data – available from many public libraries – to discover how representative these writers are.

Web analytics

With so much effort being expended on the Internet – particularly websites – it's important to undertake audience research in this area. Many media companies regularly collate statistical data and respond by providing more targeted content. For instance, the BBC has a project that monitors the use of its websites – how many people are using them – that is their *reach* – who the readers are (not personally of course), which country they are in, whether they are receiving on a mobile or computer, which pages are read, how often, and how long a reader spends on the site. There are many different web analytic tools on the market – the technical process here involves JavaScript and 'cookies' and is described in detail on the BBC Academy website (see p. 386). As with all research, the difficulty is not so much in obtaining the information, but what to do with the data produced. It tells you how much interest there is in a particular item and, therefore, how long to keep it running. It gives you a real connection with your audience and has been described as 'a tool to perform better – part of the lifeblood of any editorial process'.

Needless to say, the whole area of audience research is too big to be dealt with fully here. Further sources of information are listed under Further reading (p. 402).

Cost evaluation

What does a programme cost? Like many simple questions in broadcasting this one has myriad possible answers, depending on what you mean.

The simplest answer is to say that a programme has a financial budget of 'X' – an amount to cover the 'above the line' expenses of travel, contributors' fees, copyright, technical facilities and so on. But then what is its cost in 'people time'? Are staff salaries involved – producer, technical staff, secretarial time? Or office overheads – telephone, postage, etc.? Is the programme cost to include studio time, and is that costed by the hour to include its maintenance and depreciation? And what about transmission costs – power bills, engineering effort, capital depreciation?

Total costing will include all the management costs and all the overheads over which the individual programme maker can have little or no control. One way of looking at this is to take a station's annual expenditure and divide it by the number of hours it produces, so arriving at a cost per hour. But since this results in the same figure for all programmes, no comparisons can be made. More helpful is to allocate to a programme all cash resource costs which are directly attributable to it, and then add a share of the general overheads – including both management and transmission costs – in order to arrive at a true, or at least a truer, cost per hour figure that will bear comparison with other programmes.

Does news cost more than sport? How expensive is a well-researched documentary, or a piece of drama? How does a general magazine compare with a phone-in or a music programme? How much does a 'live' concert really cost? Of course it is not simply the actual cost of a programme that matters – coverage of a live event may result in several hours of output, which with recorded repeat capability could provide a very low cost per hour. Furthermore, given information about the size of the audience it is possible, by dividing the cost per hour by the number of listeners, to arrive at a cost per listener hour. So is this the all-important figure? No, it is an indicator among several by which a programme is evaluated.

Relatively cheap programmes that attract a substantial audience may or may not be what a station wants to produce. It might also want to provide programmes that are more costly to make and designed for a minority audience – programmes for the disabled, for a particular linguistic, religious or cultural group, or for a specific educational purpose. These will have a higher cost per listener hour but will also give a channel its public service credibility. It is important for each programme to be true to its purpose – to achieve results in those areas for which it is designed. It is also important for each programme to contribute to the station's purpose – its mission statement.

After a full evaluation, happy is the producer who is able to say:

> My programme is professionally made to a high technical standard, it meets precisely the needs of the whole audience for which it is intended, its cost per listener hour is within acceptable limits for this format, and it contributes substantially to the declared purpose of this station.

Happy, too, is the station manager.

26

Training

We learn in four main ways: by practical experience, by watching others, by studying theory, and by trying things out to see if they work. In fact, the activities are linked together. The process of observation leads us to draw conclusions about what appears to work and what doesn't; we can then test theories before launching out in practice. We monitor ourselves as we do that and the cycle starts again.

Unfortunately, most people are not equally disposed to these four ways of learning. Someone strong on observation, making judgements, reflection and theory will learn a great deal from visiting studios, watching professionals at work, asking questions and reading the literature – and they might be reluctant to try a practical exercise until they think they have mastered the theory. Someone else – the practical activist – will be

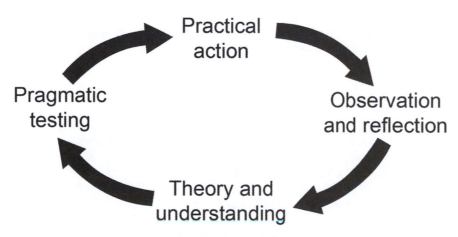

Figure 26.1   The cycle of learning styles – David Kolb

anxious to get on with 'doing it', and will be impatient of the principles. Yet another will want to read the manuals and logically plan how to go about it.

It follows that a good training scheme for producers should contain all four elements in proportions suited to the participants and, in practice, to the facilities and expertise available. There's no point in throwing someone in at the deep end ('that's the only way to learn, that's how I did it') without the essential guidelines. Nor is it sensible to insist that new producers watch someone else without any opportunity to ask questions and try things out for themselves. Giving 'on-air' experience too soon can induce in some a real sense of fear, from which recovery can be slow and painful. On the other hand, preventing people from doing anything 'for real' is likely to lead to acute frustration. It is this aspect of motivation that those responsible for a producer's training must monitor most closely. How do you find out a person's preferred method of learning? You ask them!

Is the new person enjoying learning? How do they think the training process is going? Are there signs of their being 'over-challenged', or the opposite – bored? If they are slow to finish work, is it because they are aiming at too high a standard for their present level of skill? If the work quality is low, is it because of a lack of understanding of what is required or an inability with a particular technique? Learning *how* to interview, for example, will certainly take place during practice and real interviews – but the key to *understanding* interviewing will occur during a session of expert feedback on the end result. The critical analysis of programme material by someone experienced in the craft is an essential part of the learning process. Neither is this confined to the new producer. Established producers also need to grow; their development should not be overlooked simply because they can do a day's work without supervision. New challenges, techniques, programme formats and roles will help to keep regular output producers from going stale. They may even be good enough to enthuse and train the next generation.

Triggers for training

Training used to be regarded as a one-off event experienced early on in a career. However, with continual technological and organisational change, it has to be a lifelong process. For managers and trainers, the events that trigger training include the following:

- new legislation, changes in the law;
- the organisation's own appraisal process;
- high staff turnover bringing new recruits;
- new equipment or procedures;

- a high error rate, or listener complaints;
- new programme methods;
- expansion of the broadcasting service;
- multi-skilling of staff and freelances;
- staff promotion;
- downsizing, where staff assume greater responsibilities;
- new markets, programme areas or services;
- changed organisational or departmental structures.

Trainers will be continuously aware of new training needs, but in a truly innovative organisation senior management will be the driver of visionary change, asking the training department to help implement it.

Learning objectives

What is important in all this, of course, is not the training but the learning. What is the intended effect on the learner's attitude, knowledge or skill – and

Figure 26.2   Honoré Boni training on panel layout at the MediAfrique school, Lomé, Togo (Photo: Roger Stoll)

does it happen? We should therefore start with a clear idea of what we are trying to do by writing a series of statements which describe the intentional outcomes of a training event or course.

For example, at the end of a radio news production course an already trained journalist will be able to:

- select news stories appropriate to the target audience;
- put stories in order of importance and write a programme running order;
- write bulletin material to time using accurate, clear and engaging language;
- research news leads and follow up existing stories;
- commission and brief reporters to cover news stories, including those remote from the newsroom;
- package news material, interviews and actuality in audio form in appropriately creative ways;
- produce and run a studio for a live news transmission;
- present news on-air;
- analyse and critique a news programme, giving professional feedback to contributors.

Once the trainer is sure of the skills – or competences – that have to be mastered, the course can be designed by breaking it down into individual sessions, each with its own objectives. Training outcomes, applied even to a single session, describe what the training is designed to achieve in terms of what the student will either know, or be able to do. The success of the training/learning process can then be assessed against these objectives. Even so, it is crucial that the training is closely relevant to the trainee's own environment so that the skills learned are readily transferable to the real work. In this respect the argument in favour of on-station training, rather than a remote classroom approach, is undeniable.

A special word about computer training. It is of the utmost importance that people, especially older people used to more traditional methods of programme making, are given sufficient time for training. The cost of this is an essential component of the total capital investment. It may take several weeks for staff and freelances to become familiar and comfortable with equipment of this nature. This is partly to do with the requirement for a quite different psychological approach. Understanding how things work is no longer required – as was the case, for example, with the older tape recorders. With a computer, knowing what it does and recognising its potential is the new learning. But not fully understanding a new process produces anxieties for many people, and such anxieties often inhibit learning.

Course organisation

The trainer is the enabler, developer and promoter of others. He or she is always giving away their own abilities, requiring a generosity of spirit that does not constantly show how clever they are. There are subtle differences here between the performer and the trainer. To be successful the trainer with insight sees things from the learner's point of view, knowing the needs of the trainees – their present levels of skills and knowledge – and where you want a particular training event to take them. This should help to determine whether the training is to be full-time, part-time, day-release, block-release and so on. However, whether the event is a half-day module or a three-month course, it will require five areas of attention:

1 *Aim.* The purpose of the training should be clear. What insights, skills and abilities are people to take away with them? The trainer needs a vision of what is to be achieved, and from what starting point.
2 *Logistics.* Technical, financial and other practical arrangements have to be made for the desired number of people attending the event. Teaching space, accommodation, working space (together with equipment for practical sessions), administrative effort and office support (e.g. computers, scanner, printers, software, word-processing and copying, transport, catering, visual aids – tablets, digital projector, flipcharts, books and folders, etc.) have to be predicted and provided. Even the best training should be supported by well-produced study packs, handouts and notes, preferably in audio-visual as well as printed form.
3 *Design.* What topics need to be covered, in what order? The flow should accommodate the different learning styles, the balance between theory and practical, and between individual and group work. The sessions after a midday meal should either succumb to the convention of siesta or be vigorously participative, and anything in the evening should be different again. Courses are frequently so full of input that there is little time for reflection, for students to process what is learned. Lectures should be modified with discussion 'buzz groups', mindmaps and questions, and practical work given time for debriefing and individual feedback, encouragement and critique.
4 *Lesson plans.* Each session needs its own written outline starting with its aim. What is it to achieve? How will it do it? A plan, preferably drawn up by the trainer who is to lead the session, will detail the content of the session, giving approximate timings for each section and

how it will be organised. It will list the training handouts to be distributed, equipment and software needed, aural and visual aids required: PowerPoint, CDs, videos, films or sound clips to be used, etc.

5   *Evaluation.* The initial aim will give you the success criteria for the training, but has it worked for each individual? Described in detail later, evaluation of and by trainees and tutors is part of the quality control process – and if there are standards to be met and examinations passed, these should also be with regard to the four learning styles.

The following training ideas can be adapted to suit specific conditions and represent the principles of learning by seeing–understanding–trying–doing. Observation, theoretical principle, group discussion, and working in the 'safety' of a training environment are combined with 'on-job' learning. In each case the trainer should clearly indicate what is required, by when, to what standard, with what resources.

Stretching imagination

Write a one-minute piece on a colour (see p. 232). Members of a group take colours out of a hat – black, purple, red, grey, etc. Provide access to music, poetry, Fx and compile a recording for discussion and evaluation by the group. Invite a blind person to sit in and comment.

Editorial selection

Provide each trainee with the same recording of a five-minute interview, to be edited down to two minutes. Use a transcript to mark what has been used and what cut. With a group, each person says why he or she chose certain parts and omitted others, and which parts should be rewritten as the cue. Analyse the reasons for selectivity. Does everyone stand by their own decision, or recognise that other choices may be better?

News priorities

List 12 basic stories. Choose three as the lead stories for a five-minute bulletin. With only room for nine, decide which three to drop. Analyse and discuss the reasons given:

1   Police chief presses for stricter measures against all forms of civil terrorism including up to 28 days' detention without charges being brought.

2 Farmers fear price rises of staple foods in the coming six months due to poor harvesting conditions.

3 Popular national youth movement announces plans for international rally to be held in the capital.

4 The country as a whole has attracted more tourists than ever before. Income from tourism reaches multi-million record.

5 Important political figure – in the opposition party – claims wasteful government spending on road-building programme.

6 Plane crash in desolate region involving internal flight of domestic airline with 75 people on board. Circumstances and casualties not yet known.

7 Famous local sportsman wins premier prize in an international competition.

8 Country's political leader announces new government policy for welfare facilities for the old, disabled and poor.

9 Small bomb explosion in a store in centre of the capital. Part of a continuing campaign of protest by a dissident minority leader who claims responsibility by phone call to the radio station.

10 Rural region suffers a suspected outbreak of cattle disease which threatens the destruction of livestock under a government order.

11 Industrial dispute over a pay claim threatens the shut-down of major national car plant.

12 University department of medical research announces a breakthrough in its search for a drug to alleviate arthritis in the hand and knee joints.

News exercise

An excellent 'real' news exercise is to provide a training group with the same sources as those available to the working newsroom. Alongside the professional team the trainees independently compile a five-minute bulletin for comparison with the actual output. Invite the editor to listen and comment – discuss the differences of selection and treatment.

A further exercise is to listen on the same day to bulletins from different stations or networks, analysing the reasons for the variations between them.

Voicework

When giving feedback – especially critical feedback – it is always wise to remember that the trainer is commenting on the work, *not* the person. However, with voice training the work and the person come very close together and it can be almost impossible to separate them. Listening to a trainee newsreader you might feel that he or she is acting, not being themselves. So who are they being? Do they have a mental image of a newsreader that they must live up to? Are they in effect *impersonating* a newsreader? If so, what is wrong with being him or herself? These can be difficult questions that, in the end, only the individual concerned can answer. Before going on to specific technical skills any newsreader or presenter must be comfortable with themselves. If they do not like their voice or accent they will try to change it and the whole effect may sound false; if they are too concerned with their own performance they will not have sufficient care for the listener, and communication will suffer.

The first step therefore is to record and play back some newsreading and ask the reader to comment on it. If it is very different from their normal talking voice, record this conversation and play it back comparing it with the newsreading. Why the difference? Many readers have to be assured that their normal voice – or at least something very close to it, that works for them throughout the greater part of the day – is perfectly OK for radio.

Bearing in mind the 'seven Ps' (p. 149), the object is for the trainer to 'liberate' a voice to be natural, rather than assumed, nervous or full of mannerisms and false stresses. This may take some time, for in the end it is something that only the presenter can correct. It's good then for the trainee to begin to suggest improvements that could be tried – just experiment doing things in different ways.

All voicework has something of performance about it and it is natural for broadcasters of all disciplines to want some form of professional feed-back. It follows that the opportunity for formal voice training, and for a discussion of one's personal approach to it, should be offered to both new and experienced practitioners.

Personal motivation

Using the lists in Chapter 1, write a short essay on the ideal use of the radio medium and how you see yourself being most fulfilled as a programme maker. Why and how do you want to use radio? Which of its many characteristics do you find the most attractive, and what ideally would you bring to your listener?

Vox pop

Trainees produce edited street interviews on a similar topic:

- a film or television programme;
- an aspect of industrial or agricultural policy;

- solutions to traffic problems;
- views on new buildings being erected;
- teenage dress.

Discuss – to what extent does luck play a part in the end result? How does one's personal approach affect the outcome? What is the most appropriate slot for these items?

Commentary

An initial exercise is for a trainee to be given a picture taken from a newspaper or magazine and, without showing it to the others in the group, is asked to describe it for 30 seconds. He or she then shows it to everyone, and they comment on the differences between their mind's-eye picture created by the commentary, and the real one. What was it about perspective, size or content that was distorted? The next step is to record a piece of outside scene-set or event commentary for subsequent playback (without editing) and analysis. Did it provide a coherent picture, and a credible sense of mood, movement and activity?

Drama

Develop a short piece of personal drama based on the technique outlined on pp. 302–5. Was this satisfying to do? Does the end result reflect your original feeling of the event?

Write 10 minutes of dialogue for two to four voices, with or without effects. Using actors, produce the playlet for discussion by the group. Invite the actors to give their view of how they were produced. Could they have done better? Were the producer's instructions clear? Did the end result recreate the writer's intentions?

Exercises and questions for media students

Here are some issues that arise from the subjects discussed in these chapters.

- What are the key skills and abilities you should develop to succeed in radio?
- In what ways might a council or board usefully get involved in the running of a station?
- Study carefully a journalism Code of Practice in a country other than your own.

- Who is your favourite interviewer? What are they doing that you like?
- What are the prime characteristics of a good discussion host, and what are the disadvantages of a discussion programme taking phone calls or texts?
- A programme helpline needs real off-air support. What would you suggest for your own station?
- Make a radio ad for a product you use yourself – test it on someone else.
- What do you find is the best way of generating ideas for programmes?
- How would you improve the website of your own station?

New challenges for old producers

Set out deliberately to do something personally never attempted before:

- a vox pop in an old people's home, or at a school;
- draft outline ideas for a documentary, giving research sources;
- using the clock format, construct a one-hour music sequence for a given target audience;
- list 10 new ideas for an afternoon magazine;
- write a public service message – road safety, community health;
- record some of your own broadcast output and offer it for evaluation to a senior radio manager or trainer;
- reconstruct the station's morning schedule.

Without attempting to justify the result, it should be discussed with an experienced broadcaster. It should then be done a second time, with a further appraisal.

Maintaining output

Many radio training courses with access to studio facilities will culminate with a sequence of programmes as simulated continuous output. Run in real time to a predetermined schedule, a full morning's music, news and weather, traffic reports, features, etc. are watched by professional observers who report back on programme quality, presentation style, sound levels, operational faults, the management of the studio, the producer's ability to motivate and communicate with others, and so on. Some such exercises include deliberate emergencies like equipment failure, the breaking of a major news story, the unexpected arrival of a VIP or the loss of some pre-recorded material. Such 'disasters' should not be allowed to bring everything to a halt, since the effect may be counter-productive to good morale. However, 'live' broadcasters must be encouraged to think quickly 'on their feet'.

Assessing quality

Listen to a piece of radio for subsequent discussion. What was its purpose? What effect did it have on you? Comment on content, order and presentation – the message and its style of delivery, on technique, story treatment, music selection, news values, etc. How much of the elements of 'quality' described on pp. 360–61 did it contain? If possible, invite the actual producer to comment on the discussion.

As already noted, when giving feedback – and this is a general rule – comment on the work, don't criticise the person. Critique the programme not the programme maker.

The process of discussion, analysis and evaluation is carried on continuously by professional broadcasters. It is especially important after any programme that has attracted public or government criticism. Communicators need the comments of others – and a 'gut' reaction may be just as valid as a careful intellectual assessment. All broadcasters, in training and the trained, need to maintain their own analytical reasoning in good order. Keep 'quality' on the agenda.

Training evaluation

If programmes are to be evaluated, then so must be the training process itself. Trainees can be encouraged to set their own goals at the start of any course or training period, and asked halfway through how they are doing in achieving those goals. At the end, to what extent have they succeeded in meeting their own criteria? Other 'end of course' feedback useful to trainers is obtained by questions such as:

- Which sessions did you find most useful?
- Which sessions did you find least useful?
- How did you find the balance of theory and practical? Too much theory/about right/too much practical.
- Did the training come too soon/about the right time/too late, for you?
- How far was the training relevant to the work you are doing/you expect to be doing?
- What would you like to see added to the course?

It is also useful in an end of course questionnaire to ask for 'any other comments'. This will cover training administration and group relationships, as well as course content. The trainer needs to know as much as possible about what trainees *feel* about the training experience as well as what they *think*.

Of course, radio training in all its forms is only a means to an end – better broadcasting. Real evaluation can only take place three or six months later, involving both the trainee and his or her manager. Is what has been learned being put to practical use? Are genuine results apparent from the training effort? If not, is there a mismatch between the training approach and the workplace? Training has to meet the felt needs of the programme output; and the trainer, like the programme maker, will constantly evaluate the work done in order to improve the product for the customer.

27

Back-announcement

Harry Vardon, one of the great exponents of golf, was asked why he never wrote a book setting out all that he knew about it. His reason was that when he came to put it down on paper it looked so simple that 'anyone who didn't know that much about it shouldn't be playing the game anyway!'

It looks as though it is the same with broadcasting. Is there really any more to it than: 'Have something to say, and say it as interestingly as you can?' Yet there are whole areas of output that have hardly been mentioned – educational programmes, light entertainment and comedy, programmes for young people, specialist minorities or ethnic groups. What about the special problems of short-wave or Internet broadcasting, or programmes for the listener who is culturally or geographically a long way from the broadcaster? A book, like a programme, cannot tell the whole truth. What the reader, or listener, has a right to expect is that the product is 'sold' to him in an intelligible way and then remains true to his expectations. Broadcasters talk a great deal about objectivity and balance, but even more important, and more fundamental, is the need to be fair in the relationship with the listener. A broadcasting philosophy that describes itself in programme attitudes yet ignores the listener is essentially incomplete.

What then is the purpose of being in broadcasting – the aim of it all? It is not enough to say, 'I want to communicate' – communicate what? And why? As we have seen there are several possible answers to this – to earn money, to meet the needs of the organisation, to meet your own needs, to become famous, to win all the prestigious radio awards, or to persuade others to think as you do. But the purpose of communication is surely to provide ideas and options, and a consequent freedom of action for other people – not to reduce the scope of such action by offering a half-truth or by weighting them with a personal, political or commercial bias (although, of course, there are national regimes that use the media for exactly this purpose). The reason for providing information, education and the relaxation of entertainment is to suggest alternative choices, to explain the implications of one against another, and having done so to allow for a freedom of thought and action. This assumes that people are capable of responding in a way which itself requires a regard for our fellow men and women.

Having announced your intentions and made your programme, as producer you must put your name to it. Programme credits are not there simply to feed the ego or as a reward for your labours. They are a vital element in the power which broadcasting confers on the communicator – personal responsibility for what is said. Many members of a team may contribute to a programme, but only one person can finally decide on the content. Good programmes cannot be made by committee. Group decisions inevitably contain compromise and a weakening of purpose and structure, but worse, they conceal responsibility. Communication that is not labelled or attributed is of little use to the person who receives it. And if your programme is savaged by a critic or scorned by detractors, first check it by the standards of quality referred to earlier, then remember the following comment made in a speech a dozen years before broadcasting began, by the American President, Theodore Roosevelt.

> It is not the critic who counts; not the man who points out how the strong man stumbles, or where the doer of deeds could have done them better. The credit belongs to the man who is actually in the arena; whose face is marred by dust and sweat and blood; who strives valiantly; who errs, who comes short again and again, because there is no effort without error and shortcoming; but who does actually strive to do the deeds; who knows great enthusiasms; the great devotions; who spends himself in a worthy cause; who at the best knows in the end the triumph of high achievement; and who, at the worst, if he fails, at least fails while daring greatly, so that his place shall never be with those cold and timid souls, who know neither victory nor defeat.

Programme makers face a hundred difficulties not mentioned here, but by engaging a 'professional overdrive', rather than regarding them as a personal undermining, most problems can be made to take on more the aspect of a challenge than a threat. The practicalities of production are encapsulated by the well-known Greenwich Time Signal, whose six pips must serve as their final reminder.

1   *Preparation:* state the aims, plan to meet them.
2   *Punctuality:* be better than punctual, be early.
3   *Presentation:* keep the listener in mind.
4   *Personal:* come over as yourself, be real.
5   *Punctilious:* observance of all agreed systems and procedures. If there is something you don't like, do not ignore it, change it.
6   *Professional:* putting the interests of the listener and the broadcasting organisation before your own, with the constant maintenance of an editorial judgement based on a full awareness and a competent technique.

Websites

This edition of *Radio Production* has an associated website which includes audio examples, practical exercises plus audio and video support materials: www.focalpress.com/cw/mcleish

There are far too many websites on the subject of radio to list here, but the following keywords, topics and websites are useful when searching online for further information relating to the chapters in this book.

Ch 1   Characteristics of the medium
Media in Good Governance – DFID Practice Paper

Ch 2   Structure and regulation
USA Federal Communications Commission
National Association of Broadcasters in the USA
UK Office of Communication
Australian Communications and Media Authority
Canadian Radio-television and Telecommunications Commission
Communication Authority of Kenya
Community Media Association
Canadian Campus and Community Radio Association

Ch 3   The radio studio
All manufacturers of studio equipment and software have their own websites which include operational guidance.

Ch 4   Using the Internet and social media
Social Media Today
Social Media Webtrends

Ch 5   Ethics
International Federation of Journalists
UN Human Rights Documents
UNESCO – Professional Journalistic Standards
– Code of Ethics
– Media and Good Governance

National Union of Journalists UK
Society of Professional Journalists USA
Corporation for Public Broadcasting USA
The Poynter Institute, California (Media Ethics)
Human Rights – Inter Press Service, Africa

Ch 7   **News – policy and practice**
World Press Institute
International Press Councils
Alliance of Independent Press Councils
UNESCO Handbook, Reporting without Borders
BBC Academy – Journalism
American Society of Journalists and Authors
The Guardian – media law
Nieman Lab – future of journalism

Ch 15   **Music programming**
PRS – Copyright Law UK
Copyright Law – Music UK
MCPS music libraries
Music Copyright USA, see also ASCAP

Ch 17   **Making commercials**
Radio Advertising Bureau UK
Code of Advertising Practice
All India Radio Commercial Code

Ch 23   **The work of the producer**
BBC Editorial Guidelines
Public Media Alliance

Ch 24   **The executive producer**
BBC Commissioning – Radio

Ch 25   **Programme evaluation**
Radio Joint Audience Research UK or RAJAR UK
Audience Dialogue

Ch 26   **Training**
BBC Academy – Production – radio
USA – Society of Professional Journalists
Mediawise UK
Commercial Radio – Australia
Australian Radio School
New Zealand Broadcasting

History of radio – californiahistoricalradio.com
– history of the BBC

Online radio stations – www.radio-locator.com
– www.radiofeeds.co.uk
– www.web-radio.fm
– www.live365.com

Glossary

Above-the-line cost   Expenditure under the producer's control in addition to fixed over-heads (below the line).

ACMA   Australian Communications and Media Authority. Responsible for regulating broadcasting in Australia.

Acoustic   Characteristic sound of any enclosed space due to the amount of sound reflected from its wall surfaces and the way in which this amount alters at different frequencies. See also *Reverberation time, Coloration.*

Acoustic screen   Free-standing movable screen designed to create special acoustic effects or prevent unwanted sound reaching a particular microphone. One side is soft and absorbent, the other is hard and reflective.

Actuality   'Live' recording of a real event, sounds recorded on location.

Ad   Advertisement or commercial.

A/D converter   Analogue-to-digital converter. Creates a digital output from an analogue input, e.g. a conventional analogue microphone will be converted to a digital signal when recording to a digital recorder.

Ad-lib   Unscripted announcement, 'off-the-cuff' remark.

Ad log   The daily list of advertisements to be played, against which the presenter's signature is verification.

Adobe Audition   Computer software system for recording, editing and playout of broadcast audio. Formerly Cool Edit Pro.

ADSL   Asymmetric Digital Subscriber Line. A digital system providing fast data transfer. Also known as Broadband.

Aerial   Device for transmitting or receiving radio waves at the point of transition from their electrical/electromagnetic form.

AGC   Automatic Gain Control. Amplifier circuit which compensates for variations in signal level, dynamic compression.

AM   Amplitude modulation. System of applying the sound signal to the transmitter modulation, associated with medium-wave broadcasting.

Analogue signal   An electrical signal that exactly represents the original shape of the acoustic or mechanical vibrations that caused it.

Anchor   Person acting as the main presenter in a programme involving several components.

AP   Associated Press. Syndicated news service.

App or Application   A program for information processing for a single purpose or activity, e.g. on a smartphone or tablet.

Apple and biscuit   Microphone resembling a black ball with a circular plate fixed on one side. Omni-directional polar diagram.

ASCAP   American Society of Composers, Authors and Publishers – copyright control organisation protecting musical performance rights. Collects and distributes fees due for performance in any medium to its members.

Atmosphere   Impression of environment created by use of actuality, sound effects or acoustic.

Attenuation   Expressed in decibels (dB), the extent to which a piece of equipment decreases the signal strength. Opposite of amplification.

Attenuator   Device of known or variable attenuation deliberately inserted in a circuit to reduce the signal level.

Audience figures   Expressed as a percentage of the potential audience, or in absolute terms, the number of listeners to a single programme or sequence, daily or weekly patronage, or total usage of the station. See *Patronage, Ratings, Reach.*

Audience measurement   Research into numbers and attitudes of listeners. Methods used include: 'Aided-Recall' – person to person interview; 'Diary' – the keeping of a log of programmes heard; 'Panel' – permanent representative group reporting on programmes heard; and personal meters.

Audio frequency   Audible sound wave. Accepted range 20 Hz–20 kHz.

Automatic Gain Control   See *AGC.*

Auxiliary output ('Aux')   A secondary output from a mixing desk providing a different mix independent of the main programme output, in order to send to echo, public address, foldback, etc.

Back-announcement   Where the names and details of an interview or record are given immediately after the item.

Backing track   Recording of musical accompaniment heard by a soloist while adding an additional performance.

Back-timing   The process of timing a live programme backwards from its intended closing time to ensure it ends on time.

Balance   Relative proportion of 'direct' to 'reflected' sound apparent in a microphone output. Also the relative volume of separate components in a total mix, e.g. voices in a discussion, musical instruments in an orchestra.

Balance control   Control for adjusting the relative volume of two stereo loudspeakers.

Bass cut   Device in microphone or other sound source which electrically removes the lower frequencies.

Bay   Cabinet with a frame for housing power supplies and other technical equipment used in studios or control areas. Usually arranged to match standard 19-inch rack-mounted equipment.

Bed   Instrumental backing to which words or singing are added to make a commercial or station ident.

Bias   High-frequency signal applied to the recording head of a tape machine to ensure distortion-free recording.

Bi-directional   Microphone sensitive in two directions, front and back, but completely insensitive on either side, e.g. ribbon microphone.

Bi-media   Operations involving both radio and television.

BMI   Broadcast Music Incorporated.

Board   American term for a mixing desk – control board.

Boom   Wheeled microphone support having a long arm to facilitate placement of microphone over performers, e.g. orchestra.

Boomy   Room acoustic unduly reverberant in the lower frequencies.

Boundary effect mic   Small microphone mounted on a plate with a gap between it and the plate to give a directional polar diagram. Used on-stage for opera and theatre work. Also called Pressure Zone (PZ) effect.

Break-out box   A simple input/output unit to give easy pluggable access to a recording device, e.g. a computer.

Breakthrough   Unwanted electrical interference or acoustic sound from one source or channel affecting another.

Broadband   See *ADSL*.

BSA   Broadcasting Standards Authority. Responsible for broadcast regulation in New Zealand.

Burner   Colloquial term for a CD recorder, used because of its laser process.

Byte   A string of 8, 16, 32, or more binary electrical pulses or 'bits', representing a specific piece of data.

Cans   Colloquial term for headphones.

Capacitor microphone   Microphone type based on the principle of conducting surfaces in proximity holding an electrical charge. Requires a power supply.

Card   Solid state medium for recording digital signals in an audio recorder.

Cardioid   Heart-shaped area of pick-up around a microphone.

Cassette   Enclosed reel-to-reel device of 3 mm-wide tape particularly used in domestic or miniature recording machines.

Catchline   Brief identifying title for an audio insert, e.g. news item. See also *Slug*.

CD   Compact Disc. Digital recording and playback medium.

Channel   The complete circuit from a sound source to the point in the control panel where it is mixed with others.

Check calls   A newsroom's routine phone calls to the emergency services.

Clean feed   A supply of cue programme in which a remote contributor hears all the programme elements except his own. Essential to prevent howl-round in certain two-way working conditions.

Clip   A short piece of audio extracted from a longer item and illustrative of it. See also *Sound bite*.

Cloud storage   Remote centralised computer data storage with online access via a network or the Internet.

CMA   Community Media Association

CODEC   A coder/decoder for converting analogue audio to a digital signal and vice versa. Used at the end of an ISDN communications line.

Coloration   Effect obtained in a room when one range of frequencies tends to predominate in its acoustic.

Compression   (1) Dynamic – decreasing the maximum difference between the loud and soft parts of an audio signal. (2) Digital – a process for encoding digitised audio or video signals resulting in a smaller file size. May be 'lossy' or 'lossless'. Lossy will result in a reduction in the fidelity of the signal.

Compressor   Device for narrowing the dynamic range of a signal passing through it.

Condenser microphone   See *Capacitor microphone*.

Control line   A circuit used to communicate engineering or production information between a studio and an outside source. Often also used as cue line. Not necessarily high quality – see *Music line*.

COOBE   Contributor Operated Outside Broadcast Equipment. Small suitcase of OB equipment for reporters.

Copy   Written material offered for broadcast, e.g. news copy, advertising copy.

Copyright   The legal right of ownership in a creative work invested in its author, composer, publisher or designer.

Copytaster   The first reader of copy sent to a newsroom who decides whether it should be rejected or retained for possible use.

Cough key   Switch, under the speaker's control, which cuts his microphone circuit.

Crossfade   The fading in of a new source while fading out the old.

Crossplug   The temporary transposition of two circuits, normally on a jackfield. See also *Overplugging.*

Crosstalk   Audible interference of one circuit upon another.

Cue   The pre-arranged signal to begin – visual light or gesture, verbal, musical, or scripted words.

Cue, in and **out**   The first and last words (effects or music) of a programme or item.

Cue light   A small electric light, often green, used as a cueing signal.

Cue line   A circuit used to send cue programme to a distant contributor.

Cue programme   The programme which contains a contributor's cue to start.

Cue sheet   Documentation giving technical information and introductory script for programme or insert, i.e. cue material.

Cume   Cumulating audience measurement. See *Reach.*

DAB   Digital Audio Broadcasting. Digital broadcast system for radio with accompanying text. DAB and DAB+ are two formats available mainly in Europe.

DAT   Digital Audio Tape. Sound recording and playback system in *digital mode* using a small tape cassette and rotating heads – as in a video recorder. Now a largely obsolete format.

D/A converter   Digital-to-analogue converter. Creates an analogue output from a digital input, e.g. CD playback signal is converted in order to feed conventional loudspeakers.

DAW   See *Digital audio workstation.*

dB   Decibel. Logarithmic measurement of sound intensity or electrical signal. The smallest change in level perceptible by the human ear.

Dead side of microphone   Least sensitive area.

Deferred relay   The broadcasting of a recorded programme previously heard 'live' by an audience.

Demographics   The division of the general population into precisely defined sections – by age, gender, education, job, income or ethnic origin in order to describe a 'target audience', especially for audience research.

DFID   UK Department for International Development.

Digital audio workstation   Digital equipment, usually a computer, capable of recording, editing, mixing and manipulating audio. Used for recording music and making packages as well as full programmes.

Digital effects unit   Electronic equipment capable of affecting sound quality in a variety of ways, e.g. by changing frequency response, adding coloration or reverberation. Capable of creating synthetic or 'unreal' sounds.

Digital radio   The transmission and reception of sound that has been processed using digital technology. In Europe, mainly known as DAB, but the United States has opted for a proprietary system called HD Radio technology.

Digital signal   An electrical signal that represents its original acoustic or mechanical vibrations as a series of pulses in binary code.

Din   Plug or socket manufactured to standard of Deutsche Industrie Norm.

Directional   Property of microphone causing it to be more sensitive in one direction than in others. Also applied to transmitters, receiving aerials, loudspeakers, etc. See also *Polar diagram*.

Disc jockey (DJ)   Personality presenter of music programme, generally pop music show.

Dolby system   Trade name for electronic circuitry designed to improve the signal-to-noise ratio of a programme chain.

Double-ender   Short length of audio cable with a jack plug on each end used to connect pieces of equipment or jacks on a jackfield or break-out box.

Double-headed   Style of presentation using two presenters.

Drivetime   Period of late afternoon which coincides with commuter travel and the greatest in-car listening.

Dry run   Programme rehearsal, especially drama, not necessarily in the studio, and without music, effects or movements to mic. See *Run through*.

Dub   To copy material already recorded. To make a dubbing.

Ducking unit   Automatic device providing 'voice-over' facility. See *Voice-over*.

Dynamic range   Measured in dB, the difference between the loudest and the quietest sounds.

Echo   Strictly a single or multiple repeat of an original sound. Generally refers to reverberation.

Echo plate or spring   Electro-mechanical device for artificially adding reverberation.

Edit   The rearrangement of recorded material to form a preferred order or eliminate unwanted material.

Editorial judgement   The professional philosophy that leads to decisions on programme content and treatment.

Email   Electronic mail. International means of conveying computer-generated correspondence, text or pictures via the Internet.

Engagement   The level of interest a reader shows in a website, described in terms of time spent, how many pages read and the frequency of return.

ENPS   Electronic News Production System. System initiated by Associated Press for the syndication of news material.

EQ   Equalisation or frequency control. May be applied to individual channels on a mixing desk, via an external unit or in software.

Equity   The British Actors Equity Association. Actors' union.

Facebook   Internet-based social networking service open to registered users over 13 years of age.

Fade   A decrease in sound volume (fade down or out).

Fade in   An increase in sound volume (fade up).

Fader   Volume control of a sound source used for setting its level, fading it up or down, or mixing it with other sources. See also *Pot*.

FCC   Federal Communications Commission. Body responsible for regulating broadcast operations across the USA.

Feed   A supply of programme, generally by circuit.

Feedback   See *Howl-round*.

Figure of eight   See *Bi-directional*.

Filter   Electrical circuitry for removing unwanted frequencies from a sound source, e.g. mains hum, or surface noise from an old or worn recording. Also in drama for simulating telephone or two-way radio quality, etc.

First-generation copy   A copy taken from the original recording. A copy of this copy would be a second-generation copy.

Flash card   Solid state recording medium.

Fletcher–Munson effect   The apparent decrease in the proportion of higher and lower frequencies, with respect to the middle range, as the loudspeaker listening level is decreased. Significant in correct setting of monitoring level, particularly in music balance.

Fluff   (1) Accumulation of dust on the stylus of gramophone pick-up. (2) Mistake in reading or other broadcast speech.

Flutter   Rapid variations of speed discernible in tape or disc reproduction.

FM   Frequency modulation. System of applying the sound signal to the transmitter frequency, associated with VHF broadcasting.

Focus group   A small group of people asked to give feedback on programme opinion as part of audience research.

Foldback   Means of allowing artists in the studio to hear programme elements via a loudspeaker even while studio microphones are live.

Freelance   Self-employed broadcaster of any category – producer, contributor, operator, reporter, etc. Not on permanent full-time contract. Paid by the single contribution or over a period for a series of programmes. Non-exclusive, available to work for any employer. See also *Stringer*.

Frequency   Expressed in cycles per second or hertz, the rate at which a sound or radio wave is repeated. The note 'middle A' is widely used as concert pitch in the United Kingdom and the United States with a frequency of 440 Hz. In continental Europe the frequency of A commonly varies between 440 Hz and 444 Hz. A long-wave transmitter with a wavelength of 1500 metres has a frequency of 200 kHz (200,000 cycles per second). Frequency and wavelength are always associated in the formula $F \times W =$ speed. Speed is the speed of the wave, i.e. sound or radio, and in each case remains constant.

Frequency distortion   Distortion caused by inadequate frequency response.

Frequency response   The ability of a piece of equipment to treat all frequencies within a given range in the same way, e.g. an amplifier with a poor frequency response treats frequencies passing through it unequally and so its output does not faithfully reproduce its input.

Full track   Tape recording using the whole width of the tape.

Fx   Sound effects.

Gain   Expressed in decibels (dB), the amount of amplification at which an amplifier is set. Can also refer to a receiving aerial – the extent to which it can discriminate in a particular direction, thereby increasing its sensitivity.

Gain control   The control that affects the gain of an amplifier, also loosely applied to any fader or volume control affecting the output level.

Giga   A thousand million – a 40 gigabyte hard disk holds 40,000,000,000 bytes of data.

Gramophone, gramdeck or **grams**   Turntable and associated equipment for the reproduction of records.

GTS   Greenwich Time Signal – six pips ending at a precise time.

Gun-mic   Microphone resembling a long-barrelled shotgun. Highly directional, used for nature recordings or where intelligibility is required at some distance from the sound source, e.g. OBs.

Hammocking   Scheduling term referring to the need to support a low audience or specialist programme by placing more popular material before and after it in order to maintain a strong average listening figure.

Handout   Press information or publicity sheet issued to draw attention to an event.

Hand signals   System of visual communication used through the glass window between a studio and its control area, or in a studio with a 'live' mic. See *Wind-up*.

Hard disk   The component inside a computer which stores audio data in digital form. Capacity measured in gigabytes.

Harmonic distortion   The generation of spurious upper frequencies.

Head amplifier   Small amplifier within a microphone, especially capacitor type.

Headline   Initial one-sentence summary of news event.

Hertz (Hz)   Unit of frequency, one complete cycle per second.

Hiss   Unwanted background noise in the frequency range 5–10 kHz, e.g. tape hiss.

Howl-round   Acoustic or electrical positive feedback generally apparent as a continuous sound of a single frequency. Often associated with public address systems. Avoided by decreasing the gain in the amplifying circuit, cutting the loudspeaker or, when working with a contributing studio, by using a clean feed circuit.

Hum   Low-frequency electrical interference derived from mains power supply.

Hypercardioid   A cardioid microphone having a particularly narrow angle of acceptance at its front that decreases rapidly towards the sides.

ID   Station identification or ident.

IFPI   International Federation of Phonographic Industries. International organisation of record manufacturers to control performance and usage rights.

ILR   Independent Local Radio in the UK.

In cue   The first words of a programme insert, known in advance. Can also be music.

Insert   A short item used in a programme, e.g. a 'live' insert, a tape insert.

Intercom   Local voice communication system.

Internet   Worldwide digital communication system linking similarly equipped computers. See also *Email*, *Website*.

Intro   Introduction – especially the non-vocal beginning of a song before the vocal. Timing needed for talk-overs.

ISDN   Integrated Services Digital Network. A system of conveying high-quality digital audio signals over the public telephone system. See also *CODEC*.

Jack   Socket connected to an audio circuit. Can incorporate a switch activated by insertion of jack plugs – a 'break' jack.

Jackfield or **patch panel**   Rows of jacks connected to audio sources or destinations. Provides availability of all circuits for interconnection or testing.

Jack plug or **post office jack**   Plug type used for insertion in jack socket comprising three connections, a circuit pair plus earth, known as 'ring tip and sleeve'. See also *Double-ender*.

Jingle   Short musical item used as station ID, or in advertisement.

Jingle package   The set of jingles used by a station to establish its audio logo.

Jock   See *Disc jockey*.

Key   Switch.

Kilo   Thousand. Kilohertz – frequency in thousands of cycles per second. Kilowatt – electrical power, a thousand watts.

Landline   See *Line*.

Lavalier microphone   Small microphone hung round the neck (**lanyard mic**) or fastened to clothing.

Lazy arm   Small boom-type microphone stand suitable for suspending a microphone over a talks table.

Lead sheet   Basic musical score indicating instrumentation of melody. Used for microphone control during music balance.

Lead story   The first, most important story in a news bulletin.

Leader   Inert coloured tape having the same dimensions as recording tape spliced into a recorded programme or insert to give visual indication of beginnings and endings. Before the beginning – white; intermediate spacers – yellow; after the end – red.

LED   Light Emitting Diode. Small signal light. Often arranged in a row to act as an indicator, e.g. of sound level.

Level   (1) A test prior to recording or broadcasting to check the volume of the speaker's voice – 'take some level'. (2) Expressed in dBs, plus or minus, the measurement of electrical intensity against an absolute standard, zero level (1 mW in 600 ohms).

Limiter   Device to prevent the signal level exceeding a pre-set value.

Line   Physical circuit between two points for programme or communication purposes.

Line equalisation   The process that compensates for frequency distortion at the receiving end of a landline.

Line-up   Technical setting up of circuits to conform to engineering standards. Line-up tone of standard frequency and level used to check the gain of all component parts.

Lip mic   Noise-excluding ribbon microphone designed for close working, e.g. OB commentary.

Log   Written record of station output, especially for the payment of music rights. Can also refer to recorded audio.

'M' signal   The combination of left and right stereo signals, i.e. the mono signal.

Marching box   Sound effects device comprising small box partially filled with gravel used to simulate marching feet.

MCPS   Mechanical Copyright Protection Society. Organisation that controls the copying or dubbing of copyright material.

Mega   Million. Megahertz – frequency in millions of cycles per second. Megawatt – electrical power, a million watts.

Metadata   Additional data about an item or file that is held digitally alongside or within the item.

Microphonic   Faulty piece of electronic equipment sensitive to mechanical vibration – acting like a microphone.

Middle of the road   Popular, mainstream music with general appeal. Non-extreme.

MIDI   Musical Instrument Digital Interface – A technical standard that describes a protocol, digital interface and connectors, that allows a wide variety of electronic instruments, computers and other related devices to connect and communicate with each other. The data is converted to musical sounds within MIDI software or a MIDI processor.

MiniDisc (MD)   Digital recording and playback system using a 64 mm disc in portable recorders and studio decks.

Mixer   Studio, OB or PA control desk for mixing together sound sources to the appropriate level.

Modem   Modulator/demodulator. Converts an (acoustic) analogue signal to digital signal and vice versa. Used to send computer output to a telephone line.

Modulation   Variations in a transmission or recording medium caused by the presence of programme. Often abbreviated to Mod.

Module   Interchangeable equipment component.

Montage   Superimposition of sounds and/or voices to create a composite impression.

MOR   See *Middle of the road*.

MPEG   Moving Picture Experts Group – international group responsible for developing digital standards for film and audio.

MP3   International standard for the compression of an audio signal in digital form developed by MPEG.

MU   Musicians' Union.

Multi-tracking   Two or more audio tracks are recorded separately and subsequently mixed for the final result.

Music line   High-quality landline or satellite circuit suitable for all types of programme, not only music. Compare with *Control line*.

NAB   National Association of Broadcasters. American trade organisation which secures agreement on standards of procedure and equipment, e.g. NAB spool, a professional tape reel type.

Nagra (Trade name)   Manufacturer of high-quality portable recorders.

Needle time   The total time allowed for the playing of commercially recorded music – generally expressed as minutes per day.

Noise   Extraneous sound, electrical interference, or background to a signal.

Noise gate   Device that allows a signal to pass through it only when the input level exceeds a pre-set value.

OB   Outside broadcast – US term; remote.

Ofcom   Office of Communications. UK broadcasting and communications industry regulator.

Off-mic   A speaker or other sound source working outside a microphone's most sensitive area of pick-up. Distant effect due to drop in level and greater proportion of reflected to direct sound.

Omni-directional   A microphone sensitive in all directions. Also applied to transmitters and aerials.

One-legged   'Thin' low-level quality resulting from a connection through only one wire of a circuit pair.

Open-ended   A programme without a pre-determined finishing time.

Optimod   Audio compressor to maximise the modulation of a transmitter in order to obtain optimum signal strength.

Opt in/out   Joining or leaving a programme stream from another broadcasting source.

Out cue   Final words of a contribution, known in advance, taken as a signal to initiate the following item in a sequence.

Out of phase   The decrease in level and effect on quality when two similar signals are combined in such a way as to cancel each other. Two similar signals 180° apart are said to be out of phase.

Outside source   Programme originating point remote from the studio, or the circuit connection from it.

Overload distortion   The distortion suffered by a programme signal when its electrical level is higher than the equipment can handle. When this happens non-continuously it is referred to as 'peak distortion'. Also referred to as 'squaring off'.

Overplugging   The substitution of one circuit for another by the insertion of jack plug in a break jack.

PA   (1) Press Association. News agency. (2) Public address system.

Package   Edited programme or insert offered complete with links ready for transmission.

Pan   To place a sound source to the left, or right, in a stereo sound image.

Panel   Studio mixing desk, control board or console.

Pan-pot   Panoramic potentiometer. Control on studio mixing desk which places a source to the left or right in a stereo image.

Par   Paragraph. Journalist's term often applied to news copy.

Parabolic reflector   Microphone attachment that focuses sound waves, thereby increasing directional sensitivity. Used for OBs, nature recordings, etc.

Parallel strip   An inert row of jacks mounted on a jackfield not connected to any other equipment but connected in parallel to each other. Used for joining programme sources together, connecting equipment, or multiplying outputs from a single feed. Also available on its own in the form of a 'junction box'.

P as B   Programme as broadcast. Documentation giving complete details of a programme in its final form – duration, inserts, copyright details, contributors, etc.

Patch panel   See *Jackfield.*

Patronage   See *Reach.*

PDF   Portable Document Format. A file format with all the elements of a printed document – viewed, printed or emailed.

Peak distortion   See *Overload distortion.*

Peak programme meter   Voltmeter with a slugged slow decay time, designed to indicate levels and peaks of electrical intensity for the purposes of programme control.

Phantom power   Method of providing a working voltage via a cable to a piece of equipment, e.g. a microphone, using the programme circuit and earth (ground).

Phase distortion   The effect on the sound quality caused by the imprecise combination of two similar signals not exactly in phase with each other.

Pick-up   Gramophone record reproducing components which convert the mechanical variations into electrical energy, pu-arm, pu-head, pu-shell, pu-stylus.

Pilot   Programme to test the feasibility of, or gain acceptance for, a new series or idea.

Plug   Free advertisement.

Podcast (or Netcast)   Individual programme or series, downloaded digitally through web syndication or streamed online to a computer or mobile device.

Polar diagram   Graph showing the area of a microphone's greatest sensitivity. Also applies to aerials, transmitters and loudspeakers. Directivity pattern.

Popping   Descriptive term applied to 'mic-blasting', the effect of vocal breathiness close to microphone.

Pop screen   A circle covered in layers of fine gauze placed in front of a microphone to reduce explosive vocal sounds like 'p' and 'f'.

Post office jack   See *Jack plug.*

Pot   Potentiometer. See *Fader.*

Pot cut   The cutting off of a recording during replay before it has finished by closing its fader – generally to save time. 'Instant editing'.

PPL   Phonographic Performance Ltd. Organisation of British record manufacturers to control performance and usage rights.

PPM   See *Peak programme meter.*

Prefade   The facility for hearing and measuring a source before opening its fader, generally on a studio mixing desk.

Prefade to time   The technique of beginning an item of known duration before it is required so that it finishes at a precise time, e.g. closing signature tune. Also known as back-timing.

Presence   A sense of 'realistic closeness' often on a singer's voice. Can be aided by boosting the frequencies in the range 2.8 kHz to 5.6 kHz.

Prime time   The best, most commercial hours of station output, e.g. 6.30 a.m. to 10.30 a.m.

Producer   The person in charge of a programme and responsible for it.

Promo   On-air promotion of station or programme. Also *Trail.*

PRS   Performing Right Society. Organisation of authors, composers and publishers for copyright protection.

PSA   Public Service Announcement made by the station in the public interest, or on behalf of a charity or other non-commercial body – or at its own discretion for a private individual – for which no charge is made.

Puff   See *Plug.*

PZ (Pressure Zone) mic   See *Boundary effect mic.*

Q & A   Question and answer basis of a discussion between a programme presenter and a specialist correspondent. Less formal than an interview.

QSL   Postcard or email confirmation by a station of a reception report received from a listener. Especially on Short Wave.

Radioman   Trade name of computer software system for recording, editing and playout of broadcast audio.

Radio mic   Microphone containing or closely associated with its own portable transmitter. Requires no cable connection, useful for stage work, OBs, etc.

RAJAR   Radio Joint Audience Research. Body undertaking research for all BBC and commercial stations in the UK.

Ratings   Audience measurement relating to the number of listeners to a specific programme.

RDS   Radio Data System. A data signal added to FM and digital transmissions, used for carrying text, e.g. station ident and other messages, for display by the receiver. Also provides automatic switching of car radios for local traffic information, retuning for best signal, encoding of programmes for recording, etc.

Reach   Term used in audience measurement describing the total number of different listeners to a station or service within a specified period. Most often expressed as a percentage of the potential audience. Weekly reach. See also *Market penetration*, *Patronage.*

Reduction   Playback of a multi-track music recording to arrive at a final mix. Also mix-down.

Rehearse-record   Procedure most used in music recording or drama for perfecting and recording one section before moving on to the next.

Relay   (1) Simultaneous transmission of a programme originating from another station. (2) Transmission of a programme performed 'live' in front of an audience. See also *Deferred relay.* (3) Electrically operated switch.

Residual   Artist's repeat fee.

Reverberation   The continuation of a sound after its source has stopped due to reflection of the sound waves.

Reverberation time   Expressed in seconds, the time taken for a sound to die away to one millionth of its original intensity.

Reverse talkback   Communications system from studio to control cubicle.

Ribbon microphone   A high-quality microphone using electromagnetic principle. Bi-directional polar diagram.

Rip 'n' read   News bulletin copy sent from a central newsroom designed to be read on the air without rewriting.

ROT   Recording Off Transmission. Recording made at the time of transmission, not necessarily off-air.

RSL   Restricted Service Licence. Limited licence to broadcast, e.g. for a short duration, or within a small area – university campus.

'Rubber bands'   Colloquial term for the on-screen volume or panning envelopes used in digital mixing software.

Running order   List of programme items and timings in their chronological sequence.

Run through   Programme rehearsal.

'S' signal   The difference between left and right stereo signals, i.e. the stereo component.

SABLE   Studio Automatic Barcode Logging Equipment. Computer software system for scheduling and logging music.

SADiE   Studio Audio Digital Editor. Computer software system for recording, editing and playout of broadcast audio.

Satellite studio   Small outlying studio, perhaps without permanent staff but capable of being used as a contribution point via a link with the main studio/station centre.

SB   Simultaneous Broadcast. Relay of programme originating elsewhere. Conveyed from point to point by system of permanent SB lines, or taken 'off-air'.

Script   Complete text of a programme or insert from which the broadcast is made.

Segue   The following of one item immediately after another without an intervening pause or link. Especially two pieces of music.

SEO   Search Engine Optimisation – making web pages search friendly in order to attract new traffic.

SESAC   Society of European Stage Authors and Composers.

Share   Audience measurement term describing the amount of listening to a specified station or service expressed as a percentage of the total listening to all services heard in that area.

Sibilance   An emphasis on the 's' sounds in speech. May be accentuated or reduced by type and position of microphone.

Sig tune   Signature tune. Identifying music at the beginning and end of a programme or regular insert.

Simulcasting   Simultaneous broadcasting of one programme by two separate channels, e.g. AM and FM, or radio and TV.

Slug   Short identifying title given to a short item, particularly a news insert. See also *Catchline*.

Solid state   Transistorised or integrated circuitry as opposed to that containing valves. Also refers to solid state storage in memory cards.

Soundbite   A short piece of audio said to sum up a particular truth or point of view, able to stand alone.

Spin-doctor   Organisation person hired to promote positive aspects of policy and events, and to suppress the negative.

Spot Fx   Practical sound effects created live in the studio.

Squelch   Means of suppressing unwanted noise in the reception of a radio signal. See *Noise gate*.

Sting   Single music chord, used for dramatic effect.

Stock music   In-house library of recorded music.

Streaming   Sending a station's live output to the Internet.

Stringer   Freelance contributor paid by the item. Generally a journalist at outlying place not covered by staff.

Sustaining programme   Programme supplied by a syndicating source or elsewhere to maintain an output for a station making its own programmes for only part of the day.

Sweep   The process of audience survey for a particular station or service within a given timescale.

Sync output   Programme replayed from the record heads of a multi-track recording machine heard by performers while they record further tracks.

Talkback   Voice communication system from control cubicle to studio or other contributing point.

Talks table   Specially designed table for studio use, often circular with an acoustically transparent surface and a hole in the middle to take a microphone.

TBU   Telephone Balance Unit. Interface device used in conjunction with calls on a phone-in programme. Isolates public telephone equipment from broadcaster's equipment.

Telex   Teleprinter system of written communication.

Tie line   Any circuit pair connecting two programme areas, especially within the same building.

TOC   Table of Contents. Non-audio data held on a MiniDisc governing the track order and labelling. Affected by editing.

Tone   A test or reference signal of standard frequency and level. For example, 1 kHz at 0 dB.

Top and tail   A shortened programme rehearsal where only the opening and closing of inserts are played. Also, adding the opening and closing to a package.

Traffic   Station department responsible for scheduling and billing commercial advertising.

Trail   Broadcast item advertising forthcoming programme. On-air promotion or 'promo'.

Transcript   The text of a broadcast as transmitted, often produced from an off-air recording.

Transcription   A high-quality recording of a programme, often intended for reproduction by another broadcasting service.

Transducer   Any device that converts one form of energy into another, e.g. mechanical to electrical, acoustic to electrical, electrical to magnetic, etc.

Transient response   The ability of a microphone or other equipment to respond rapidly to change of input or brief energy states.

Tri-media   Description of a newsroom or function involving all three television, radio and Internet media.

Tweet   Message posted on the Twitter website, limited to 140 characters.

Twitter   Online social networking and micro-blogging service that enables users to send and read short messages – 'tweets'.

Two-way   Discussion or interview between two studios remote from each other. Also an interview of a specialist correspondent by a programme presenter. See also *Q & A*.

UHF   Ultra High Frequency. Radio or television transmission in the range of frequencies from 300 MHz to 3000 MHz.

Uncapped   Describing an Internet website to which access may be limited or reduced, but never severed.

UPI   United Press International. Syndicated news service.

USB   Universal Serial Bus. A type of computer connection.

USB stick   A memory device with a USB connection.

VCS DiRA   Video Computer Systems Digital Radio System. Station software system for recording, editing and playout.

VHF   Very High Frequency. Radio or television transmission in the range of frequencies from 30 MHz to 300 MHz.

ViLoR   Virtualisation of Local Radio. The BBC system of centralising the availability of news and music for local stations.

Voice-over Voiced announcement superimposed on lower level material, generally music.

Voice report Broadcast news piece in the reporter's own voice.

Vox pop 'The voice of the people'. Composite recording of 'street' interviews.

VU meter Volume Unit meter calibrated in decibels measuring signal level, especially as a recording level indicator.

Warm-up Initial introduction and chat, generally by a programme presenter or producer, designed to make an audience feel at home and create the appropriate atmosphere prior to a live broadcast or recording.

Wavelength Expressed in metres, the distance between two precisely similar points in adjacent cycles in a sound or radio wave. The length of one cycle. Used as the tuning characteristic or 'radio address' of a station. See also *Frequency*.

Website An organisation's or individual's location on the Internet for promotional publicity, information, sales, etc. Not generally the address for correspondence.

Wild or **wild track** Term borrowed from film to describe the recording of atmosphere, actuality or effects at random without a precise decision on how they are to be used in a programme.

Windshield Protective cover of foam rubber, plastic or metal gauze, designed to eliminate wind noise from a microphone. Essential for outdoor use or close vocal work.

Wind-up Signal given to broadcaster to come to the end of his programme contribution. Often by means of index finger describing slow vertical circles, or by flashing cue light.

Wire Service News agency supplying information by 'wire' – line feed to computer.

Wrap A short piece of actuality audio 'wrapped around' by a vocal introduction and a back-announcement. Frequently used in news bulletins.

WWW World Wide Web. The global network of information provided over the Internet.

Zero level tone A standard reference level signal (0 dB or 1 mW in 600 ohms at a frequency of 1 kHz) used to line up broadcasting equipment.

Zoom (trade name) Japanese handheld digital audio recorder.

Further reading

Historical survey and reference

Albarran, A. and Pitts, G., *The Radio Broadcasting Industry*. Allyn & Bacon, 2000

Barnard, S., *On the Radio: Music Radio in Britain*. Open University Press, 1989

Barnouw, E., *A History of Broadcasting in the United States, vols 1–3. A Tower in Babel, to 1933; The Golden Web, 1933–53; The Image Empire, 1950 Onward*. New York: Oxford University Press, 1966–70

Boemer, M. L., *The Children's Hour: radio programs for children 1929–1956*. Scarecrow, 1989

Briggs, A., *The History of Broadcasting in the United Kingdom, vols 1–4. The Birth of Broadcasting; The Golden Age of Wireless; The War of Words; Sound and Vision*. Oxford University Press, 1961–79

Briggs, A., *The BBC: the first fifty years*. Oxford University Press, 1985

Cain, J., *The BBC: 70 years of broadcasting*. BBC, 1992

Carpenter, H., *The Envy of the World: 50 years of the BBC Third Programme and Radio 3*. Phoenix-Orion, 1997

Chapman, R., *Selling the Sixties: the pirates & pop music radio*. Routledge, 1992

Crissell, A., *An Introductory History of British Broadcasting*. Routledge, 2002

Douglas, G., *The Early Days of Radio Broadcasting*. McFarland & Co (USA), 2001

Douglas, S., *Listening In: radio and the American imagination*. Times Books, 2000

Dunning, J., *On The Air: the encyclopaedia of old-time radio*. Oxford University Press, 1998

Dyke, G., *Inside Story*. HarperCollins, 2004

Fang, I., *A History of Mass Communication*. Focal Press, 1997

Gifford, D., *The Golden Age of Radio: an illustrated companion*. Batsford, 1985

Gilbert, S., *World Radio, TV Handbook 2005: the directory of international broadcasting*. WRTH Publications, 2004

Harmon, J., *The Great Radio Heroes*. McFarland & Co (USA), 2001

Hilmes, M., *Radio Voices: American broadcasting, 1922–1952*. University of Minnesota Press, 1997

Hilmes, M., *Radio Reader: essays in the cultural history of radio*. Routledge, 2001

Hind, J. and Mosco, S., *Rebel Radio*. Pluto, 1985

Hines, M., *The Story of Broadcasting House – Home of the BBC*. Merrell, 2008

Lackmann, R., *The Encyclopaedia of American Radio*. Checkmark Books, 2000

List, D., *Audience Survey Cookbook*. Australian Broadcasting Corp., 1997

McIntyre, I., *The Expense of Glory: a life of John Reith*. HarperCollins, 1993

Mansell, G., *Let Truth be Told: 50 years of BBC external broadcasting*. Weidenfeld, 1982

Martin-Jenkins, C., *Ball by Ball: the story of cricket broadcasting*. Grafton, 1990

Mickelson, S., *America's Other Voice: Radio Free Europe and Radio Liberty*. Praeger, 1983

Milne, A., *The Memoirs of a British Broadcaster*. Hodder & Stoughton, 1988

Milner, R., *Reith and the BBC years*. Mainstream, 1983

Mytton, G., *Handbook on Radio & Television Audience Research*. BBC, UNICEF and UNESCO, 1999

Nicholson, M., *A Measure of Danger: memoirs of a British war correspondent*. HarperCollins, 1991

Partner, P., *Arab Voices: the BBC Arabic Service 1938–1988*. BBC, 1988

Pegg, M., *Broadcasting and Society, 1918–39*. Croom Helm, 1983

Reeves, G., *Communications and the Third World*. Routledge, 1993

Reid, C., *Action Stations: a history of Broadcasting House*. Robson Books, 1987

Robinson, J., *Learning over the Air: sixty years of partnership in adult learning*. BBC, 1982

Rusling, P., *The Lid off Laser 558*. Pirate, 1984

Seymour, C., *The British Press and Broadcasting Since 1945*. Blackwell, 1996

Smith, E., *Road Map to Public Service Broadcasting*. Pub Asia-Pacific Broadcasting Union, 2012

Sterling, C. and Kittross, J., *Stay Tuned: a history of American broadcasting* (3rd edn). Routledge, 2001

Terrace, V. (ed.), *Radio's Golden Years: the encyclopaedia of radio programmes 1930–60*. Tantivy, 1981

Took, B., *Laughter in the Air: an informal history of radio comedy* (revised edn). Robson Books, 1982

Toye, J., *The Archers: family ties 1951–1967*. BBC, 1998

Tusa, J., *A World in Your Ear*. Broadside Books, 1992

Wedell, G. (ed.), *Making Broadcasting Useful: the African experience*. Manchester University Press, 1986

Whitburn, V., *The Archers: the official inside story*. Virgin, 1992

Whitehead, K., *The Third Programme: a literary history*. Clarendon, 1989

Wolfe, K., *The Churches and the BBC 1922–1956: the politics of broadcast religion*. SCM Press, 1984

Wood, J., *History of International Broadcasting*. Peter Peregrinus, 1992

Radio programmes

Bennett, A., *Talking Heads: six classic monologues*. BBC, 1998

Bennett, A., *Telling Tales*. BBC Worldwide Ltd, 2000

Best Radio Plays of 1986. Methuen/BBC, 1987.

Causley, C., *Poetry Please*. BBC, 1999

Cooke, A., *Letter From America 1946–2004*. Alfred A. Knopf, 2004

Curtis, M., *Asian Auntie-Ji: life with the BBC Asian Network*. Troubador, 2015

Donovan, P., *All Our Todays: 40 years of 'Today'*. BBC, 2013

Grant, T., *From Our Own Correspondent: the best of 50 years*. BBC, 2010

Hall, L., *Spoonface Steinberg & Other Plays from Radio 4*. BBC, 1997

Halper, D., *Full-service Radio: programming for the community*. Focal Press, 1991

Humphrys, J., *Devil's Advocate*. Hutchinson, 1999, Arrow Books, 2000

MacGregor, S., *Woman of Today*. Headline Publications, 2002

Purves, L., *Radio: A True Love Story*. Hodder & Stoughton, 2002, Coronet Paperback, 2003

Simpson, J., *News From No Man's Land: reporting the world*. Pan, 2003

Simpson, J., *The Wars Against Saddam: taking the hard road to Baghdad*. Pan, 2004

Smethurst, W., *The Archers: the history of radio's most famous programme*. Michael O'Mara Books, 2000

Thompson, J., *Good Morning! A decade of thoughts for the day*. SPCK, 2003

Journalism and ethics

Berry, D., *Ethics and Media Culture: practices and representations*. Focal Press, 1999

Boyd, A., *Broadcast Journalism: techniques of radio and TV news* (5th edn). Focal Press, 2000

Chantler, P. and Harris, S., *Local Radio Journalism*. Focal Press, 1996

Chantler, P. and Stewart, P., *Basic Radio Journalism*. Focal Press, 2003

Christians, C., Fackler, M., Rotzoll, K. and McKee, K., *Media Ethics* (6th edn). Longman, 2001

Crone, T., *Law and the Media* (4th edn). Focal Press, 2005

Curran, J. and Gurevitch, M., *Mass Media and Society* (3rd edn). Arnold, 2000

Dodd, M. and Hanna, M., *McNae's Essential Law for Journalists* (22nd edn). Oxford University Press, 2014

Douglas, L. and Kinsey, M., *A Guide to Commercial Radio Journalism*. Focal Press, 1999

Harcup, T., *Journalism: principles & practice*. Sage Books, 2003

Hargreaves, I., *Journalism: Truth or Dare?* Oxford University Press, 2003

Herbert, J., *Journalism in the Digital Age*. Focal Press, 1999

Herbert, J., *Practising Global Journalism*. Focal Press, 2000

Kieran, M. (ed.), *Media Ethics*. Routledge, 1998

Kovacs, B. and Rosenthal, T., *The Elements of Journalism*. Atlantic/Guardian Books, 2003

McBride, K. and Rosenstiel, T., *The New Ethics of Journalism: principles for the 21st century*. Poynter Institute, 2013

Morrison, J., *Essential Public Affairs for Journalists* (3rd edn). Oxford University Press, 2013

Quinn, F., *Law for Journalists* (3rd edn). Longman, 2011

Quinn, S., *Digital Sub-Editing and Design*. Focal Press, 2001

Raine, M., *Editorial Guidelines* (4th edn). Pub. Commonwealth Broadcasting Association, 2010

Rudin, R. and Ibbotson, T., *Introduction to Journalism*. Focal Press, 2002

Smith, J. and Butcher, J., *Essential Reporting, NCTJ Guide for Trainee Journalists*. Sage, 2007

Spark, S., *Investigative Reporting*. Focal Press, 1999

Ward, M., *Journalism Online*. Focal Press, 2005

Programme production

Abel, J. and Glass, I., *Radio: an illustrated guide*. WBEZ Alliance Inc., 1999

Ash, W., *The Way to Write Radio Drama*. Elm Tree Books, 1985

Bartlett, B. and Bartlett, J. *Recording Music on Location* (2nd edn). Focal Press, 2014

Beaman, J., *Interviewing for Radio*. Routledge, 2000

Beaman, J., *Programme Making for Radio*. Routledge, 2006

Beck, A., *Radio Acting*. A & C Black, 1997

Connelly, D. *Digital Radio Production* (2nd edn). Waveland Press, 2012

Crook T., *Radio Drama*. Routledge, 1999

Diem, P., Mytton, G. and Van Dam, P., *Handbook on Media Research*. Sage 2015. Renamed *Media Audience Research: a guide for professionals*. Sage 2016.

Ditingo, V., *The Remaking of Radio*. Focal Press, 1995

Fleming, C., *The Radio Handbook* (2nd edn). Routledge, 2002

Geller, V., *Beyond Powerful Radio* (2nd edn). Focal Press, 2011

Gooch, S., *Writing a Play* (2nd edn). A & C Black, 1995

Gross, L., Gross, B. and Perebinossoff, P., *Programming for TV, Radio, & The Internet*. Focal Press, 2005

Hall, R., *Writing Your First Play*. Focal Press, 1998

Hand, R. and Traynor, M. *Radio in Small Nations: production, programmes, audiences*. University of Wales Press, 2012

Hilliard, R. and Keith, M., *The Broadcast Century and Beyond* (4th edn). Focal Press, 2004

Hausman, C., Benoit, P., Messere, F. and O'Donnell, L., *Announcing: broadcast communication today* (5th edn). Wadsworth Publishing, 2003

Hausman, C., Benoit, P., Messere, F. and O'Donnell, L., *Modern Radio Production: production, programming, and performance* (6th edn). Wadsworth Publishing, 2003

Hendricks, J. and Mims, B., *Keith's Radio Station: broadcast, internet, and satellite* (9th edn). Focal Press, 2013

Huber, D., *Modern Recording Techniques* (6th edn). Focal Press, 2005

Johnson, A., *Radio Programmer: core skills*. Saland Publishing, 2010

Keith, M., *The Radio Station* (6th edn). Focal Press, 2004

Lellis, C., *Music Production: recording*. Focal Press, 2013

MacLoughlin, S., *Writing for Radio*. How to Books, 1998

McCoy, Q. and Crouch, S., *No Static: a guide to creative radio programming*. Backbeat Books, 2002

McCrum, S. and Hughes, I., *Interviewing Children*. Save the Children, 1997

Media In Governance, The. Department for International Development. UK Government, 2008

Mills, J., *The Broadcast Voice*. Focal Press, 2004

Mitchell, C. (ed.), *Women and Radio: airing differences*. Routledge, 2000

O'Day, D., *101 Ways to Make Your Radio Station Invincible*. Website: www.danoday.com

Price-Davies, E. and Tacchi, J., *Community Radio in a Global Context: a comparative analysis*. Community Media Association, 2001

Priestman, C., *Web Radio*. Focal Press, 2001

Reese, D. and Gross, L., *Radio Production Worktext* (3rd edn). Focal Press, 1998

Schultz, B., *Sports Broadcasting*. Focal Press, 2001

Senior, M., *Recording Secrets for the Small Studio*. Focal Press, 2014

Starkey, G., *Radio in Context* (2nd edn). Macmillan, 2013

Stephenson, A., *Broadcast Announcing Worktext* (2nd edn). Focal Press, 2004

Stewart, P., *Essential Radio Skills* (2nd edn). A&C Black, 2006

Trewin, J., *Presenting on TV and Radio*. Focal Press, 2003

Utterback, A., *Broadcast Voice Handbook* (3rd edn). Bonus Books, 2000

Warren, S., *Radio: the book*. Focal Press, 2004

Weinberger, M., Campbell, L. and Brody, B., *Effective Radio Advertising*. Lexington Books, 1994

Weiss, A., *Experimental Sound and Radio*. MIT Press, 2001

Index

absolute privilege 86
accidents, reporting of 97, 266
accommodation 251–2
accountability 5, 17, 341; journalists 81;
   politicians 111; producers 336
accuracy: Codes of Practice 21, 65; in
   news reporting 91, 92–3; objectivity 66;
   programme evaluation 361, 362;
   vox pops 134
acoustics: drama 309–11, 317–18; live
   music 274, 275; location interviews
   127–8
actors: commercials 235; documentaries
   326; drama 307–9, 317, 318, 379;
   eminence 361
actuality 92, 267, 323, 325, 327, 328; *see
   also* background noise
ad-libbing 78, 147, 203, 208, 318; *see
   also* spontaneity
administration 340–1
advertising 6, 23, 228–44; ACMA
   regulation 22; Codes of Practice 21,
   65; copy policy 229–30; ethics 68–9;
   humour 242–4; music 238–41; premise
   231–2; stereo 241–2; target audience
   230–1; testing 244; writing copy 232–5
advice programmes 14, 172, 177,
   179–81, 184
advisory bodies 19
Africa 12
agenda setting 13, 86–7
AIDS 3, 91, 181, 192, 231, 237–8
All India Radio 230
alterations to scripts 154
AM 39
ambiguity, avoiding 76–7, 94–5, 147

American Society of Composers, Authors,
   and Publishers (ASCAP) 194–5
anonymity 185
apologies 189, 355
appraisals 345, 372
appropriateness 360, 362
apps 53–4, 103, 139
archival policy 357–8
assistants 173–4, 177, 181, 340
atmosphere: actuality 92; documentaries
   324, 325, 326; drama 314, 316, 317;
   feature programmes 319; live music
   271, 276; location interviews 127; sound
   effects 241; sport events 262; studio
   336, 337; vox pops 138
audience research 3, 346, 362–9
audiences, studio 288, 291
audio labels 142
Australian Communications and Media
   Authority (ACMA) 22
automated playout systems 9–10
automatic gain control (AGC) 130–1, 138
auxiliary output 30, 32, 33, 37

'back-announcements' 143, 145, 146, 147
background listening 6–7
background noise: cues 146;
   documentaries 325; drama 314;
   interviews 127–9, 130; live music 275;
   vox pops 136–7, 141; *see also* actuality
backtiming 225
balance 66, 146; discussions 162–3;
   phone-ins 182–3; *see also* objectivity
Baraka FM 208
BBC: Broadcasting Code 21;
   commissioning programmes 351;